Ground Based Synthetic Aperture Radar

Ground based synthetic aperture radar (GB-SAR) is used to effectively mitigate natural disasters and monitor social infrastructure such as bridges, dams, and airport pavement surfaces. This book explains the fundamentals of radar technology, the principles of synthetic aperture radar (SAR) image generation, and interferometric SAR (InSAR) processing for observing small ground surface deformation less than 1 mm. More advanced multiple-input multiple-output (MIMO) radar for ground surface observation is introduced. The authors also provide examples of GB-SAR used for monitoring landslide and vegetation to show its potential and limitations. Understanding this advanced technology and its applications will help readers plan and install GB-SAR systems in real-life projects.

Features:

- Introduces GB-SAR, an advanced tool that measures in-millimeter ground surface displacement.
- Explains how the GB-SAR system can be installed for landslide monitoring.
- Provides a new radar technology that monitors vibrations of infrastructure remotely.
- Discusses the advanced radar technology related to polarimetry and interferometry.
- Includes several case studies applying the tools and techniques discussed to natural disasters, such as landslides, volcanoes, glaciers, and so on.

This book is intended for civil professionals who deal with disaster mitigation and infrastructure monitoring and those in electrical engineering, including radar technology. It is also an excellent resource for upper-level undergraduate and graduate students taking courses in remote sensing and photogrammetry, geography, geodesy, information science, engineering, and geology, as well as researchers and scientists interested in learning the techniques and technologies for collecting, analyzing, managing, and visualizing geospatial data sets.

SAR Remote Sensing

Series Editor: Jong-Sen Lee
Naval Research Laboratory, Washington DC

Synthetic Aperture Radar (SAR) is an indispensable and highly capable Earth remote sensing instrument. It is a day-night and all-weather sensor for all aspects of environmental monitoring, disaster (earthquake, tsunami, forest fire, etc.) assessment, and military applications. Since the 1980s, a plethora of space-borne SAR systems have been launched and are in operation, and new satellites are launched every day. Many SAR imaging modes have been developed and tailored to specific applications. SAR is becoming a vast and continuously evolving field of remote sensing technology and applications. This book series provides timely information on advancements and encompasses innovative SAR scattering theory, processing techniques, and applications in all environments. Books in this series serve and benefit professionals, academics, and students in the remote sensing community.

Spatial Analysis for Radar Remote Sensing of Tropical Forests
Gianfranco D. De Grandi and Elsa Carla De Grandi

Radar Scattering and Imaging of Rough Surfaces: Modeling and Applications with MATLAB®
Kun-Shan Chen

Polarimetric SAR Imaging: Theory and Applications
Yoshio Yamaguchi

Moon-Based Synthetic Aperture Radar: A Signal Processing Prospect
Zhen Xu and Kun-Shan Chen

Imaging Radar Polarimetric Rotation Domain Interpretation: Theory and Applications
Si-Wei Chen

Ground Based Synthetic Aperture Radar
Motoyuki Sato, Weike Feng, Yuta Izumi, and Amila Karunathilake

For more information about this series, please visit:
www.routledge.com/SAR- Remote-Sensing/book-series/CRCSRS

Ground Based Synthetic Aperture Radar

Motoyuki Sato, Weike Feng, Yuta Izumi,
and Amila Karunathilake

CRC Press
Taylor & Francis Group
Boca Raton London New York

CRC Press is an imprint of the
Taylor & Francis Group, an **informa** business

Designed cover image: © Motoyuki Sato, Weike Feng, Yuta Izumi and Amila Karunathilake

First edition published 2025
by CRC Press
2385 NW Executive Center Drive, Suite 320, Boca Raton FL 33431

and by CRC Press
4 Park Square, Milton Park, Abingdon, Oxon, OX14 4RN

CRC Press is an imprint of Taylor & Francis Group, LLC

ISBN: 978-1-032-32706-8 (hbk)
ISBN: 978-1-032-32707-5 (pbk)
ISBN: 978-1-003-31631-2 (ebk)

DOI: 10.1201/9781003316312

Typeset in Times
by Apex CoVantage, LLC

Contents

Preface...ix
About the Authors...xi

Chapter 1 Introduction 1

    1.1 GB-SAR for Microwave Remote Sensing..................................1
    1.2 History...5
    1.3 Radar Configuration 6
        1.3.1 Conventional GB-SAR 6
        1.3.2 MIMO Radar 6
    1.4 Applications 10
    1.5 Construction of This Book...12
    References 13

Chapter 2 Fundamentals of GB-SAR 20

    2.1 SAR Principle 20
    2.2 GB-SAR Signal Model 22
        2.2.1 SFCW and FMCW 22
        2.2.2 Geometry and Signal Model 26
    2.3 GB-SAR Imaging Algorithms 30
        2.3.1 Back-Projection Algorithm 30
        2.3.2 Range Migration Algorithm 31
        2.3.3 Far-Field Pseudo-Polar Format Algorithm.............33
    2.4 Interferometry 36
        2.4.1 GB-SAR Interferometry....................................36
        2.4.2 Interferogram Forming and Pixel Selection.............37
        2.4.3 Noise Filtering and Phase Unwrapping.................43
        2.4.4 Atmospheric Phase Compensation 46
        2.4.5 Displacement Estimation and Geocoding 46
    2.5 Polarimetry..47
        2.5.1 Scattering Matrix..47
        2.5.2 Coordinate System..48
    References 49

Chapter 3 Installation and Operation of GB-SAR............................53

    3.1 Site Modeling...53
        3.1.1 3D GIS Model Design for GB-SAR Illumination.........53
        3.1.2 Environmental Modeling in the Spatial Domain 53
        3.1.3 3D Model Design..55

3.1.4   LiDAR Survey 56
3.1.5   Digital Elevation Model 57
3.2   Estimation of the Illumination Area 58
3.2.1   Antenna Radiation Pattern 58
3.2.2   Vector Grid Design 62
3.2.3   Estimation of Illumination Pixels 63
3.3   Case Study of Minami–Aso 65
3.3.1   Site Description 65
3.3.2   Selection of the GB-SAR Site 66
3.3.3   Optimization of the GB-SAR Configuration 67
3.3.4   Field Validation 68
3.4   System Installation 69
3.4.1   Hardware Installation 71
3.4.2   Software Parameter Settings 74
3.5   Long-Term Observation 77
3.5.1   Automatic Mapping 80
3.5.2   Surface Displacement Observation 83
3.5.3   Landslide Early Warning System 97
References 98

Chapter 4   Atmospheric Phase Delay 99

4.1   Introduction 99
4.2   Theoretical Background 100
4.2.1   Refractivity Change Effect in the
Propagation Phase 100
4.2.2   Stochastic Modeling 101
4.3   Compensation Methods 103
4.3.1   Model-Based Approach 103
4.3.2   Geostatistical Approach 108
References 112

Chapter 5   Applications 115

5.1   Imaging of Vegetation Using Polarimetric GB-SAR 115
5.1.1   System Specification 115
5.1.2   Algorithm for 3D Imaging 115
5.1.3   Imaging of Three Different Trees 117
5.2   Rice Paddy Field Monitoring Using Polarimetric GB-SAR ...126
5.2.1   Observation of Vegetation by SAR 126
5.2.2   Measurement Configuration in a Rice Field 126
5.2.3   Ear Emergence Detection 129
5.2.4   Rice Growing until Ear Emergence 130
5.2.5   Summary 135
5.3   Soil Erosion Estimation 137
5.3.1   Landslide Monitoring in Arato-Zawa 137

		5.3.2	Erosion Factor	139
		5.3.3	Soil-Loss Assessment	144
		5.3.4	Cross-Correlation between Displacement and Precipitation	212
		5.3.5	Early Warning System	214
		References		218

Chapter 6 MIMO Radar Systems 220

	6.1	Theoretical Background		220
		6.1.1	Introduction	220
		6.1.2	MIMO Radar Development	221
		6.1.3	Discussion	253
	6.2	Block Sparsity-Based Imaging Method		256
		6.2.1	Introduction	256
		6.2.2	Target Azimuth Dependency	256
		6.2.3	Block Sparsity and Under-Sampling	260
		6.2.4	Cross-Correlation–Based Fusion	263
		6.2.5	Experiments and Results	265
		6.2.6	Discussion	275
	6.3	Tensor CS-Based Imaging Method		276
		6.3.1	Introduction	276
		6.3.2	3D Pseudopolar Coordinate System	277
		6.3.3	Tensor-Based IAA	280
		6.3.4	Experiments and Results	283
		6.3.5	Discussion	286
	6.4	Summary		288
	References			289

Index 291

Preface

Natural disasters such as earthquakes, landslides, and floods often occur in many countries all over the world. Although many types of disaster mitigation technologies have been developed, natural disasters such as volcanic activities and earthquakes are inevitable. Sensors such as an inclinometer, strain meter, or global navigation satellite system (GNSS) have been used for landslide monitoring and observation of ground surface deformation; however, these sensors can monitor separate positions, and installation of the sensor requires accurate prediction of the position of displacement beforehand.

The ground based synthetic aperture radar (GB-SAR) discussed in this book is a radar system that can be fixed on the ground and can continuously measure an area of several square kilometers from a location more than 1 km away. Furthermore, using interferometric SAR technology, we can estimate the displacement of the ground surface with 1-mm accuracy in real time and with a spatial resolution of several meters. GB-SAR has been put into practical use in the last decade. The GB-SAR system has a relatively simple configuration and appears not to be very difficult to operate compared with spaceborne SAR systems. However, without understanding the specific problems that GB-SAR suffers from, practical use of the system is not easy. We can find many journal papers on GB-SAR, but no textbook has described the practical procedure of GB-SAR monitoring.

GB-SAR is used at a fixed point to observe a fixed area. Based on this feature, GB-SAR is very useful for the long-term, continuous monitoring of ground surface deformation, which could be caused by landslides or volcanic activities. In addition, monitoring of social infrastructure has recently garnered interest. GB-SAR is also useful for this purpose.

This book begins with the mathematical formulations required for GB-SAR signal processing and imaging. Next, we describe the hardware construction of the system. On the basis of our experience installing a GB-SAR system for landside monitoring, we describe the procedure. We believe that this is a unique topic that will combine theory and practice.

Based on basic research on GB-SAR equipment conducted by the authors, this book shows the social implementation of long-term landslide monitoring in Miyagi and Kumamoto prefectures, both in Japan, and its effectiveness.

We describe the atmospheric correction method required for the high-precision measurement of ground surface displacement and propose the use of radar polarimetry and multiple frequencies to obtain more advanced ground surface information. In addition, while the current GB-SAR system moves the radar transceiver device on a fixed rail to acquire data, the next-generation system, multiple-input multiple-output (MIMO) radar, is introduced, which uses multiple fixed transmitter/receiver antennas. MIMO radar has excellent environmental resistance, and we demonstrate its performance.

We believe that this book will be useful for professionals working in civil engineering for disaster mitigation and monitoring infrastructure and for upper-level undergraduate and graduate students taking courses in radar technology, remote sensing, and photogrammetry in geography, geodesy, information science, engineering, and geology.

Microwave remote sensing technology has made remarkable progress in the last three decades. For example, environmental monitoring and natural disaster hazard detection using synthetic aperture radar mounted on satellites are widely used. Many people think of remote sensing as a technology that uses sensors mounted on satellites to observe the surface of the Earth over a wide area. However, in reality, in addition to satellites, radar sensors mounted on aircraft, vehicles, and ground based radar systems are also used.

The authors have been developing environmental measurement technology using electromagnetic waves at the Center for Northeast Asian Studies, Graduate School of Environmental Studies, and Faculty of Engineering at Tohoku University. We are particularly interested in advanced radar technologies, including radar polarimetry and SAR interferometry. We used JERS-1, ALOS, and ALOS-2, which are spaceborne SARs operated by the Japan Aerospace Exploration Agency (JAXA), and Pi-SAR and Pi-SAR2, which are airborne SARs operated by the National Institute of Information and Communications Technology (NICT), Japan. We have studied soil moisture measurement, vegetation classification, and natural disaster monitoring using GPR and spaceborne and airborne SAR. As we conducted "ground truth" on the ground surface with spaceborne and airborne SAR data, we realized that comparison with GB-SAR would provide extremely useful scientific data. We noticed that GB-SAR has led to many unique studies that go beyond "ground truth" and to practical applications, such as landslide and infrastructure monitoring. This book summarizes the studies on GB-SAR conducted with many graduate and undergraduate students.

Motoyuki Sato,
Cavtat, Croatia, April 2024

About the Authors

Motoyuki Sato is a professor emeritus at Tohoku University, Japan, and CEO of ALISys Co., Ltd, Japan. He received his B.E., M.E., and Dr. Eng. in information engineering from Tohoku University, Sendai, Japan, in 1980, 1982, and 1985, respectively. From 2007 to 2011 he was a distinguished professor at Tohoku University and from 2009 to 2013 was the director of the Center for Northeast Asian Studies. He was a visiting researcher at the Federal German Institute for Geoscience and Natural Resources (BGR) in Hannover, Germany, in 1988–1989. His current interests include transient electromagnetics and antennas, radar polarimetry, ground-penetrating radar (GPR), borehole radar, electromagnetic induction sensing, and ground based synthetic aperture radar (GB-SAR) and multiple-input multiple-output (MIMO) radar systems. He developed GPR sensors for humanitarian demining, and they are used in mine-affected countries, including Cambodia and Ukraine. He received the 2014 Frank Frischknecht Leadership Award from society of exploration Geophysicists (SEG) for his sustained and important contributions to near-surface geophysics in the field of ground-penetrating radar. He received the The Institute of Electronics, Information and Communication Engineers (IEICE) Best Paper award in 2017 (Kiyasu Award) and 2020, Achievement Award in 2019, IEEE GRSS Education Award in 2012, and IEEE Ulrich L. Rohde Innovative Conference Paper Awards on Antenna Measurements and Applications in 2017. He was a visiting professor at Jilin University, China; Delft University of Technology, the Netherlands; and the Mongolian University of Science and Technology. Dr. Sato was the general chair of the IEEE International Geoscience and Remote Sensing Symposium (IGARSS) in 2011 in Sendai–Vancouver.

Weike Feng is an associate professor at Air Force Engineering University, Shaanxi, China, and the leader of a youth innovation team of Shaanxi universities. He received his B.E. from Air Force Engineering University, Shaanxi, China, in 2013 and his Ph.D. from Tohoku University, Sendai, Japan, in 2019. Dr. Feng received the Young Researcher Award from the IEEE GRSS All Japan Joint Chapter, the Student Award from IEEE AP-S Japan, and the Excellent Paper Award from IET International Radar Conference in 2018, as well as the Professor Fujino Incentive Award from Tohoku University and the Best Student Paper Award from PIERS in 2019.

Yuta Izumi received his B. Eng and M. Eng from Chiba University, Japan, in 2016 and 2018, respectively, and his Ph.D. in radar remote sensing from Tohoku University, Japan, in 2021. From 2018 to 2020, he was a JSPS research fellow DC1. From 2019 to 2020, he stayed at the Swiss Federal Institute of Technology (ETH), Zurich, Switzerland. From 2021 to 2022, he was a JSPS research fellow (postdoctoral) at the Institute of Industrial Science, University of Tokyo. He is currently an assistant professor at the Muroran Institute of Technology. His research interests include SAR interferometry and polarimetry and their applications for disaster mitigation and infrastructure health monitoring. Dr. Izumi is a recipient of the Student

Paper Award at the ISRS in 2017, President Award from Chiba University in 2018, Dean Award from the Graduate School of Advanced Integration in 2018, Science Young Researcher Award from the IEEE Geoscience and Remote Sensing Society All Japan Joint Chapter in 2019, and Student Poster Award at the IEICE-AP in 2020.

Amila Karunathilake is a senior researcher at Advanced Technologies Research Laboratory, Asia Air Survey Co., LTD., Kawasaki, Japan. He received his B.Sc. and M.Sc. from the University of Peradeniya, Kandy, Sri Lanka, in 2011 and 2014, respectively, and his Ph.D. in environmental studies from Tohoku University, Sendai, Japan, in 2017. He was a post-doctoral research fellow with Tohoku University in 2018. He has shown his expertise by involvement in multi-participant disaster prevention research projects under the Cross-Ministerial Strategic Innovation Promotion Program (SIP) from 2019 to 2020. He collaborated with the Ministry of Land, Infrastructure, Transport and Tourism (MLIT) from 2021–2022 in Japan by consulting SAR remote sensing, spatial data processing, and analysis. His research interests include 3D spatial modeling, LiDAR-based mobile mapping systems (MMSs), new algorithms, application development for spaceborne and ground based synthetic aperture radar, interferometric and polarimetric SAR analysis for infrastructure monitoring, and disaster prevention and mitigation. Dr. Karunathilake was a recipient of the Young Researcher Award from the IEEE GRSS Japan Chapter and the best Ph.D. Student Award from Graduate School of Environmental Studies, Tohoku University, in 2017.

1 Introduction

1.1 GB-SAR FOR MICROWAVE REMOTE SENSING

After a theoretical investigation by James Clerk Maxwell, electromagnetic waves were experimentally demonstrated by Heinrich Hetz. Practical use of electromagnetic waves is based on these fundamental studies, and its applications have been extended to various fields, including communication, sensing, and energy transmission. Radar technology is one of the typical applications of sensing using electromagnetic waves.

Nowadays, radar is used in various applications, such as tracking of airplanes and ships, anti-collision radar for automobiles, measurement of precipitation/wind velocity, monitoring of post-landslide slopes, and generation of digital elevation models (DEMs) of the Earth [1–2]. Compared to other remote sensing tools, such as optical systems, due to the cloud and smoke penetration capacity of electromagnetic waves, radar can be used during the day and night in all weather conditions. Also, there are some advanced radar systems that can penetrate material such as foliage, soil, and walls and are used for foliage penetration (FOPEN) radar, ground-penetrating radar (GPR), and through-the-wall radar (TWR) [3–5], resulting in additional advantages over conventional remote sensing techniques. For example, GPR is widely used for the detection of buried facilities, including pipes and cables, archeological surveys, and the detection of landmines and unexploded ordnance.

RADAR, the acronym for RAdio Detection and Ranging, is initially used to detect and locate targets by measuring the travel time of the electromagnetic wave reflected by the targets. This is the one-dimensional range information of the target. By changing the antenna direction, we can add the azimuth information of the targets, and a two-dimensional or three-dimensional location of the targets can be achieved. This is equivalent to "imaging" targets, and it is a simple way to realize imaging radar.

Imaging radar is an important function of modern radar, which mainly includes high-range-resolution (HRR) imaging, real-aperture-radar (RAR) imaging, and synthetic-aperture-radar (SAR) imaging [6–11]. HRR imaging may be the simplest radar imaging technique. The range resolution of a radar is determined by its frequency bandwidth [6]. Assuming that the bandwidth is B, the range resolution is given by $\delta_r = c/2B$, where c is the speed of light. With a wide frequency bandwidth, a high-resolution range profile, that is, a one-dimensional (1D) image of the target, can be obtained. By analyzing this range profile, that is, the projection of the three-dimensional (3D) target on the line of sight (LOS) direction of the radar system, the target information (such as type, shape, and size) can be extracted.

In addition to the range resolution, in many applications, a high azimuth resolution (or angular resolution) is also desired to obtain more details of the target. Using an antenna with a large aperture or a phased array with many antenna elements, an RAR imaging technique has been developed to obtain a two-dimensional (2D) image

DOI: 10.1201/9781003316312-1

of the target with a high azimuth resolution. Assuming that the antenna aperture or the phased array length is D and the wavelength of the transmitted signal is λ, the angular resolution of RAR can be approximated by $\delta_\theta = \lambda / D$ [7]. Therefore, when the target distance is r, the azimuth resolution is given by $\delta_\alpha = r\lambda / D$. If the wavelength λ is 20 mm (corresponding to a carrier frequency of 15 GHz) and a 1-m azimuth resolution is required for a target at the target distance $r = 10$ km, the antenna aperture D should be 200 m, which is difficult to implement in practice, especially on a moving platform, considering the size, weight, and cost. In addition, to cover a large imaging area, beam scanning is necessary for RAR. Typically, mechanically steering the narrow radiation beam of a large antenna or electronically shifting the phase of the signal emitted from each element of a phased array is used to perform beam scanning, which is either time consuming or complicated and expensive.

By exploiting the relative movement between the radar platform and the target, SAR can achieve a large synthetic aperture without using many antenna elements, as in a phased array, while the antenna size is not necessarily large [11]. As shown in Figure 1.1, where the "stop-go" assumption is used, during the movement of the radar platform, the reflected signal from the target is measured and stored at each antenna position. Then, a synthetic aperture of length L can be generated by coherently processing all the returned signals via the matched filtering-based techniques. The azimuth resolution of SAR imaging is thus given by $\delta_\alpha = r\lambda / 2L$: the difference (a factor of 2) from RAR ($\delta_\alpha = r\lambda / D$) is introduced by the two-way target distance. As the target enters the radar radiation beam, its reflection is recorded. As the target moves away from the radar radiation beam, its reflection stops being recorded. Therefore, according to Figure 1.1, the maximum synthetic aperture length is about $L = r\lambda / D$, giving the azimuth resolution of the target at distance r as $\delta_\alpha = D / 2$. It can be seen that when a maximum synthetic aperture can be generated, the azimuth resolution of the SAR is independent of the target distance and signal wavelength.

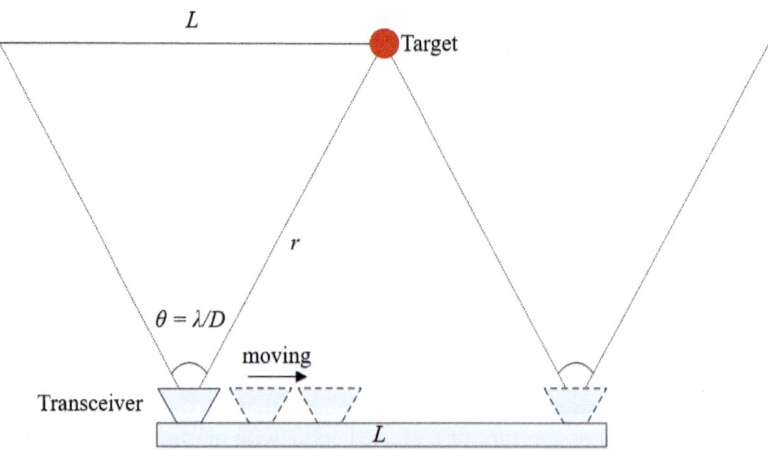

FIGURE 1.1   Principle of synthetic aperture radar (SAR) in stripmap mode.

Therefore, a 2-m (instead of 200-m) antenna aperture is sufficient to obtain a 1-m azimuth resolution.

It is worth noting that, similar to SAR imaging, when the radar platform is stationary while the target is moving, their relative motion can still be exploited to generate a high-resolution 2D image of the moving target, which is known as inverse SAR (ISAR) imaging [12–13]. In addition, there are different modes for SAR imaging, such as stripmap mode, spotlight mode, and scan mode [14]. In this book, SAR imaging with stripmap mode for stationary targets is mainly considered. Note that the velocity of electromagnetic waves in air is $c = 3 \times 10^8\, m/s$ and is much faster than the physical movement of antennas and targets. Therefore, we use the term "stop-go" for antenna scanning in ground based synthetic aperture radar (GB-SAR); the antenna can move without stopping if the travel time of the radar signal is much shorter than the data acquisition interval.

The phase information of the radar image provides in-depth target information. The phase difference between two radar images at the same pixels contains unique information related to surface elevation and motion, which is known as radar interferometry. Radar interferometry is thus employed for topographic mapping [15], moving target velocity estimation [16], and land displacement monitoring [17].

One pair of SAR images, which will be used for radar interferometry, can be acquired by two different ways. Interferometric SAR (InSAR) can be achieved using SAR data sets acquired by 2 separated antennas on a platform or two or more platforms working simultaneously, and height information of the ground surface can be obtained. Differential InSAR (DInSAR) can be achieved using two or more SAR images in a single SAR system acquired at different times, which generates time-lapse information of the ground surface or targets on the ground [18–19]. Radar polarimetry is another sophisticated SAR data analysis technique, and polarimetric SAR (PolSAR) can retrieve more information about targets by combining different polarizations [19–20]. Combining the interferometric and polarimetric SAR techniques, polarimetric InSAR (PolInSAR) and polarimetric DInSAR (PolDInSAR) provide rich information about the radar targets, which include Earth-surface topography mapping; terrain subsidence monitoring; research on earthquakes, landslides, and volcanic activities; classification of land cover; and forest height estimation.

Electromagnetic scattering from objects is strongly dependent on the size and shape of the targets and the wavelength of the electromagnetic wave. The propagation characteristics of electromagnetic waves in air, which include attenuation and phase shift, are also dependent on the wavelength. Based on these physical requirements, most of the current imaging radar systems used for terranean observation use microwaves, which correspond to electromagnetic waves between 300 MHz and 300 GHz.

Microwave remote sensing is a sensing technique that is independent of the time of day and under all weather conditions [1]. In particular, an active microwave sensor allows sensing of the Earth's surface under almost all conditions because the microwave can penetrate clouds and (to some extent) rain. Radar remote sensing measures the backscattered microwave that is related to the electrical and geometric properties of the observing targets. Therefore, the target's electrical and surface properties can be retrieved from the measured data by solving the inverse problem. SAR uses the phase information of the echo signal to achieve high azimuth resolution and generates 2D complex images (amplitude and phase). Therefore, signal processing rather than image processing provides further analysis of SAR images. In addition to sensing capability in all weathers and times, SAR provides a complementary and useful addition to optical sensors [21].

The polarization properties of electromagnetic waves are also exploited for imaging radar observation. In particular, a radar system that can acquire radar signals at different polarizations is called polarimetric radar, and polarimetric radar observation provides additional biophysical and geophysical parameters for targets biophysical and geophysical parameters. These interferometric and polarimetric techniques broaden radar remote sensing applications and enhance target characterization.

According to the adopted platform, there are different types of SAR imaging systems, such as spaceborne SAR, airborne SAR, unmanned aerial vehicle (UAV)-borne SAR, car- or truck-based SAR, and ground based SAR (GB-SAR) [22–26]. For DInSAR applications, the spaceborne SAR system may be the most powerful and popular. It can observe the largest area and provide high-accuracy and high-resolution target displacement estimation based on the permanent scatter (PS) technique and the SqueeSAR technique [27–28]. However, spaceborne SAR has several limitations in practical applications. For example, a digital elevation model is normally required for DInSAR processing to compensate for the phase error introduced by the terrain. In addition, a long revisiting period (normally from several days to tens of days) makes spaceborne SAR unsuitable for real-time displacement measurements, such as stability monitoring of a post-landslide slope or open-pit mine slope. Furthermore, because of long spatial/temporal baselines, atmospheric and environmental changes, and system thermal noise, the decorrelation problem of spaceborne SAR should be well overcome. Finally, limited by the orbital trajectory and the LOS direction, spaceborne SAR cannot provide high displacement estimation sensitivity for all parts in the observed area.

In contrast, although it can only cover a local area (a few square kilometers), GB-SAR is a more compact and flexible system that enjoys several advantages [26, 29–30]:

1. zero baseline, making the phase error caused by the terrain ignorable;
2. high data sampling rate, that is, short revisiting time (from several seconds to tens of minutes), making real-time displacement monitoring and early warning with high coherence more feasible;
3. high mobility and accessibility to almost any location at any time, making the displacement estimation of any area of interest (AoI) easy to configure from an optimal angle.

Hence, GB-SAR has been viewed as an effective complement to spaceborne and airborne SAR.

As indicated, conventional SAR imaging can only provide 2D information (range and azimuth parameters) of the target (although the InSAR technique can provide height estimation, no resolution in the elevation direction can be obtained; thus it cannot distinguish targets in the same range-azimuth pixel). To overcome the spatial ambiguity problem (i.e., targets with the same range and azimuth but different elevations will be projected into the same pixel of the focused SAR image) and retrieve the real 3D spatial parameters of the target, advanced 3D radar imaging techniques have been further developed, such as curvilinear SAR imaging [31], circular SAR imaging [32], multi-baseline tomography SAR (TomoSAR) inversion [33], and MIMO-SAR imaging [34].

By exploiting acquisitions carried out with multiple baselines, forming a 2D synthetic aperture by complex moving trajectories, or using a real aperture in one direction and a synthetic aperture in another direction, in addition to the range and azimuth directions, high resolution can also be obtained in the elevation direction by these techniques, which permits an accurate reconstruction of the observation scene in the 3D spatial domain. Moreover, by using interferometric approaches, 3.5D radar imaging is also possible; the target motion can be estimated along the time direction. For MIMO-SAR imaging, the advanced technology uses the multiple-input-multiple-output (MIMO) radar technique, which originated from the communication society, to form a virtual array with a reduced number of transmitters and receivers. In Chapter 6, MIMO radar 2D and 3D imaging are studied to reduce the data acquisition period of GB-SAR.

To summarize, radar imaging (1D, 2D, and 3D) is an important and innovative remote sensing technique that has been used for a broad range of applications.

1.2 HISTORY

From the late 1990s, several experiments were conducted for displacement measurement by ground based SAR. The first displacement study by GB-SAR in a controlled environment was performed in [35], followed by field studies for dam displacement monitoring [36], pedestrian bridge monitoring [37], and landslide monitoring [38]. Compared with spaceborne and airborne SAR, GB-SAR can perform stationary observations and achieve a much shorter revisit time. Therefore, a series of radar images can be generated in near real time or real time.

The use of three-dimensional and polarimetric GB-SAR for the observation of vegetation has been reported [39]. Although this was a short-range (less than 50 m) observation, a high potential of GB-SAR for the characterization of targets was shown.

Radar interferometry allows the monitoring of displacement in the radar line-of-sight direction. The measured displacement is thus a projected displacement of the real displacement to the LOS [30]. Hence, some displacement phenomena that move nearly orthogonal to the LOS show very low sensitivity. However, GB-SAR can maximize sensitivity by selecting the appropriate radar location. This is possible because the ground-based radar monitoring configuration is flexible compared to air- and spaceborne SAR owing to its portability. In addition, interferometric repeat-pass observation can be performed with a zero-baseline configuration in GB-SAR for displacement measurement. Thus, the topographic phase component related to terrain elevation can be neglected, while it must be corrected by a digital elevation model in spaceborne SAR interferometry.

Some campaigns employed the polarimetric GB-SAR system and used the polarimetric information for their operation [20, 40–43]. GB-SAR can obtain full polarimetric information without any sacrifice of swath width, whereas full-polarimetric spaceborne SAR acquisition compromises the reduction of swath width. In [20, 43], temporal polarimetric analyses were conducted to describe the better temporal behavior of an urban area [20] and glacier [43]. Another reported application of the full polarimetric data set is polarimetric coherence optimization to enhance the interferometric coherence and obtain an accurate interferometric phase [40, 42].

1.3 RADAR CONFIGURATION

1.3.1 CONVENTIONAL GB-SAR

The simplest and most popular GB-SAR system configuration consists of a radar system and a linear rail (guide). In this configuration, the radar system collects the received signals along the linear rail to realize synthetic aperture processing by employing antennas with wider azimuth beamwidth (generally horn antenna). The azimuth resolution of this type of radar is degraded as the azimuth angle increases. This radar type is most frequently used for radar imaging applications [44].

Unlike linear scanning, arc-scanning SAR (ArcSAR) generates an SAR image by moving the antennas along an arc [45–48]. In this configuration, the antennas are fixed to a rotating arm to realize arc scanning. By rotating the arm 360°, SAR images with the entire azimuth angle can be obtained with a constant azimuth resolution.

1.3.2 MIMO RADAR

All the previous radar types require mechanical movement of the antennas. On the other hand, the recent development of multi-input-multi-output GB-SAR technology does not require the mechanical movement of the antennas and, therefore, can achieve fast sampling acquisition of 2D SAR images realized by the array antennas. In particular, the virtual array concept that allows fewer antennas (sparse array) than that of an equivalent physical array with proper antenna position design reduces the system complexity and cost [49–52]. Monitoring examples with MIMO GB-SAR have been increasing because of these advantages.

Inspired by the MIMO technique used to increase the spectral efficiency in the wireless communication field by using advanced space–time coding methods [53–54], MIMO radar has drawn considerable interest since the beginning of this century [55–56]. According to the arrangement of antennas, that is, the position of the transceivers relative to the observed targets, MIMO radar can be divided into two categories [57–58]: (1) distributed MIMO radar (a.k.a. statistical MIMO radar), for which the transceivers are widely separated from each another to obtain different observation angles of the target, and (2) collocated MIMO radar (a.k.a. coherent MIMO radar), for which the transceivers are collocated and thus have the same looking angle of the target at far range for coherent processing. In imaging and displacement estimation applications, the collocated MIMO mode is generally adopted [52] [59–61] and is considered in Chapter 6. In the following, MIMO radar is used to indicate collocated MIMO radar for simplicity.

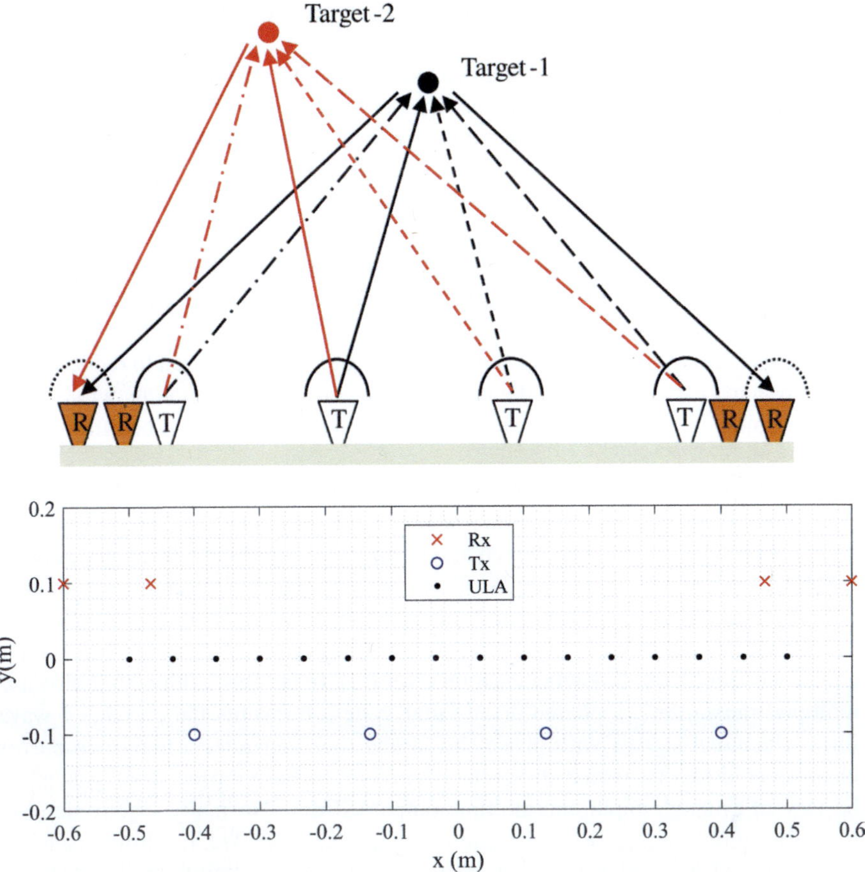

FIGURE 1.2 Geometry of the MIMO radar with two targets and four pairs of transceivers. In Figure 1.2, the top sub-figure, both transmitter and receiver have wide beamwidths. In Figure 1.2, a height difference between the transmitting and receiving arrays is introduced to clearly show the equivalent virtual uniform linear array (ULA).

Compared with conventional radar systems, one of the key advantages of MIMO radar is its high-angle resolution capacity, which can be obtained by forming a virtual array with a much reduced number of transceivers than would be required for an equivalent physical array, as shown in Figure 1.2.

For MIMO radar, assuming that there are M independent transmitters with orthogonal waveforms and N receivers with a uniform weighting vector, a virtual uniform linear array with MN channels can be created. The configuration of the virtual array elements is determined by the spatial convolution of the transmitting and receiving patterns of the MIMO radar. Using $x_{T/m}$ and $x_{R/n}$ to denote the positions of the m-th transmitter and the n-th receiver in the x direction and $y_{T/m}$ and $y_{R/n}$ to denote their positions in the y direction, the position of each element in the generated virtual array is given by $(x_{T/m} + x_{R/n})/2$ and $(y_{T/m} + y_{R/n})/2$, where $m = 1, 2, \ldots, M$, $n = 1, 2, \ldots,$ N, and the far-field assumption is used.

For example, in Figure 1.2, there are four pairs of transceivers, giving a uniform linear array with 16 elements, where the first element of the uniform linear array is generated by the first transmitter and the first receiver, while the second element is generated by the first transmitter and the second receiver. According to this, it can be found that two main advantages of MIMO radar over traditional radar are:

1. the angular resolution can be easily improved with an increased aperture length of the virtual array, while the number of physical antennas does not need to be large.
2. a sparse transmitting array and/or receiving array can be used to reduce the system cost and mitigate the impact of antenna mutual coupling, while the angular resolution and the sidelobe/grating lobe level will not be affected.

It should be also noted that, for MIMO radar, all transmitters have wide beamwidths. Since the transmitting waveforms are orthogonal to each other, no beamforming on the transmission is performed, which may be the most significant difference between MIMO radar and phased-array radar. Therefore, all targets in the entire area of interest can be illuminated at the same time by MIMO radar. Then, target reflections corresponding to different transmitters are separated in reception using, for example, a matched filter bank following each receiver [61]. In contrast, although it also uses multiple transmitters, a phased-array radar system transmits the correlated waveforms (identical signals with phase shifts) and thus generates a high-gain narrow beam steered to a particular direction. Therefore, a phased-array radar system can be viewed as single input and multiple output (SIMO). Normally, electronic scanning is required for phased-array radar to search the entire AoI, resulting in a longer period than MIMO radar. However, because its directivity (gain) is higher than that of MIMO radar, phased-array radar can achieve a higher target signal-to-noise ratio (SNR) [62]. In addition, because of the orthogonal waveforms, additional processing complexities are normally introduced to MIMO radar.

To create a complete virtual array and achieve the maximum benefit of the MIMO concept, the transmitting waveforms should be distinguished by the reception to separate different transmitter–receiver paths, which is realized by transmitting orthogonality [63–64]. There are different ways to achieve transmitting orthogonality for MIMO radar, such as using different frequencies, polarizations, or phase coding methods. For the applications considered in Chapter 6, the observed targets are normally stationary. Therefore, the time division multiplexing (TDM) approach, that is, different transmitters emit the same waveform sequentially, is adopted to guarantee transmitting orthogonality. Compared with other methods, the TDM method provides the best signal separation capacity, reduces the complexity of signal reception, and eliminates cross-correlations caused by environmental interactions. The disadvantage of the TDM method is that it normally requires a longer data sampling period than other methods that simultaneously transmit orthogonal waveforms, such as the code division multiplexing (CDM) method [64]. For example, if there are M transmitters and N receivers that can record the target reflections simultaneously, M times the data acquisition period is needed for the TDM method to form the same virtual array as other methods. For moving target detection, this will degrade the performance of the MIMO radar and reduce the unambiguous Doppler range. However, the required time of TDM-based MIMO radar to generate one focused image of stationary targets is typically much shorter than that of GB-SAR and is sufficient for most potential applications considered in Chapter 6 (such as the bridge vibration measurement). Therefore, the TDM method is adopted by the most MIMO systems developed to replace/update GB-SAR.

Based on the MIMO technique, many advanced radar imaging techniques have been developed recently. For example, by using super-resolution spectral estimation methods, e.g., APES and IAA, the 2D imaging (range-Doppler and range-angle) performance of MIMO radar has been proven to be superior to that of traditional radar [65–66]. In addition, with a narrow-band signal, a perpendicular MIMO array is used to create a virtual planar array for high-resolution target 2D imaging [67]. Furthermore, target 3D imaging can be achieved using a 2D cross-MIMO array with a wide frequency bandwidth 68]. The MIMO-ISAR imaging technique has also been proposed for moving target imaging with a reduced integration time [69], and the airborne MIMO-SAR imaging technique has been developed for ground stationary target 3D imaging [70]. The drawbacks of the conventional GB-SAR technique, such as the limited aperture length, heavy and precise mechanical rail, high transport-ability constraint, high purchase and maintenance price, and low data acquisition speed, can also be overcome by MIMO imaging radar. As mentioned before, a longer virtual array can be easily generated by the MIMO technique with a relatively small number of transceivers, and thus the aperture length will not be limited to obtain a higher spatial resolution. No heavy and accurate mechanical rail is required by MIMO radar, making system transportation easy and lowering the system cost. The temporal resolution can be improved by the MIMO radar technique, facilitating vibration measurements of civil structures.

During the last decade, to exploit the benefits of the MIMO technique, many MIMO systems have been developed to reduce the weight, cost, and data acquisition period of GB-SAR. For example: (1) the Melissa system fabricated in JRC to improve the performance of their Lisa system (as mentioned in Section 1.1.2) has been introduced in [59] and [60], where the system structure, arrangement of antenna positions, and potential applications have been carefully studied. It is shown that an image refreshing time (time interval between the formation of two adjacent images) smaller than 4 ms and an LOS displacement estimation accuracy better than 10 μm can be achieved by Melissa. (2) With the consideration of antenna size, non-collinear transmitting and receiving arrays have been adopted by the MIMO-SAR system designed by Beijing Institute of Technology (CN) in 2017 [61]. To satisfy the imaging requirements, advanced design methods that consider the array height difference and grating lobe level have been proposed. On the basis of this system, the 2D generalized ambiguity function (GAF) of MIMO imaging radar in the far field is discussed in [72], which can be used to analyze the resolution and sidelobe properties of a MIMO imaging radar system. (3) To further reduce the production cost and simplify the installation procedure, an X-band

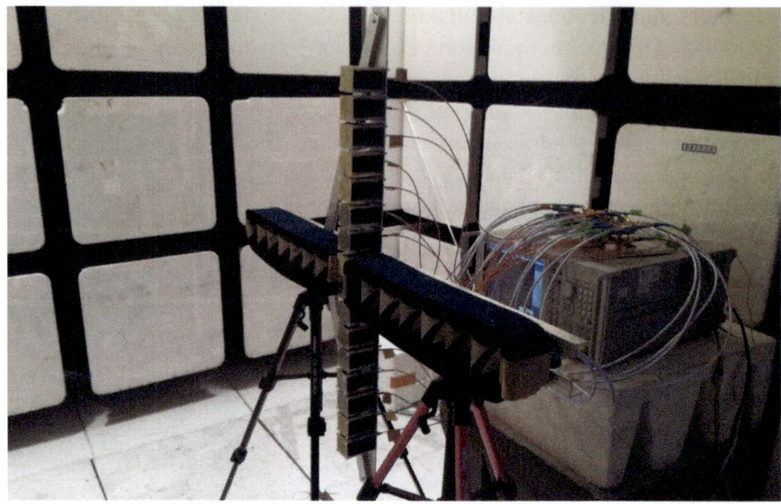

FIGURE 1.3 Cross-MIMO radar system developed by Tohoku University for 3D target imaging.

MIMO radar system that has a modular and integrated architecture [73], which is the same as [50], called SPARX, was developed by the IDS GeoRadar company in 2019. (4) Different from the previous radar systems, a compressive sensing (CS)–based MIMO radar system from the University of Florence (IT) for bridge monitoring was presented in [52]. By exploiting the sparsity of the scene, a better angular resolution than that of the conventional MIMO radar system with the same number of transceivers can be achieved. (5) Finally, as shown in [69], a cross-MIMO radar system was developed by Tohoku University (JP) in 2019, as shown in Figure 1.3, where a virtual planar array can be generated, providing 3D target imaging capacity rather than the 2D imaging capacity of the previously mentioned systems. In addition, by using a tensor-CS–based imaging method, the spatial sampling points (i.e., the number of transceivers) of the system can be further reduced, and high-level artifacts can be effectively suppressed.

1.4 APPLICATIONS

With the advantages mentioned in Section 1.1, many GB-SAR applications have been reported since 1997 [44]. Most of the publications aimed at displacement estimation, which can be roughly categorized into the following specific purposes: landslide monitoring/prediction, open pit mining monitoring, artificial structure monitoring, and glacier/snow monitoring. In addition, few polarimetric applications for vegetation monitoring exist.

Landslide monitoring is the most popular application (see publication ratio in [44]). The main objective of the monitoring is to understand the landslide mechanism and predict the potential risk of slope failure [30, 73]. Inclinometers, extensometers, and GPS networks have been conventionally employed for landslide monitoring [74]. Such point-wise measurements must be installed directly onto the landslide surface. On the other hand, ground-based radar presents a denser and higher spatial extent than conventional measurements and allows monitoring of inaccessible areas. Therefore, many varieties of landslide phenomena and fields have been observed using radar techniques [38, 30, 73, 75–81]. Both fast-moving and slow-moving landslides were observed. For fast-moving landslides, continuous observation has been carried out to follow the rapid variation of temporal behavior [10, 75, 81–83]. In this case, the radar is left on the field, acquiring the data in near real time [84]. Monitoring campaigns as an early warning need continuous observation [85]. On the other hand, GB-SAR observations of slow-moving landslides have generally been performed several months apart (depending on the rate of displacement) [30, 79, 80]. In this case, radar can be installed for each observation. A few campaigns for volcano deformation monitoring have been reported in [85–92] as applications of slope failure detection and early warning.

Open-pit Mine fields and quarries monitoring by GB-SAR also aims at slope failure detection and prediction, as reported in [93]. Such monitoring produces an early warning to minimize the potential risk of operators in the field. Ground-based radar monitoring is an alternative method to conventional point-wise mine field measurement with spatially distributed prisms and a total station [93]. Wider azimuth angle observation is preferable for monitoring the entire mining field. The IBIS-ArcSAR developed by IDS GeoRadar can produce SAR images with full 360° horizontal coverage in less than 1 min [47].

The application of artificial structure monitoring by GB-SAR has also been widely performed globally. Monitoring examples include buildings [94], urban areas [30, 41, 94, 95], dams [96], bridges [37, 55, 97–99], towers [100, 101], cultural heritage sites [102, 103], and monuments [104]. Structure monitoring assesses the potential risk of collapse and stability and understands the dynamic response [103]. Those very high acquisition rate measurements have generally been made by real-time monitoring, which can be realized by the real-aperture mode without any 2D image generation (taking into account only range profile). Nevertheless, the recent advanced MIMO radar technique allows a very high rate of 2D displacement monitoring (real-time monitoring) of the structure [30]. Most of the structural measurements by ground-based radar have been performed in continuous operational mode, while the urban measurement campaign in [30, 41] aimed at slow land deformation monitoring in discontinuous mode. Some studies have reported displacement measurements of the bollard pulled by ships [106], cruise ships (Costa Concordia) [115], and quay-subsided areas affected by the 2011 Tohoku Earthquake and associated tsunami [82].

Other key applications include glacier and snow cover monitoring. Glacier surface velocity measurement provides essential information for studying its variation as a response to climate change [108] and detects the potential risk of glacier failures as an early warning [109]. Related articles are in [109–115]. Because of the fast-moving and decorrelated characteristics of the glacier, the measurements should be performed in continuous operational mode with a short temporal baseline for interferometric purposes. Cases of surface velocity estimation of glaciers using the offset tracking method in amplitude information have also been reported [116, 117]. The application of snow avalanche detection for publishing avalanche bulletins has emerged [118, 119]. Furthermore, some campaigns have shown the applicability of snow cover depth measurement by ground-based radar [120, 121].

Some publications have reported vegetation monitoring by polarimetric ground-based radar [39, 122–124]. In addition, seasonal changes in the branches and leaves of trees were revealed by polarimetric analysis [39]. In [122], long-term rice paddy phenology monitoring was performed inside a controlled environment. Vegetation sample measurements were performed in the large anechoic chamber of the European Microwave Signature Laboratory (EMSL), JRC, Ispra, Italy [123, 124]. Vegetation monitoring has generally been performed for research purposes to investigate the signature of vegetation morphology using multi-polarimetric data analysis and has not been practically applied to operational use.

1.5   CONSTRUCTION OF THIS BOOK

This book has six chapters. In Chapter 2, we describe the technical and theoretical fundamentals of GB-SAR. Understanding the basic theory of radar systems is significantly important for the application and interpretation of radar signals and images. In Chapter 3, we demonstrate practical approaches for designing GB-SAR monitoring, installation of the radar system, and operation of the radar system. This chapter is based on our experience installing the GB-SAR system in Minami Aso, Japan. Interferometric analysis of GB-SAR data for quantitative displacement measurement based on interferometry is a very important application of GB-SAR. This information is strongly influenced by the atmospheric conditions of space, where the radar signal is propagating. Chapter 4 describes a method to compensate for this effect. In Chapter 5, we show several practical applications of GB-SAR. Conventional GB-SAR has a radar unit equipped with Tx-Rx antennas, which moves on a rail to acquire data. MIMO radar is a new approach replacing this conventional type of GB-SAR system. We explain the theoretical basis of MIMO radar and its applications to GB-SAR measurements in Chapter 6.

This book is based on research conducted in my laboratory at Tohoku University, Japan. More than 50 undergraduate and graduate students of the Engineering School and the Graduate School of Environmental Studies, have contributed. Some parts of this book are edited from M.Sc. and Ph.D. theses of these students. Some of the contents of their theses are included in each chapter as follows:

Chapter 1 Weike Feng and Yuta Izumi
Chapter 2 Weike Feng and Yuta Izumi
Chapter 3 Amila Karunathilake
Chapter 4 Yuta Izumi
Chapter 5 Zheng-Shu Zhou, Masayoshi Matsumoto, and Delima Canny Valentine
Chapter 6 Weike Feng

REFERENCES

[1] F. T. Ulaby, D. J. Long, W. J. Blackwell, et al., Microwave radar and radiometric remote sensing, vol. 4, no. 5, Ann Arbor: University of Michigan Press, 2014.

[2] J. Richards, Remote sensing with imaging radar, New York, USA: Springer, 2009.

[3] M. E. Davis, Foliage penetration radar: Detection and characterization of objects under trees, Raleigh, NC, USA: SciTech, 2012.

[4] H. M. Jol, Ground penetrating radar theory and applications, Amsterdam, Netherlands: Elsevier, 2008.

[5] M. Amin, Through-the-wall radar imaging, Boca Raton, FL: CRC Press, 2010.

[6] M. Cheney and B. Borden, Fundamentals of radar imaging, Philadelphia, PA, USA: SIAM Press, 2009.

[7] D. Wehner, High resolution radar, Norwood, MA, USA: Artech House, 1987.

[8] A. J. Fenn, D. H. Temme, W. P. Delaney, et al., "The development of phased-array radar technology," Lincoln Lab. J., vol. 12, no. 2, pp. 321–340, 2000.

[9] W. D. Wirth, Radar techniques using array antennas, London, UK: IEE, 2001.

[10] B. D. Steinberg and H. M. Subbaram, Microwave imaging techniques, New York, NY, USA: Wiley, 1991.

[11] M. Soumekh, Synthetic aperture radar signal processing with MATLAB algorithms, New York, NY, USA: Wiley, 1999.

[12] C. Özdemir, Inverse synthetic aperture radar imaging with MATLAB algorithms, New York, NY, USA: Wiley, 2012.

[13] V. C. Chen and M. Martorella, Inverse synthetic aperture radar imaging: Principles, algorithms and applications, Raleigh, NC, USA: SciTech, 2014.

[14] A. Moreira, P. Prats-Iraola, M. Younis, et al., "A tutorial on synthetic aperture radar," IEEE Geosci. Remote Sens. Mag., vol. 1, no. 1, pp. 6–43, Mar. 2013.

[15] L. C. Graham, "Synthetic interferometer radar for topographic mapping," Proc. IEEE, vol. 62, no. 6, pp. 763–768, 1974.

[16] H. A. Zebker and R. M. Goldstein, "Topographic mapping from interferometric synthetic aperture radar observations," J. Geophys. Res., vol. 91, no. B5, pp. 4993–4999, 1986.

[17] A. K. Gabriel, R. M. Goldstein, and H. A. Zebker, "Mapping small elevation changes over large areas: Differential radar interferometry," J. Geophys. Res., vol. 94, no. B7, pp. 9183–9191, 1989.

[18] S. R. Cloude and K. P. Papathanassiou, "Polarimetric SAR interferometry," IEEE Trans. Geosci. Remote Sens., vol. 36, no. 5, pp. 1551–1565, Sep. 1998.

[19] L. Pipia, X. Fabregas, A. Aguasca, et al., "Polarimetric differential SAR interferometry: First results with ground-based measurements," IEEE Geosci. Remote Sens. Lett., vol. 6, no. 1, pp. 167–171, Dec. 2008.

[20] J. Lee and E. Pottier, Polarimetric imaging: From basics to applications, Boca Raton, FL, USA: CRC Press, 2009.

[21] I. G. Cumming and F. H. Wong, Digital processing of synthetic aperture radar data: Algorithms and implementation, Norwood, MA, USA: Artech House, 2004.

[22] C. Elachi, T. Bicknell, R. L. Jordan, et al., "Spaceborne synthetic-aperture imaging radars: Applications techniques and technology," Proc. IEEE, vol. 70, no. 10, pp. 1174–1208, Oct. 1982.

[23] G. Fornaro, "Trajectory deviations in airborne SAR: Analysis and compensation," IEEE Trans. Aerosp. Electron. Syst., vol. 35, no. 3, pp. 997–1009, Jul. 1999.

[24] W. Wang, Q. Peng, and J. Cai, "Waveform-diversity-based millimeter-wave UAV SAR remote sensing," IEEE Trans. Geosci. Remote Sens., vol. 47, no. 3, pp. 691–700, Mar. 2009.

[25] O. Frey, C. L. Werner, U. Wegmuller, et al., "A car-borne SAR and InSAR experiment," Proc. IEEE Int. Geosci. Remote Sens. Symp., pp. 93–96, Jul. 2013.

[26] O. Monserrat, M. Crosetto, and G. Luzi, "A review of ground-based SAR interferometry for deformation measurement," ISPRS J. Photogramm. Remote Sens., vol. 93, pp. 40–48, Jul. 2014.

[27] A. Ferretti, C. Prati, and F. Rocca, "Permanent scatterers in SAR interferometry," IEEE Trans. Geosci. Remote Sens., vol. 39, no. 1, pp. 8–20, Jan. 2001.

[28] A. Ferretti, A. Fumagalli, F. Novali, et al., "A new algorithm for processing interferometric data-stacks: SqueeSAR," IEEE Trans. Geosci. Remote Sens., vol. 49, no. 9, pp. 3460–3470, May 2011.

[29] D. Leva, G. Nico, D Tarchi, et al., "Temporal analysis of a landslide by means of a ground-based SAR interferometer," IEEE Trans. Geosci. Remote Sens., vol. 41, no. 4, pp. 745–752, Jun. 2003.

[30] R. Iglesias, A. Aguasca, X. Fabregas, et al., "Ground-based polarimetric SAR interferometry for the monitoring of terrain displacement phenomena-Part I: Theoretical description," IEEE J. Sel. Top. Appl. Earth Obs., vol. 8, no. 3, pp. 994–1007, Mar. 2015.

[31] J. Li, Z. Bi, Z.-S. Liu, et al., "Use of curvilinear SAR for three-dimensional target feature extraction," IEE Proc. Radar Sonar Nav., vol. 144, no. 5, pp. 275–283, Oct. 1997.

[32] O. Ponce, P. Prats-Iraola, M. Pinheiro, et al., "Fully polarimetric high-resolution 3-D imaging with circular SAR at L-band," IEEE Trans. Geosci. Remote Sens., vol. 52, no. 6, pp. 3074–3090, Jun. 2014.

[33] X. Zhu and R. Bamler, "Superresolving SAR tomography for multidimensional imaging of urban areas," IEEE Signal Process. Mag., vol. 31, no. 4, pp. 51–58, Jul. 2014.

[34] G. Krieger, "MIMO-SAR: Opportunities and pitfalls," IEEE Trans. Geosci. Remote Sens., vol. 52, no. 5, pp. 2628–2645, May 2014.

[35] D. Tarchi, E. Ohlmer, and A. Sieber, "Monitoring of structural changes by radar interferometry," Res. Nondestruct. Eval., vol. 9, no. 4, pp. 213–225, 1997.

[36] D. Tarchi, H. Rudolf, G. Luzi, et al., "SAR interferometry for structural changes detection: A demonstration test on a dam," in International Geoscience and Remote Sensing Symposium (IGARSS), 1999.

[37] M. Pieraccini, D. Tarchi, H. Rudolf, et al., "Structural static testing by interferometric synthetic radar," NDT E Int., vol. 33, no. 8, pp. 565–570, Dec. 2000.

[38] D. Tarchi, N. Casagli, R. Fanti, et al., "Landslide monitoring by using ground-based SAR interferometry: an example of application to the Tessina landslide in Italy," Eng. Geol., vol. 68, pp. 15–30, Feb. 2003.

[39] Z. Zhou, W. Boerner, and M. Sato, "Development of a ground-based polarimetric broadband SAR system for noninvasive ground-truth validation in vegetation monitoring," IEEE Trans. Geosci. Remote Sens., vol. 42, no. 9, pp. 1803–1810, Sept. 2004.

[40] L. Pipia, X. Fabregas, A. Aguasca, et al., "Polarimetric coherence optimization for interferometric differential applications," in International Geoscience and Remote Sensing Symposium (IGARSS), 2009.

[41] L. Pipia, X. Fabregas, A. Aguasca, et al., "Polarimetric differential SAR interferometry : First results with ground-based measurements," IEEE Geosci. Remote Sens. Lett., vol. 6, no. 1, pp. 167–171, 2009.

[42] R. Iglesias, D. Monells, X. Fabregas, et al., "Phase quality optimization in polarimetric differential SAR interferometry," IEEE Trans. Geosci. Remote Sens., vol. 52, no. 5, pp. 2875–2888, 2014.

[43] S. Baffelli, O. Frey, and I. Hajnsek, "Polarimetric analysis of natural terrain observed with a ku-band terrestrial radar," IEEE J. Sel. Top. Appl. Earth Obs. Remote Sens., vol. 12, no. 12, pp. 5268–5288, 2019.

[44] M. Pieraccini and L. Miccinesi, "Ground-based radar interferometry: A bibliographic review," Remote Sens., vol. 11, no. 9, 2019.

[45] H. Lee, J. H. Lee, K. E. Kim, et al., "Development of a truck-mounted arc-scanning synthetic aperture radar," IEEE Trans. Geosci. Remote Sens., vol. 52, no. 5, pp. 2773–2779, 2014.

[46] Y. Luo, H. Song, R. Wang, et al., "Arc FMCW sar and applications in ground monitoring," IEEE Trans. Geosci. Remote Sens., vol. 52, no. 9, pp. 5989–5998, 2014.

[47] F. Viviani, A. Michelini, L. Mayer, et al., "IBIS-ARCSAR: An innovative ground-based SAR system for slope monitoring," in International Geoscience and Remote Sensing Symposium (IGARSS), 2018.

[48] M. Pieraccini and L. Miccinesi, "ArcSAR: Theory, simulations, and experimental verification," IEEE Trans. Microw. Theory Tech., vol. 65, no. 1, pp. 293–301, 2017.

[49] P. F. Sammartino, D. Tarchi, and C. J. Baker, "MIMO radar topology: A systematic approach to the placement of the antennas," in Proceedings—2011 International Conference on Electromagnetics in Advanced Applications, ICEAA'11, 2011.

[50] X. Zhuge and A. G. Yarovoy, "Sparse multiple-input multiple-output arrays for high-resolution near-field ultra-wideband imaging," IET Microw. Antennas Propag., vol. 5, no. 13, pp. 1552–1562, 2011.

[51] Y. Liu, X. Xu, and G. Xu, "MIMO radar calibration and imagery for near-field scattering diagnosis," IEEE Trans. Aerosp. Electron. Syst., vol. 54, no. 1, pp. 442–452, 2018.

[52] M. Pieraccini and L. Miccinesi, "An interferometric MIMO radar for bridge monitoring," IEEE Geosci. Remote Sens. Lett., vol. 16, no. 9, pp. 1383–1387, 2019.

[53] A. J. Paulraj, "An overview of MIMO communications-a key to Gigabit wireless," Proc. IEEE, vol. 92, no. 2, pp. 198–218, Feb. 2004.

[54] A. B. Gershman and N. D. Sidiropoulous, Space-time processing for MIMO communications, London, UK: Wiley, 2005.

[55] D. Rabideau, "Ubiquitous MIMO digital array radar," Proc. 37th Asilomar Conf. Signals Syst. Compu., pp. 1057–1064, Nov. 2003.

[56] D. W. Bliss and K. W. Forsythe, "Multiple-Input Multiple-Output (MIMO) radar and imaging: Degrees of freedom and resolution", Proc. 37th Asilomar Conf. Signals Syst. Comput., vol. 1, pp. 54–59, Nov. 2003.

[57] A. M. Haimovich, R. S. Blum, and L.J. Cimini, "MIMO radar with widely separated antennas," IEEE Signal Process. Mag., vol. 25, no. 1, pp. 116–129, Jan. 2008.

[58] J. Li and P. Stoica, "MIMO radar with colocated antennas," IEEE Signal Process. Mag., vol. 24, no. 5, pp. 106–114, Sep. 2007.

[59] D. Tarchi, F. Oliveri, and P. F. Sammartino, "MIMO radar and ground based SAR imaging systems: Equivalent approaches for remote sensing," IEEE Trans. Geosci. Remote Sens., vol. 51, no. 1, pp. 425–435, Jan. 2013.

[60] C. Hu, J. Wang, W. Tian, et al., "Design and imaging of ground-based Multiple-Input Multiple-Output Synthetic Aperture Radar (MIMO SAR) with non-collinear arrays," Sensors, vol. 17, no. 3, pp. 598–616, Mar. 2017.

[61] C.-Y. Chen, "Signal processing algorithms for MIMO radar," Ph.D. dissertation, California Institute of Technology, USA, 2009.

[62] M. S. Davis, "MIMO radar: Signal processing, waveform design, and applications to synthetic aperture imaging," Ph.D. dissertation, Georgia Institute of Technology, USA, 2015.

[63] H. Sun, F. Brigui, and M. Lesturgie, "Analysis and comparison of MIMO radar waveforms", in Proceedings of IEEE International Radar Conference, pp. 1–6, 2014.

[64] L. Lo Monte, B. Himed, T. Corigliano, et al., "Performance analysis of time division and code division waveforms in co-located MIMO," Proceedings of IEEE International Radar Conference, pp. 794–798, May 2015.

[65] L. Xu, J. Li, and P. Stoica, "Radar imaging via adaptive MIMO techniques," Proceedings 14th European Signal Processing Confnference, Sept. 2006.

[66] W. Roberts, P. Stoica, J. Li, et al., "Iterative adaptive approaches to MIMO radar imaging," IEEE J. Sel. Topics Signal Process., vol. 4, no. 1, pp. 5–20, Feb. 2010.

[67] D. Wang, X. Ma, and Y. Su, "Two-dimensional imaging via a narrowband MIMO radar system with two perpendicular linear arrays", IEEE Trans. Image Process., vol. 19, no. 5, pp. 1269–1279, May 2010.

[68] W. Feng, J.-M. Friedt, G. Nico, et al., "Three-dimensional ground based imaging radar based on C-band cross-MIMO array and tensor compressive sensing," IEEE Geosci. Remote Sens. Lett., vol. 16, no. 10, pp. 1585–1589, Oct. 2019.

[69] Y. Zhu, Y. Su, and W. Yu, "An ISAR imaging method based on MIMO technique," IEEE Trans. Geosci. Remote Sens., vol. 48, no. 8, pp. 3290–3299, Aug. 2010.

[70] K. Han, Y. Wang, W. Tan, et al., "Efficient pseudopolar format algorithm for down-looking linear-array SAR 3-D imaging," IEEE Geosci. Remote Sens. Lett., vol. 12, no. 3, pp. 572–576, Mar. 2015.

[71] C. Hu, J. Wang, W. Tian, et al., "Generalized ambiguity function properties of ground-based wideband MIMO imaging radar," IEEE Trans. Aerosp. Electron. Syst., vol. 55, no. 2, pp. 578–591, Apr. 2018.

[72] A. Michelini, F. Coppi, A. Bicci, et al., "SPARX, a MIMO array for ground-based radar interferometry," Sensors, vol. 19, no. 2, p. 252, Feb. 2019.

[73] C. Atzeni, M. Barla, M. Pieraccini, et al., "Early warning monitoring of natural and engineered slopes with ground-based synthetic-aperture radar," Rock Mech. Rock Eng., vol. 48, pp. 235–246, 2015.

[74] M. G. Angeli, A. Pasuto, and S. Silvano, "A critical review of landslide monitoring experiences," Eng. Geol., vol. 55, pp. 133–147, 2000.

[75] D. Leva, G. Nico, D. Tarchi, et al., "Temporal analysis of a landslide by means of a ground-based SAR interferometer," IEEE Trans. Geosci. Remote Sens., vol. 41, no. 4, Part I, pp. 745–752, 2003.

[76] D. Tarchi, G. Antonello, N. Casagli, et al., On the use of ground-based SAR interferometry for slope failure early warning: The cortenova rock slide (Italy), Landslides, Berlin, Germany: Springer Berlin Heidelberg, 2005, pp. 337–342.

[77] F. Bozzano, I. Cipriani, P. Mazzanti, et al., "Displacement patterns of a landslide affected by human activities: Insights from ground-based InSAR monitoring," Nat. Hazards, vol. 59, pp. 1377–1396, 2011.

[78] K. Takahashi, M. Matsumoto, and M. Sato, "Continuous observation of natural-disaster-affected areas using ground-based SAR interferometry," IEEE J. Sel. Top. Appl. Earth Obs. Remote Sens., vol. 6, no. 3, pp. 1286–1294, 2013.

[79] L. Noferini, M. Pieraccini, D. Mecatti, et al., "Long term landslide monitoring by ground-based synthetic aperture radar interferometer," Int. J. Remote Sens., vol. 27, no. 10, pp. 1893–1905, 2006.

[80] L. Noferini, et al., "Using GB-SAR technique to monitor slow moving landslide," Eng. Geol., vol. 95, no. 3–4, pp. 88–98, 2007.

[81] W. Frodella, M. Pieraccini, D. Mecatti, et al., "A method for assessing and managing landslide residual hazard in urban areas," Landslides, vol. 15, no. 2, pp. 183–197, 2018.

[82] G. Antonello, A. Ciampalini, F. Bardi, et al., "Ground-based SAR interferometry for monitoring mass movements," Landslides, vol. 1, no. 1, pp. 21–28, 2004.

[83] M. Pieraccini, N. Casagli, G. Luzi, et al., "Landslide monitoring by ground-based radar interferometry: A field test in Valdarno (Italy)," Int. J. Remote Sens., vol. 24, no. 6, pp. 1385–1391, 2003.

[84] O. Monserrat, M. Crosetto, and G. Luzi, "A review of ground-based SAR interferometry for deformation measurement," ISPRS J. Photogramm. Remote Sens., vol. 93, pp. 40–48, 2014.

[85] N. Casagli, F. Catani, C. Del Ventisette, and G. Luzi, "Monitoring, prediction, and early warning using ground-based radar interferometry," Landslides, vol. 7, pp. 291–301, 2010.

[86] F. Di Traglia, T. Nolesini, L. Solari, et al., "Lava delta deformation as a proxy for submarine slope instability," Earth Planet. Sci. Lett., vol. 488, pp. 46–58, Apr. 2018.

[87] F. Di Traglia, T. Nolesini, A. Ciampalini, et al., "Tracking morphological changes and slope instability using spaceborne and ground-based SAR data," Geomorphology, vol. 300, pp. 95–112, Jan. 2018.

[88] F. Di Traglia, E. Intrieri, T. Nolesini, et al., "The ground-based InSAR monitoring system at Stromboli volcano: Linking changes in displacement rate and intensity of persistent volcanic activity," Bull. Volcanol., vol. 76, no. 2, pp. 1–18, 2014.

[89] E. Intrieri, F. Di Traglia, C. Del Ventisette, et al., "Flank instability of Stromboli volcano (Aeolian Islands, Southern Italy): Integration of GB-InSAR and geomorphological observations," Geomorphology, vol. 201, pp. 60–69, Nov. 2013.

[90] S. Kuraoka, Y. Nakashima, R. Doke, et al., "Monitoring ground deformation of eruption center by ground-based interferometric synthetic aperture radar (GB-InSAR): A case study during the 2015 phreatic eruption of Hakone volcano," Earth Planets Sp., vol. 70, no. 1, 2018.

[91] T. Nolesini, F. Di Traglia, C. Del Ventisette, et al., "Deformations and slope instability on Stromboli volcano: Integration of GBInSAR data and analog modeling," Geomorphology, vol. 180–181, pp. 242–254, Jan. 2013.

[92] G. Wadge, D. G. Macfarlane, D. A. Robertson, et al., "AVTIS: A novel millimetre-wave ground based instrument for volcano remote sensing," J. Volcanol. Geotherm. Res., vol. 146, no. 4, pp. 307–318, Sep. 2005.

[93] J. Severin, E. Eberhardt, L. Leoni, et al., "Development and application of a pseudo-3D pit slope displacement map derived from ground-based radar," Eng. Geol., vol. 181, pp. 202–211, 2014.

[94] A. Montuori, G. Luzi, C. Bignami, et al., "The interferometric use of radar sensors for the urban monitoring of structural vibrations and surface displacements," IEEE J. Sel. Top. Appl. Earth Obs. Remote Sens., vol. 9, no. 8, pp. 3761–3776, 2016.

[95] M. Pieraccini, L. Noferini, D. Mecatti, et al., "Integration of radar interferometry and laser scanning for remote monitoring of an urban site built on a sliding slope," IEEE Trans. Geosci. Remote Sens., vol. 44, no. 9, pp. 2335–2342, 2006.

[96] D. Tarchi, H. Rudolf, G. Luzi, et al., "SAR interferometry for structural changes detection: A demonstration test on a dam," pp. 1522–1524, 2003.

[97] M. Pieraccini, "Monitoring of civil infrastructures by interferometric radar: A review," Sci. World J., vol. 2013, p. 786961, 2013.

[98] M. Pieraccini, M. Fratini, F. Parrini, et al., "Dynamic monitoring of bridges using a high-speed coherent radar," IEEE Trans. Geosci. Remote Sens., vol. 44, no. 11, pp. 3284–3288, 2006.

[99] T. A. Stabile, A. Perrone, M. R. Gallipoli, et al., "Dynamic survey of the Musmeci bridge by joint application of ground-based microwave radar interferometry and ambient noise standard spectral ratio techniques," IEEE Geosci. Remote Sens. Lett., vol. 10, no. 4, pp. 870–874, 2013.

[100] G. Nico, G. Prezioso, O. Masci, et al., "Dynamic modal identification of telecommunication towers using ground based radar interferometry," Remote Sens., vol. 12, no. 1211, 2020.

[101] C. Atzeni, A. Bicci, D. Dei, et al., "Remote survey of the leaning tower of pisa by interferometric sensing," IEEE Geosci. Remote Sens. Lett., vol. 7, no. 1, pp. 185–189, 2010.

[102] D. Tarchi, H. Rudolf, M. Pieraccini, et al., "Remote monitoring of buildings using a ground-based SAR: Application to cultural heritage survey," Int. J. Remote Sens., vol. 21, no. 18, pp. 3545–3551, 2000.

[103] F. Pratesi, T. Nolesini, S. Bianchini, et al., "Early warning GBInSAR-based method for monitoring volterra (Tuscany, Italy) city walls," IEEE J. Sel. Top. Appl. Earth Obs. Remote Sens., vol. 8, no. 4, pp. 1753–1762, 2015.

[104] M. Pieraccini, M. Betti, D. Forcellini, et al., "Radar detection of pedestrian-induced vibrations on Michelangelo's David," PLoS One, vol. 12, no. 4, 2017.

[105] G. Nico, G. Prezioso, O. Masci, et al., "Monitoring strategies of displacements and vibration frequencies by ground-based radar interferometry," in International Workshop on R3 in Geomatics: Research, Results and Review, Cham, Switzerland: Springer, 2020, pp. 363–374.

[106] G. Nico, G. Cifarelli, G. Miccoli, et al., "Measurement of pier deformation patterns by ground-based SAR interferometry: Application to a bollard pull trial," IEEE J. Ocean. Eng., vol. 43, no. 4, pp. 822–829, 2018.

[107] J. Broussolle, V. Kyovtorov, M. Basso, et al., "MELISSA, a new class of ground based InSAR system: An example of application in support to the Costa Concordia emergency," ISPRS J. Photogramm. Remote Sens., vol. 91, pp. 50–58, 2014.

[108] L. D. Euillades, P. A. Euillades, N. C. Riveros, et al., "Detection of glaciers displacement time-series using SAR," Remote Sens. Environ., vol. 184, pp. 188–198, 2016.

[109] S. Baffelli, O. Frey, and I. Hajnsek, "Geostatistical analysis and mitigation of the atmospheric phase screens in Ku-Band terrestrial radar interferometric observations of an Alpine Glacier," IEEE Trans. Geosci. Remote Sens., vol. 58, no. 11, pp. 7533–7556, Apr. 2020.

[110] G. Luzi, M. Pieraccini, D. Mecatti, et al., "Monitoring of an alpine glacier by means of ground-based SAR interferometry," IEEE Geosci. Remote Sens. Lett., vol. 4, no. 3, pp. 495–499, 2007.

[111] N. Dematteis, G. Luzi, D. Giordan, F. Zucca, et al., "Monitoring Alpine glacier surface deformations with GB-SAR," Remote Sens. Lett., vol. 8, no. 10, pp. 947–956, 2017.

[112] H. Han and H. Lee, "Motion of Campbell Glacier, East Antarctica, observed by satellite and ground-based interferometric synthetic aperture radar," in 2011 3rd International Asia-Pacific Conference on Synthetic Aperture Radar (APSAR), 2011, pp. 2–5.

[113] L. Noferini, D. Mecatti, G. Macaluso, et al., "Monitoring of belvedere glacier using a wide angle GB-SAR interferometer," J. Appl. Geophys., vol. 68, no. 2, pp. 289–293, 2009.

[114] P. Riesen, T. Strozzi, A. Bauder, et al., "Short-term surface ice motion variations measured with a ground-based portable real aperture radar interferometer," J. Glaciol., vol. 57, no. 201, pp. 53–60, 2011.

[115] D. Voytenko, T. H. Dixon, C. Werner, et al., "Monitoring a glacier in southeastern Iceland with the portable Terrestrial Radar Interferometer," in International Geoscience and Remote Sensing Symposium (IGARSS), 2012.

[116] S. Conzett, A. Wieser, and J. A. Butt, "Exploiting amplitude of terrestrial radar data for feature tracking in areal monitoring of inhomogeneous surfaces," in Proc. 3rd Joint International Symposium on Deformation Monitoring (JISDM), 2016.

[117] A. Chapuis, C. Rolstad, and R. Norland, "Interpretation of amplitude data from a ground-based radar in combination with terrestrial photogrammetry and visual observations for calving monitoring of Kronebreen, Svalbard," Ann. Glaciol., vol. 51, no. 55, pp. 34–40, 2010.

[118] A. Martinez-Vazquez and J. Fortuny-Guasch, "A GB-SAR processor for snow avalanche identification," IEEE Trans. Geosci. Remote Sens., vol. 46, no. 11, pp. 3948–3956, 2008.

[119] C. Lucas, I. Hajnsek, A. Marino, et al., "Investigation of snow avalanches with ground based ku-band radar," in Proceedings of the European Conference on Synthetic Aperture Radar, EUSAR, 2016.

[120] A. Martinez-Vazquez, "Monitoring of the snow cover with a ground-based synthetic aperture radar," in EARSeL eProceedings, 2005, pp. 171–178.

[121] S. Leinss, H. Löwe, M. Proksch, et al., "Anisotropy of seasonal snow measured by polarimetric phase differences in radar time series," Cryosphere, vol. 10, pp. 1771–1797, 2016.

[122] Y. Izumi, S. Demirci, M. Z. bin Baharuddin, et al., "Analysis of dual- and full-circular polarimetric SAR modes for rice phenology monitoring: An experimental investigation through ground-based measurements," Appl. Sci., vol. 7, no. 4, p. 368, 2017.

[123] J. M. Lopez-Sanchez, J. Fortuny, S. R. Cloude, et al., "Indoor polarimetric radar measurements on vegetation samples at l, s, c and x band," J. Electromagn. Waves Appl., vol. 14, no. 2, pp. 205–231, 2000.

[124] J. M. Lopez-Sanchez, D. Ballester-Berman, and J. Fortuny-Guasch, "Indoor wideband polarimetric measurements on maize plants: A study of the differential extinction coefficient," IEEE Trans. Geosci. Remote Sens., vol. 44, no. 4, pp. 758–767, 2006.

2 Fundamentals of GB-SAR

2.1 SAR PRINCIPLE

Imaging is an important function of modern radar, which mainly includes high-range-resolution (HRR) imaging, real-aperture-radar (RAR) imaging, and synthetic-aperture-radar (SAR) imaging [1–6]. HRR imaging may be the simplest radar imaging technique. The range resolution of radar is determined by its frequency bandwidth [1]: assuming the bandwidth is B, the range resolution is given by $\delta_r = c/2B$, where c is the speed of light. With a wide frequency bandwidth, a high-resolution range profile, that is, a one-dimensional (1D) image of the target, can be obtained. By analyzing this range profile, that is, the projection of the three-dimensional (3D) target on the line of sight (LOS) direction of the radar system, the target information (such as type, shape, and size) can be extracted.

In addition to the range resolution, in many applications, a high azimuth resolution (or angular resolution) is also desired to obtain more details of the target. Using an antenna with a large aperture or a phased array with many antenna elements, an RAR imaging technique has been developed to obtain a two-dimensional (2D) image of the target with a high azimuth resolution. Assuming that the antenna aperture or the phased array length is D and the wavelength of the transmitted signal is λ, the angular resolution of RAR can be approximated by $\delta\theta = \lambda/D$ [2]. Therefore, when the target distance is r, the azimuth resolution is given by $\delta_a = r\lambda/D$. If the wavelength is 20 mm (corresponding to a carrier frequency of 15 GHz) and a 1-m azimuth resolution is required for a target at 10 km, the antenna aperture should be 200 m, which is

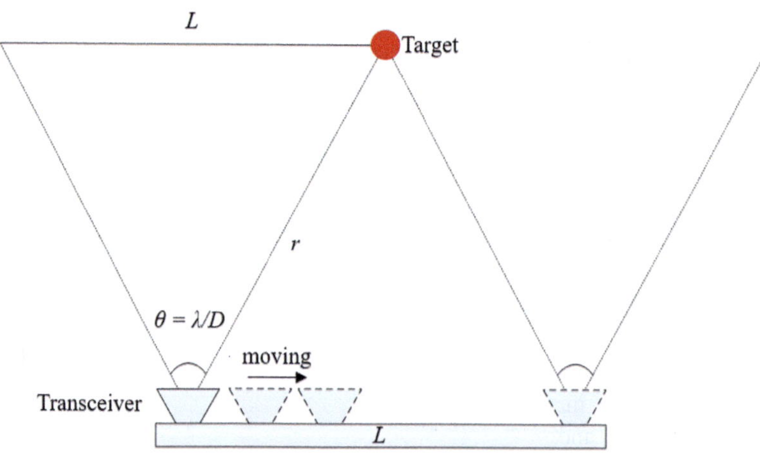

FIGURE 2.1 Principle of synthetic aperture radar in stripmap mode.

DOI: 10.1201/9781003316312-2

difficult to implement in practice, especially on a moving platform, considering the size, weight, and cost. In addition, to cover a large imaging area, beam scanning is necessary for RAR. Typically, mechanically steering the narrow radiation beam of a large antenna or electronically shifting the phase of the signal emitted from each element of a phased array is used to perform beam scanning, which is either time consuming or complicated and expensive.

By exploiting the relative movement between the radar platform and the target, SAR can achieve a large synthetic aperture without using many antenna elements, as in a phased array, while the antenna size is not necessarily large [6]. As shown in Figure 2.1, where the "stop-go" assumption is used, during the movement of the radar platform, the reflected signal from the target is measured and stored at each antenna position. Then, a synthetic aperture of length L can be generated by coherently processing all the returned signals via the matched filtering-based techniques. The azimuth resolution of SAR imaging is thus given by $\delta_a = r\lambda/2L$: the difference (a factor of 2) from RAR ($\delta_a = r\lambda/D$) is introduced by the two-way target distance. As the target enters the radar radiation beam, its reflection is recorded. As the target moves away from the radar radiation beam, its reflection stops being recorded. Therefore, according to Figure 2.1, the maximum synthetic aperture length is approximately $L = r\lambda/D$, giving the azimuth resolution of the target at distance r as $\delta_a = D/2$. It can be seen that when a maximum synthetic aperture can be generated, the azimuth resolution of SAR is independent of the target distance and signal wavelength. Therefore, a 2-m (instead of 200-m) antenna aperture is sufficient to obtain a 1-m azimuth resolution. It is worth noting that, similar to SAR imaging, when the radar platform is stationary while the target is moving, their relative motion can still be exploited to generate a high-resolution 2D image of the moving target, which is known as inverse SAR (ISAR) imaging [7–8]. In addition, there are different modes for SAR imaging, such as stripmap mode, spotlight mode, and scan mode [9]. In this book, SAR imaging with stripmap mode for stationary targets is mainly considered.

Based on 2D complex (amplitude and phase) images generated by SAR imaging with a spatial baseline (two antennas) or a temporal baseline (repeated passes), interferometric SAR (InSAR) or differential InSAR (DInSAR) techniques have been developed [10–11], which can be viewed as 2.5D radar imaging techniques. Using the InSAR technique, in addition to the range and azimuth parameters, the height (cross-track) and motion (along-track) of the target can also be reconstructed. Using the DInSAR technique, the displacement of the imaging scene along the time direction can be estimated with a centimeter or millimeter accuracy (depending on the wavelength, the characteristics of targets, and the target to radar distances). Furthermore, by exploiting the polarimetric signatures of targets [12–14], polarimetric SAR (PolSAR), polarimetric InSAR (PolInSAR), and polarimetric DInSAR (PolDInSAR) techniques have also been proposed in the past to extract more scattering properties of the targets. With the development of these radar imaging techniques, the applications of radar have been significantly expanded, such as Earth surface topography mapping; terrain subsidence monitoring; research on earthquakes, landslides, and volcanic activities, classification of land cover; and forest height estimation. In this book, the displacement estimation capacity provided by the DInSAR technique is considered.

According to the adopted platform, there are different types of SAR imaging systems, such as spaceborne SAR, airborne SAR, unmanned aerial vehicle–borne SAR, car- or truck-based SAR, and ground based SAR (GB-SAR) [15–19]. For DInSAR applications, the spaceborne SAR system may be the most powerful and popular. It can observe the largest area and provide high-accuracy and high-resolution target displacement estimation based on the permanent scatter (PS) technique and the SqueeSAR technique [20–21]. However, spaceborne SAR has several limitations in practical applications. For example, a digital elevation model (DEM) is normally required for DInSAR processing to compensate for the phase error introduced by the terrain. In addition, a long revisiting period (normally from several days to tens of days) makes spaceborne SAR unsuitable for real-time displacement measurements, such as stability monitoring of a post-landslide slope or open-pit mine slope. Furthermore, because of long spatial/temporal baselines, atmospheric and environmental changes, and system thermal noise, the decorrelation problem of spaceborne SAR should be well overcome. Finally, limited by the orbital trajectory and the LOS direction, spaceborne SAR cannot provide high displacement estimation sensitivity for all parts in the observed area. On the contrary, although it can only cover a local area (few square kilometers), GB-SAR is a more compact and flexible system that enjoys several advantages [19, 22–23]: (1) zero baseline, making the phase error caused by terrain ignorable; (2) high data sampling rate, that is, short revisiting time (from several seconds to tens of minutes), making real-time displacement monitoring and early-warning with high coherence more feasible; and (3) high mobility and accessibility to almost any location at any time, making the displacement estimation of any area of interest (AoI) easy to configure from an optimal angle. Hence, GB-SAR has been viewed as an effective complement to spaceborne and airborne SAR.

2.2 GB-SAR SIGNAL MODEL

2.2.1 SFCW AND FMCW

Before introducing the geometry and signal model of GB-SAR, two waveforms that are commonly used by GB-SAR, stepped frequency continuous waveform (SFCW) and frequency modulated continuous waveform (FMCW), are reviewed.

As shown in Figure 2.2, the SFCW signal in the time domain can be expressed as

$$s_0(t) = \sum_{q=1}^{Q} rect\left[\frac{t-(q-1)T}{T}\right] \exp\left\{-j2\pi\left[f_0 + (q-1)\Delta f\right]t\right\} \qquad (2.1)$$

where T is the pulse width for each frequency, $q = 1, 2, \ldots, Q$ is the frequency index, f_0 is the start frequency, Δf is the frequency step, and $rect(x)$ is unity when $0 \leq x \leq 1$ and is zero otherwise. For a target at R_0, the time delay is given by $\tau_0 = 2R_0/c$, where c is the speed of light, and the reflected signal is given by

$$s(t) = \sum_{q=1}^{Q} rect\left[\frac{t-\tau_0-(q-1)T}{T}\right] \exp\left\{-j2\pi\left[f_0 + (q-1)\Delta f\right](t-\tau_0)\right\} \qquad (2.2)$$

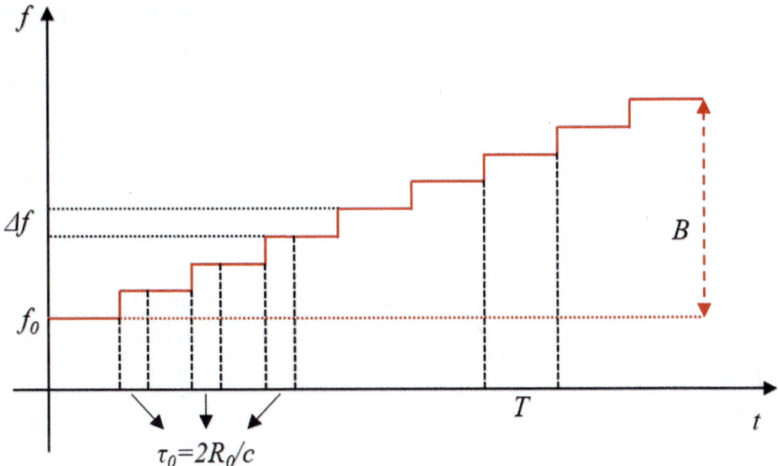

FIGURE 2.2 Stepped frequency continuous waveform.

where A is the target reflection coefficient. If the delay τ_0 is smaller than the pulse width T, the received signal for each frequency can be demodulated as follows:

$$s(f_q) = A \exp\{-j2\pi f_0 \tau_0\} \exp\{-j2\pi(q-1)\Delta f \tau_0\} \tag{2.3}$$

According to (2.3), the time delay and amplitude of the target can be estimated by the inverse fast Fourier transform (IFFT), resulting in the following complex range profile.

$$s(\tau) = A \exp\{-j2\pi f_0 \tau_0\} \exp\{j\pi(Q-1)\Delta f(\tau-\tau_0)\} \frac{\sin\left[\pi B(\tau-\tau_0)\right]}{Q\sin\left[\pi B(\tau-\tau_0)/Q\right]} \tag{2.4}$$

where τ is the time delay and $B = Q\Delta f$ is the frequency bandwidth. It can be seen from (2.4) that, when $\tau = m/\Delta f + \tau_0$, where m is an integer, $s(\tau)$ can always get its maximum amplitude A. Therefore, the unambiguous range of the SFCW signal, which should be larger than the range of the farthest target, is given by

$$R_{um} = c/2\Delta f \geq R_{max} \tag{2.5}$$

In addition, the time resolution of (2.4) can be approximated by

$$\delta_\tau = 0.88/B \simeq 1/B \tag{2.6}$$

Therefore, the range resolution of the SFCW signal is given by

$$\delta_R = \delta_\tau c/2 = c/2B = c/2Q\Delta f \tag{2.7}$$

Furthermore, when the target has a displacement of d, the range profile can be expressed as

$$s'(\tau) \simeq \exp\left\{-j4\pi\left[f_0 + (Q-1)\Delta f / 2\right]d / c\right\} s(\tau) \tag{2.8}$$

Therefore, the target displacement can be estimated by the interferometry process as follows:

$$\tilde{d} = -\arg\left[s'(\tau)s^*(\tau)\right] \cdot c / 4\pi f_c = -\Delta\phi \cdot \lambda / 4\pi \tag{2.9}$$

where $f_c = f_0 + (Q-1)\Delta f/2$ is the central frequency, $(\cdot)^*$ denotes the complex conjugate operation, $\arg[\cdot]$ denotes the argument of a complex value, and $\Delta\phi = \arg[s'(\tau)s^*(\tau)] = \phi' - \phi$ is the phase difference between the two range profiles. It can be found from (2.9) that the displacement estimation is dependent on the wavelength, which should be considered in the implementation of GB-SAR.

The advantage of SFCW is that by increasing the frequency step by step, it can easily achieve a wide frequency bandwidth with a low hardware requirement. However, since the pulse width T should be long enough to receive the reflection of the farthest target, $T \geq 2R_{max}/c$, the main drawback of the SFCW signal is that its sweep time is too long for some applications [24]. For example, given the maximum target range as 1.5 km and the number of frequencies as 5000, the total time needed to generate an SFCW signal with a 500 MHz bandwidth is 50 ms. Then, assuming a 2-m synthetic aperture length and a 5-mm moving step are used for azimuth compression, the minimum data acquisition period for GB-SAR is 20 s, which will be further increased in practical implementations considering data acquisition and antenna moving.

As shown in Figure 2.3, the FMCW signal in the time domain can be expressed as follows:

$$s_0(t) = \mathrm{rect}\left[\frac{t + T/2}{T}\right]\exp\left[j2\pi(f_c t + \gamma t^2 / 2)\right] \tag{2.10}$$

where f_c is the carrier frequency, $\gamma = B/T$ is the chirp rate, T is the sweep time, and B is the frequency bandwidth. The instantaneous frequency of the FMCW signal is thus given by $f = f_c + \gamma t$, where $-T/2 \leq t \leq T/2$. For a target with time delay $\tau_0 = 2R_0/c$, the reflected signal can be expressed as follows:

$$s(t) = A\,\mathrm{rect}\left[\frac{t - \tau_0 + T/2}{T}\right]\exp\left\{j2\pi\left[f_c(t-\tau_0) + \gamma(t-\tau_0)^2 / 2\right]\right\} \tag{2.11}$$

On the basis of the transmitted signal in (2.10), the complex conjugate of the received signal mixed with the transmitted signal is given by

$$s^*(t) = A\,\mathrm{rect}\left[\frac{t - \tau_0 + T/2}{T}\right]\exp\left\{j2\pi\left[f_c\tau_0 - \gamma\tau_0^2 / 2\right]\right\}\exp\left\{j2\pi\gamma\tau_0 t\right\} \tag{2.12}$$

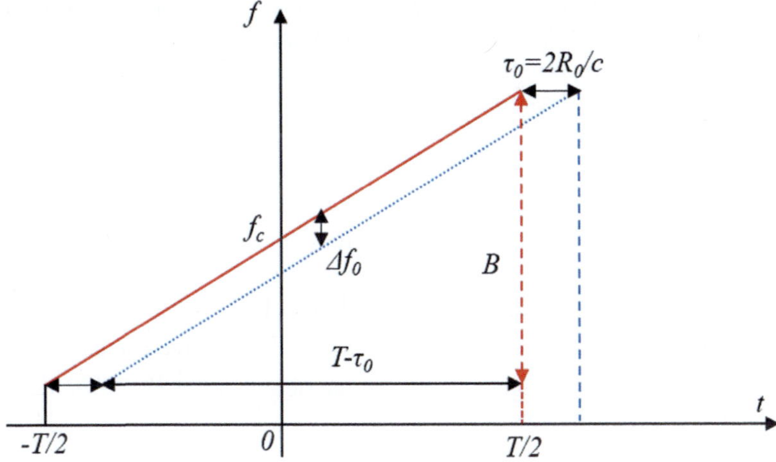

FIGURE 2.3   Frequency modulated continuous waveform.

According to (2.12), the time delay and amplitude of the target can be estimated by FFT, giving the following complex range profile:

$$s(f_r) = AT \exp\left\{ j2\pi \left[f_c\tau_0 - \gamma\tau_0^2/2 \right] \right\} \mathrm{sinc}\left[T(f_r - \Delta f_0) \right]$$

$$\exp\left\{ -j2\pi (f_r - \Delta f_0)\tau_0 \right\} \tag{2.13}$$

where sinc $(x) = \sin(\pi x)/\pi x$, f_r is the frequency in the range direction and $\Delta f_0 = \gamma\tau_0$ is the target range frequency. It can be found from (2.13) that when $f_r = \Delta f_0$, the range profile $s(f_r)$ gets its maximum amplitude AT. Therefore, the target distance can be estimated as follows:

$$\tilde{R}_0 = \tilde{f}_r c / 2\gamma \tag{2.14}$$

In addition, according to the properties of the *sinc* function in (2.13), the range frequency resolution is about $1/T$, and the range resolution of the FMCW signal is given by

$$\delta_R = c/2\gamma T = c/2B \tag{2.15}$$

More strictly, the frequency resolution is determined by the effective sweep time $T_0 = T - \tau_0$, as shown in Figure 2.3. The range resolution of the FMCW signal is thus given by $\delta_R = c/2\gamma T_0 \geq c/2B$, which is degraded with an increase of the target distance. Therefore, the dechirp-on-receive process is always conducted with a reference signal having a sweep time of $T_{ref} \geq T$ and a delay of $\tau_{ref} = 2R_{ref}/c$, where R_{ref} is the reference distance (normally the distance of the center of an imaging scene) [25–26] to increase the effective sweep time and reduce the target range frequency for data sampling rate and memory burden reduction. Furthermore, because the signal needs to be sampled in the digital form, the ambiguity range frequency is determined by the data sampling frequency f_s. Therefore, the unambiguous range of the FMCW signal can be calculated as follows:

$$R_{um} = f_s c / 2\gamma \geq R_{max} \tag{2.16}$$

When the target has a displacement of d, the range profile can be expressed as

$$s'(f_r) \simeq AT \exp\left\{j4\pi\left[f_c d / c\right]\right\} s(f_r) \tag{2.17}$$

where the term of $\gamma d^2/c^2$ is ignored and the interferometry process can be used to estimate the target displacement as follows:

$$\tilde{d} = \arg\left[s'(f_r)s*(f_r)\right]\cdot c / 4\pi f_c = \Delta\phi \cdot \lambda / 4\pi \tag{2.18}$$

Similar to the SFCW signal, the sweep time of the FMCW signal should be sufficient to receive the reflection of the farthest target. Since the total signal generation time is the same as the sweep time, FMCW signal-based GB-SAR can be more efficient [27]. For example, giving the maximum target range as 1.5 km and the data sampling frequency as 2 MHz, a 2.5-ms sweep time can be used to generate an FMCW signal with a 500-MHz bandwidth. For GB-SAR high azimuth resolution imaging with 400 antenna positions, the minimum data acquisition period is only 1 s, which is much smaller than the SFCW signal-based GB-SAR. However, because the SFCW signal can be easily generated, the SFCW signal is used for establishing the signal model and deriving the imaging algorithms in the following.

2.2.2   GEOMETRY AND SIGNAL MODEL

The geometry for GB-SAR imaging is shown in Figure 2.4, where the "stop-go" approximation is used. The transceiver is sled along a rail to generate a synthetic aperture L with a constant step Δx while a set of frequencies are transmitted and the target scattered echoes are received and stored at each transceiver position. For the p-th transceiver position (i.e., the p-th spatial sampling point) at $(x_p, 0)$ with $p = 1, 2, \ldots, P$ and the q-th frequency as $f_q = f_0 + (q-1)\Delta f$ with $q = 1, 2, \ldots, Q$, the demodulated received signal from the target at (x_0, y_0) can be written as

$$S_0(p,q) = \sigma(x_0, y_0)\exp\left\{-j4\pi f_q R_p(x_0, y_0) / c\right\} + N(p,q) \tag{2.19}$$

where $\sigma(x_0, y_0)$ is the target reflection coefficient, $R_p(x_0, y_0) = \sqrt{(x_p - x_0)^2 + y_0^2}$ is the target distance, and $N(p, q)$ denotes the thermal noise.

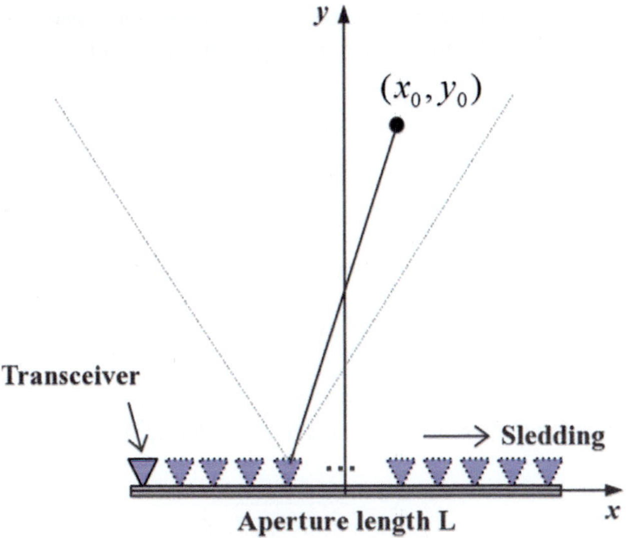

FIGURE 2.4   Basic imaging geometry of GB-SAR.

For the *p*-th spatial sampling point, the range profile can be obtained by IFFT as follows:

$$S_0(p,\tau) = \sigma(x_0, y_0)\exp\{-j4\pi R_p(x_0, y_0)/\lambda\}$$
$$\mathrm{sinc}\left[B\left(\tau - 2R_p(x_0, y_0)/c\right)\right] + N(p,\tau) \qquad (2.20)$$

where τ is the time delay and the phase term $\exp\{j\pi(Q-1)\Delta f\tau\}$ has been compensated. According to (2.20), the phase history of the target, that is, the target phase change against the transceiver position x_p, can be expressed as follows:

$$\varphi(x_p) = -4\pi\sqrt{(x_p - x_0)^2 + y_0^2}\,/\lambda \qquad (2.21)$$

For targets in the far field [28–29], the phase history can be approximated by

$$\varphi(x_p) \simeq -4\pi\left[r_0 - x_p\sin\theta_0\right]/\lambda \qquad (2.22)$$

where $r_0 = \sqrt{x_0^2 + y_0^2}$ and $\sin\theta_0 = x_0/r_0$. Therefore, the spatial frequency in the x direction can be obtained from the derivative of (2.22) with respect to x_p as follows:

$$f_x = \frac{1}{2\pi}\frac{d\varphi(x_p)}{dx_p} = \frac{2}{\lambda}\sin\theta_0 \qquad (2.23)$$

According to the Nyquist theory, the maximum unambiguous spatial frequency f_x^{am}, which is determined by the transceiver moving step Δx, should satisfy the following condition:

$$\frac{f_x^{am}}{2} = \frac{1}{2\Delta x} \geq \frac{2}{\lambda}\left|\sin\theta_0\right| \tag{2.24}$$

Therefore, to unambiguously resolve the target, considering the antennas are non-directional, the transceiver moving step Δx should satisfy

$$\Delta x \leq \frac{\lambda}{4} = \frac{\lambda}{4\left|\sin\theta_0\right|_{\max}} \tag{2.25}$$

When directional antennas are used, the condition in (2.25) can be relaxed to

$$\Delta x \leq \frac{\lambda}{4\left|\sin\left(\theta_{BW}/2\right)\right|} \tag{2.26}$$

where θ_{BW} is the beam width of antenna. In addition, based on (2.23), the following relationship can also be derived:

$$\delta f_x = \frac{\partial f_x}{\partial\theta_0}\delta\theta_0 = \frac{2\cos\theta_0}{\lambda}\delta\theta_0 \tag{2.27}$$

Assuming that the synthetic aperture length is L, the resolution of the spatial frequency f_x is given by $\delta f_x = 1/L$. Hence, the angular resolution of GB-SAR can be calculated by

$$\delta\theta_0 = \frac{\lambda}{2L\cos\theta_0} \tag{2.28}$$

It can be seen that, for linear scanning-based GB-SAR, the angular resolution will be degraded with the increase of azimuth angle, which is one of its drawbacks [30]. In addition, according to the range resolution in (2.7) and the angular resolution in (2.28), the spatial resolution cell of GB-SAR in the x-y coordinate is illustrated as in Figure 2.5, which can be used to discretize the observation scene for GB-SAR imaging.

Based on this analysis, the basic parameters of GB-SAR, such as wavelength λ, bandwidth B, frequency step Δf, pulse width T, antenna moving step Δx, and synthetic aperture length L, can be determined by the system performance requirements, such as the spatial resolution and maximum detection distance and angle. For real applications, all targets in the entire illuminated area are superimposed to give the following received signal:

$$S(p,q) = \iint_\Omega \sigma(x,y)\exp\left\{-j4\pi f_q R_p(x,y)/c\right\}dxdy + N(p,q) \tag{2.29}$$

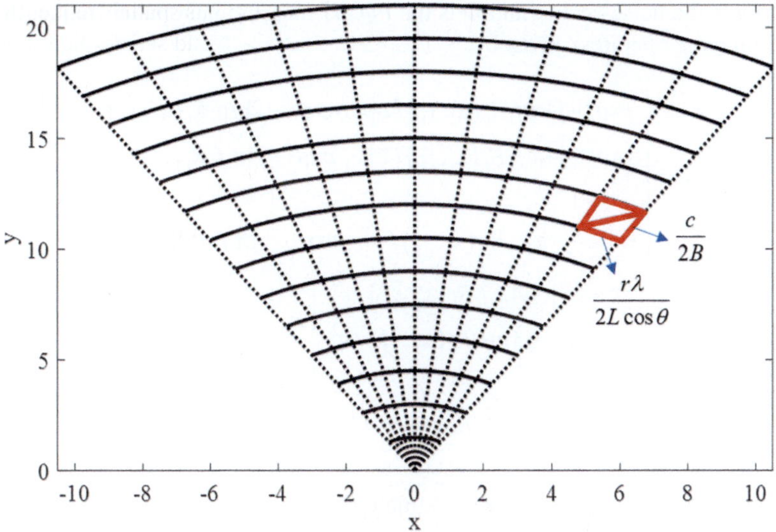

FIGURE 2.5   Sketch of the spatial resolution cell of GB-SAR.

where $R_p(x, y)$ is the distance between the p-th spatial sampling point and the target at (x, y), $\sigma(x, y)$ is the reflection coefficient of the target at (x, y), and Ω denotes the illuminated area.

Conventionally, for imaging, the scene is discretized into I grids in the x direction and into J grids in the y direction. Assuming that the targets are located exactly at these grids, the received signal at the p-th spatial sampling point and the q-th frequency is given by

$$S(p,q) = \sum_{i=1}^{I}\sum_{j=1}^{J}\sigma\left(x_i, y_j\right)\exp\left\{-j4\pi f_q R_p\left(x_i, y_j\right)/c\right\} + N(p,q) \tag{2.30}$$

Let

$$\sigma = \left[\sigma\left(x_1, y_1\right),...,\sigma\left(x_i, y_j\right),...,\sigma\left(x_I, y_J\right)\right] \in C^{IJ \times 1} \tag{2.31}$$

be the vectorized reflection coefficients of targets and

$$s = \left[S(1,1),...,S(1,Q),...,S(P,1),...,S(P,Q)\right] \in C^{PQ \times 1} \tag{2.32}$$

be the vectorized received signal for all frequencies and spatial sampling points; we can obtain the following linear equation:

$$s = \Phi\sigma + n \tag{2.33}$$

where $\boldsymbol{n}$ is the noise vector, and $\boldsymbol{\Phi}$ is the $PQ{\times}IJ$ measurement matrix (a.k.a. the forward imaging operator), given by

$$
\boldsymbol{\Phi} = \begin{bmatrix}
\exp\left\{-\text{j}4\pi f_1 R_1\left(x_1,y_1\right)/c\right\} & \exp\left\{-\text{j}4\pi f_1 R_1\left(x_2,y_1\right)/c\right\} \\
\exp\left\{-\text{j}4\pi f_2 R_1\left(x_1,y_1\right)/c\right\} & \exp\left\{-\text{j}4\pi f_2 R_1\left(x_2,y_1\right)/c\right\} \\
\vdots & \vdots \\
\exp\left\{-\text{j}4\pi f_Q R_P\left(x_1,y_1\right)/c\right\} & \exp\left\{-\text{j}4\pi f_Q R_P\left(x_2,y_1\right)/c\right\}
\end{bmatrix}
$$

$$
\begin{matrix}
\cdots & \exp\left\{-\text{j}4\pi f_1 R_1\left(x_I,y_J\right)/c\right\} \\
\cdots & \exp\left\{-\text{j}4\pi f_2 R_1\left(x_I,y_J\right)/c\right\} \\
\ddots & \vdots \\
\cdots & \exp\left\{-\text{j}4\pi f_Q R_P\left(x_I,y_J\right)/c\right\}
\end{matrix}
\tag{2.34}
$$

The goal of GB-SAR imaging is to estimate $\boldsymbol{\sigma}$ accurately and effectively on the basis of (2.33). In the following section, some conventional GB-SAR imaging algorithms are introduced.

2.3 GB-SAR IMAGING ALGORITHMS

2.3.1 BACK-PROJECTION ALGORITHM

One of the conventional imaging methods for SFCW-based GB-SAR imaging is the frequency-domain back projection (FDBP) method [31], which coherently sums the received signals from different spatial sampling points and frequencies to estimate the reflection coefficient of the target at $(x_i,\,y_j)$ as follows:

$$
\tilde{\sigma}_{FDBP}\left(x_i,y_i\right) = \sum_{p=1}^{P}\sum_{q=1}^{Q} S\left(p,q\right)\exp\left\{+\text{j}4\pi f_q R_p\left(x_i,y_i\right)/c\right\}
\tag{2.35}
$$

Therefore, the vectorized reflection coefficients of all targets can be estimated by

$$
\tilde{\sigma}_{FDBP} = \boldsymbol{\Phi}^{\text{H}}\boldsymbol{s} = \boldsymbol{\Phi}^{\text{H}}\boldsymbol{\Phi}\boldsymbol{\sigma} + \boldsymbol{\Phi}^{\text{H}}\boldsymbol{n}
\tag{2.36}
$$

where $(\cdot)^{\text{H}}$ denotes the conjugate transpose operator.

It can be seen from (2.36) that the FDBP method uses $\boldsymbol{\Phi}^{\text{H}}$ to approximate the inverse of $\boldsymbol{\Phi}$ to solve (2.33). If the columns of $\boldsymbol{\Phi}$ are orthogonal (or near orthogonal) to each other, that is, $\boldsymbol{\Phi}^{\text{H}}\boldsymbol{\Phi}\approx\boldsymbol{I}$, where $\boldsymbol{I}$ is the unit matrix, and the targets are located exactly at the discretized grids, (2.36) can be used to obtain a well-focused GB-SAR image. However, in such a case, to obtain good orthogonality, the spatial grid size of the measurement matrix should approximately be the range and angular resolutions, that is, $c/2B$ and $\lambda/2L$. If the grid size is decreased to more accurately locate the targets, that is, IJ becomes (much) bigger than PQ, the columns of $\boldsymbol{\Phi}$ are no longer orthogonal but correlated, resulting in high-level sidelobes in the GB-SAR image. Therefore, it can be concluded that the FDBP-based imaging method suffers from limited resolution and high-level sidelobes. The frequency bandwidth B and aperture size L of the GB-SAR system are limited in practice. Thus, the spatial resolution of the FDBP method is bounded. Sidelobes of strong targets may make weak targets undetectable, and the sidelobes of multiple targets produce spurious peaks, resulting in negative effects on the subsequent processing, that is, displacement estimation.

Another critical problem that limits the practical applications of the FDBP method is its significantly high computation complexity, which, in terms of complex multiplication number, is $PQIJ$. Correspondingly, the time-domain implementation of the BP method (TDBP) is commonly used to solve this problem. With the received signal vector $s(p)$ at each spatial sampling point, the formulation of TDBP can be written as

$$\tilde{\sigma}_{TDBP}\left(x_i,y_j\right) = \Sigma_{p=1}^{P} s_t\left[p, 2R_p\left(x_i,y_j\right)/c\right] \tag{2.37}$$

where $s_t[p, 2R_p(x_i, y_j)/c]$ is calculated by interpolating the range profile $s_t(p, \tau)$ obtained by the inverse Fourier transform of $s_{zp}(p)$, which is the zero-padded vector of $s(p)$. The computational complexity of TDBP is $P[Q_{zp}\log_2(Q_{zp}) + IJ]$, which is much lower than that of FDBP. Note that zero padding is necessary to compensate for the phase difference of different spatial sampling points, which can normally be conducted as follows:

$$s_{zp}\left(p\right) = \left[s_z^1, s\left(p\right), s_z^2\right]^{T} \in C^{N_{IFFT}\times 1} \tag{2.38}$$

where $s_z^1 = 0 \in C^{Q1\times 1}$ and $s_z^2 = 0 \in C^{Q2\times 1}$ are two all-zero vectors with $Q_1 = [f_1/\Delta f]$, $[\cdot]$ as the nearest integral value, $Q_2 = N_{IFFT} - Q - Q_1$, and N_{IFFT} as the IFFT length that is typically set to a power of 2 to improve efficiency.

If only s_z^2 is used, the formulation of TDBP should be modified to

$$\tilde{\sigma}_{TDBP}\left(x_i,y_j\right) = \sum_{p=1}^{P} s_t'\left(p, 2R_p\left(x_i,y_j\right)/c\right)\exp\left\{+j4\pi f_0 R_p\left(x_i,y_j\right)/c\right\} \tag{2.39}$$

Finally, it should be pointed out that, instead of the zero-padded FFT and interpolation, type-2 non-uniform FFT (NUFFT) can be employed to directly calculate the range compression result at non-uniform delays to increase the accuracy.

2.3.2 Range Migration Algorithm

Another precise algorithm usually used to image a local observation area with fine spatial grids is the range migration algorithm (RMA) (a.k.a. the f-k algorithm), which will be introduced in the following. First, without considering thermal noise, the received signal reflected by the entire observation scene in the wavenumber domain can be written as

$$S\left(p,q\right) = \iint_{\Omega} \sigma\left(x,y\right)\exp\left\{-j2k_q R_p\left(x,y\right)\right\} dx dy \tag{2.40}$$

where $k_q = 2\pi f_q/c$ denotes the q-th wavenumber.

Then, by the Fourier transform of $S(p, q)$ with respect to x_p, we obtain

$$S\left(k_p, k_q\right) = \int_{x_p} S\left(p, q\right) \exp\left\{-jk_p x_p\right\} dx_p = \iint_\Omega \sigma\left(x, y\right) S_1\left(k_p, k_q\right) dxdy \qquad (2.41)$$

where k_p is the spatial wavenumber in the x direction and

$$\begin{aligned} S_1\left(k_p, k_q\right) &= \int \exp\left\{-j2k_q R_p\left(x, y\right)\right\} \exp\left\{-jk_p x_p\right\} dx_p \\ &= \int \exp\left\{-j\left[2k_q\sqrt{\left(x_p - x\right)^2 + y^2} + k_p x_p\right]\right\} dx_p \end{aligned} \qquad (2.42)$$

On the basis of the principle of stationary phase (PSP) [32], $S_1(k_p, k_q)$ can be approximated by

$$S_1\left(k_p, k_q\right) \simeq \exp\left\{-j\left[\sqrt{4k_q^2 - k_p^2 y} + k_p x\right]\right\} \qquad (2.43)$$

By substituting (2.43) into (2.41), we derive

$$S\left(k_x, k_y\right) = \iint_\Omega \sigma\left(x, y\right) \exp\left\{-j\left[k_x x + k_y y\right]\right\} dxdy \qquad (2.44)$$

where $k_x = k_p$ and $k_y = \sqrt{4k_q^2 - k_x^2}$.

It can be found from (2.44) that the target reflection coefficient can be estimated by the 2D Fourier transform of $S(k_x, k_y)$. To use FFT to increase the computing speed [33], the interpolation process is required to get a signal matrix uniformly sampled in the k_x-k_y domain given by

$$\tilde{S}\left(k_x, k_y\right) = S\left(k_x, k_q = \sqrt{k_x^2 + k_y^2}\ /2\right) \Leftarrow S\left(k_x, k_q\right) \qquad (2.45)$$

where the arrow represents interpolation. With $k_q \in [0.5, 1]$ and $k_x \in [-1, 1]$, the interpolation scheme is illustrated in Figure 2.6, where the circles denote the raw data and the dots denote the interpolated data. According to (2.45), the GB-SAR image can be obtained by 2D FFT as

$$\tilde{\sigma}_{RMA}\left(x, y\right) = F_x^H\left[\tilde{S}\left(k_x, k_y\right)\right] F_y^* \qquad (2.46)$$

where $\boldsymbol{F}_x$ and $\boldsymbol{F}_y$ are the Fourier transform matrices in the x and y directions, respectively.

Similar to back-projection algorithm (BPA)–based imaging, Fourier transform–based RMA imaging also suffers from limited resolution and high-level sidelobes in the x and y directions [34]. Advanced spectral estimation algorithms, such as multiple signal classification (MUSIC) and amplitude and phase estimation (APES) algorithms [34–35], can be used to improve the imaging performance.

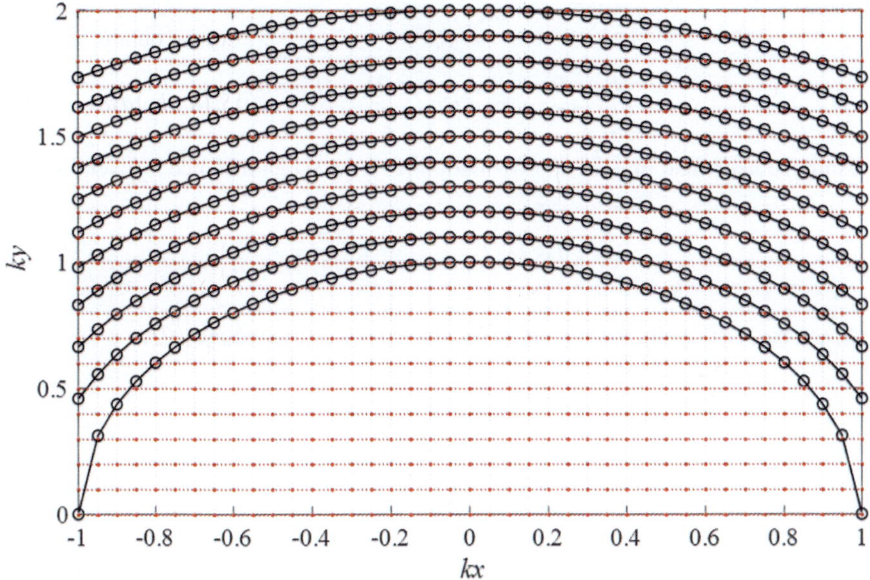

FIGURE 2.6 RMA interpolation scheme for GB-SAR imaging with $k_q \in [0.5,1]$ and $k_x \in [-1,1]$.

The computational complexity of RMA is $QP_{zp}\log_2(P_{zp}) + 3Q_yP_{zp} + Q_yP_{zp}\log_2(Q_yP_{zp})$, where P_{zp} is the number of samples in the x direction, and Q_y is the number of wavenumbers in the y direction that can be much larger than Q as $B_{ky} = 2k_{max} - 2k_{min}\cos(\phi_{max}) > 2B_k$, where $\sin(\phi_{max}) = k_{max}/2k_{min}$. For short-range local area imaging applications, RMA is normally more effective than BPA. However, RMA is not suitable for imaging a large observation scene because zero padding is necessary in the azimuth direction; that is, large P_{zp} and Q_y are needed in such a case, resulting in many samples in the azimuth direction and a time-consuming 2D interpolation in the k_x-k_y domain.

2.3.3 Far-Field Pseudo-Polar Format Algorithm

The previously introduced BPA and RMA can be viewed as general GB-SAR imaging algorithms; that is, they can be used for targets at different positions without limitation. When the target is in the far field, some approximations can be used to reduce the computational complexity. The far-field pseudo-polar format algorithm (FPFA) is one of the commonly used GB-SAR imaging methods for targets in the far field [28] and is introduced in the following. First, instead of the Cartesian coordinate (x, y), the formulation of the FDBP method in the polar coordinate (r, θ) can be expressed as follows:

$$\tilde{\sigma}_{FDBP}(r,\theta) = \sum_{P=1}^{P}\sum_{q=1}^{Q}S(p,q)\exp\left\{+j4\pi f_q R_p(r,\theta)/c\right\} \tag{2.47}$$

where

$$R_p(r,\theta) = \sqrt{\left(x_p - r\sin\theta\right)^2 + (r\cos\theta)^2} \tag{2.48}$$

Using the near-field Fresnel approximation [29], the target distance is given by

$$R_p(r,\theta) \simeq r - x_p \sin\theta + \frac{\cos^2\theta}{2r} x_p^2 \tag{2.49}$$

On the basis of (2.49), the target coefficient can be estimated by

$$\tilde{\sigma}(r,\theta) = \sum_{P=1}^{P}\sum_{q=1}^{Q} S(p,q)\exp\left\{j2\pi\left[f_q 2r/c - x_p 2\sin\theta/\lambda\right]\right\}$$
$$\exp\left\{-j\varphi_1\right\}\exp\left\{-j\varphi_2\right\} \tag{2.50}$$

where

$$\varphi_1 = 4\pi x_p \hat{f}_q \sin\theta/c = 4\pi x_p\left(f_q - f_c\right)\sin\theta/c \tag{2.51}$$

and

$$\varphi_2 = 2\pi x_p^2 f_q \cos^2\theta/rc \tag{2.52}$$

The upper bound of $|\varphi_2|$ is given by

$$|\varphi_2|_{\max} = 2\pi\left|x_p\right|_{\max}^2 \cdot \left|f_q\right|_{\max}/c \cdot \cos^2\theta_{\min}/r$$
$$= \pi L^2 \cos^2\theta_{\min}/2r\lambda_{\min} \leq \pi L^2/2r\lambda_{\min} \tag{2.53}$$

where $\theta_{\min}$ is the minimal target angle, $|x_p|_{\max} = L/2$, $\lambda_{\min} = c/|f_q|_{\max}$ is the smallest wavelength, and $|f_q|_{\max} = f_Q$ is the highest frequency. If the maximum absolute value of ϕ_2 is smaller than $\pi/4$ [36], the last exponential in (2.50) can be ignored. Therefore, it can be derived that

$$\pi L^2/2r\lambda_{\min} \leq \pi/4 \rightarrow r \geq 2L^2/\lambda_{\min} \tag{2.54}$$

which is a commonly used condition for far-field approximation. Similarly, the upper bound of $|\phi_1|$ is given by

$$|\varphi_1|_{\max} = 4\pi\left|x_p\right|_{\max} \cdot \left|\hat{f}_q\right|_{\max}/c \cdot \sin\theta_{\max} = \pi LB\sin\theta_{\max}/c \leq \pi L/2\delta_R \tag{2.55}$$

where θ_{max} is the maximum target angle and $|f_q - f_c|_{max} = B/2$. Considering the defocusing effect on the GB-SAR image, if the following condition can be satisfied [36], the second exponential in (2.50) can also be ignored:

$$\pi L / 2\delta_R \leq \pi / 2 \rightarrow L / \delta_R \leq 1 \qquad (2.56)$$

which implies that the synthetic aperture length should be comparable to the range resolution. For spaceborne or airborne SAR, this condition cannot be satisfied. However, for GB-SAR imaging, this condition can be used to simplify image formation. For example, for a GB-SAR with $L = 1$ synthetic aperture length and $B = 150$ MHz frequency bandwidth, $L/\delta_R = 1$. In addition, as shown in (2.51), since the mean value of x_p is zero and due to the factor of $\sin\theta$, $|\phi_1|$ is always much smaller than that given in (2.55). Therefore, for GB-SAR imaging, the simplified imaging method for a target at the far field can be obtained as follows:

$$\tilde{\sigma}_{FPFA}(\alpha,\beta) = \sum_{p=1}^{P}\sum_{q=1}^{Q} S(p,q)\exp\left\{j2\pi\left[f_q\alpha - x_p\beta\right]\right\} = \boldsymbol{F}_{\alpha}^{H}[S(p,q)]\boldsymbol{F}_{\beta} \qquad (2.57)$$

where $\boldsymbol{F}_{\alpha}$ and $\boldsymbol{F}_{\beta}$ are the Fourier transform matrices with respect to f_q and x_p, respectively. $\alpha = 2r/c$ and $\beta = 2\sin\theta/\lambda$ are two variables in a pseudo-polar coordinate system, and the corresponding grid in the Cartesian coordinate system can be calculated by

$$\begin{cases} x = c\lambda\alpha\beta / 4 \\ y = c\alpha \cos[\arcsin(\lambda\beta / 2)] / 2 \end{cases} \qquad (2.58)$$

The relationship between the pseudo-polar coordinate and the Cartesian coordinate is shown in Figure 2.7, where r is from 75 m to 90 m with a 1.5-m step and $\sin\theta$ is from -0.3 to 0.3 with a 0.03 step. The computational complexity of (2.57) is only $PQ\log_2(PQ)$, which is much lower than that of BPA and RMA for imaging a large observation scene.

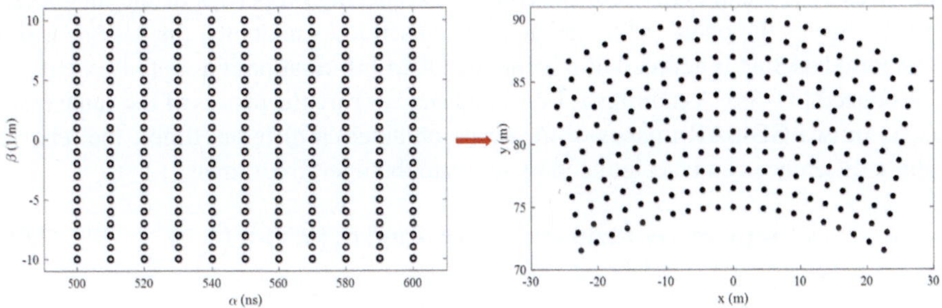

FIGURE 2.7   Uniform pseudopolar grids and the corresponding Cartesian grids.

To obtain a more accurate imaging result, the second exponential in (2.50) is not discarded but can be approximated by its Taylor series expansion as [28]

$$\exp\{-j\varphi_1\} = \sum_{o=1}^{o} \frac{1}{(o-1)!}\left[-j4\pi x_p \hat{f}_q \sin\theta / c\right]^{o-1} \tag{2.59}$$

where O is the expansion order, which should be selected to balance imaging accuracy and computational efficiency. The FPFA can then be reformulated as

$$\tilde{\sigma}_{FPFA}(\alpha,\beta) = \sum_{o=1}^{o} \frac{1}{(o-1)!}\left[-j2\pi\beta / \hat{f}_c\right]^{o-1}\left\{\boldsymbol{F}_\alpha^H\left[S(p,q)\cdot\left(x_p f_q\right)^{o-1}\right]\boldsymbol{F}_\beta\right\} \tag{2.60}$$

It should be noted that the larger the expansion order O, the more accurate the GB-SAR image that can be obtained. The larger the ratio L/δ_R, the larger the expansion order O required to guarantee the convergence of the Taylor series expansion [28]. The computational complexity of (2.60) is $OPQlog_2(PQ) + O(O-1)PQ$. Since parallel processing can be used to generate different image series, the computing cost of (2.60) will not be significantly increased based on 2D FFT than (2.57). In real applications, for far-field target imaging, the formulation in (2.57), that is, the first-order FPFA, can always be used to reduce the image refreshing period. Compared with BPA and RMA, the advantage of FPFA is its reduced computational time for large-area imaging, whereas the cost is the fact that FPFA can only be used for targets in the far field. In addition, because the Fourier transform is used, the spatial resolution of FPFA is limited and high-level sidelobes will be generated. In fact, because the signal model of FPFA imaging is range-azimuth decoupled, the 2D high-resolution spectral estimation algorithms can be employed to obtain better imaging performance [34].

2.4 INTERFEROMETRY

2.4.1 GB-SAR Interferometry

On the basis of the aforementioned imaging algorithms, a complex (amplitude and phase) GB-SAR image of the observation scene can be obtained. The amplitude of the SAR image can be used to interpret the scattering properties of the monitored area, whereas the phase of the image can be used to estimate the displacement if a temporal baseline is adopted or to generate DEM information if a spatial baseline is introduced [37–38]. Assuming $\varphi_1(x,y)$ and $\varphi_2(x,y)$ are the phases of the same pixel (x, y) in two GB-SAR images of the scene obtained at different times, the relative phase difference can be calculated to generate the interferogram as

$$\psi(x,y) = \varphi_2(x,y) - \varphi_1(x,y) = -\arg\left[\sigma_2(x,y)\sigma_1^*(x,y)\right] \tag{2.61}$$

where $\sigma(x,y)$ is the estimated complex amplitude of the target. Because the calculated phase in (2.61) is always wrapped into $[-\pi, \pi]$, phase ambiguity occurs. The absolute phase difference (i.e., the unwrapped phase difference) is given by

$$\phi(x,y) = \psi(x,y) + 2\pi k \tag{2.62}$$

where k is an unknown integral value. In practice, the absolute interferometric phase can be expressed as the sum of the following four terms [19, 39]

$$\phi(x, y) = \phi_{defo} + \phi_{atm} + \phi_{geom} + \phi_{noise} \tag{2.63}$$

where ϕ_{defo} and ϕ_{atm} are the phase changes caused by the displacement and the atmospheric effects; ϕ_{geom} denotes the geometric phase difference between two acquisitions, which is assumed to be zero in this book; and ϕ_{noise} is the noise phase component generated by the changes of target scattering properties among different data acquisitions and by other noise sources, such as signal processing noise and thermal noise. It should be mentioned that in (2.63), the topographic phase difference is not considered because no spatial baseline is considered in this book. Based on (2.63), the displacement can be estimated according to the following relationship:

$$\frac{4\pi d}{\lambda} = \phi_{defo} = \psi(x, y) + 2\pi k - \phi_{atm} - \phi_{geom} - \phi_{noise} \tag{2.64}$$

where d is the displacement during the interval between the two data acquisitions.

Assuming no phase unwrapping is conducted and the phase changes caused by other effects are ignored, the maximum unambiguous displacement that can be estimated by (2.64) is bounded by a quarter of the wavelength as follows:

$$d_{max} = \frac{\lambda |\psi(x, y)|_{max}}{4\pi} = \frac{\lambda}{4} \tag{2.65}$$

To more accurately estimate the displacement (probably larger than a quarter of the wavelength), additional processing is required to compensate for the displacement-unrelated phase components in (2.64). The main processing stages, such as pixel selection, interferogram filtering, phase unwrapping, and atmospheric phase compensation, are introduced.

2.4.2 Interferogram Forming and Pixel Selection

2.4.2.1 Image Co-Registration

In the previous derivation of displacement estimation, the same pixel (x, y) in two different GB-SAR images is assumed to represent the same target. In practice, because of the position difference of the GB-SAR system between two different acquisitions, this assumption may not be fulfilled, especially in the case where the discontinuous model is adopted [40]. Therefore, SAR image co-registration is required in some specific applications to increase image coherence and reduce phase error [41]. In addition, when two or more GB-SAR systems are used simultaneously to estimate the 2D/3D target displacement vector, high-accuracy SAR image co-registration may be mandatory. For most measurements in this book, GB-SAR or MIMO radar with continuous working mode is used for short-range or middle-range target imaging during the short observation time. Therefore, the image co-registration stage can be avoided.

2.4.2.2 Interferogram Forming

For the following processing stages of displacement estimation, interferograms and coherence images are the main inputs. After obtaining a series of co-registered GB-SAR images, the interferograms are generated by

$$\psi\left(x,y\right) = -\arg\left[\sigma_{s}\left(x,y\right)\sigma_{r}^{*}\left(x,y\right)\right] \tag{2.66}$$

where $\sigma_r(x,\ y)$ and $\sigma_s(x,\ y)$ denote the reference and secondary GB-SAR images, respectively.

Furthermore, for O GB-SAR images, according to the interferogram network, which determines how many times a specific GB-SAR image is acted as the reference or secondary image, there are different ways to generate the interferograms and coherence images. Figure 2.8 describes three types of image networks defined based on connectivity and the number of links.

The network in Figure 2.8 (a) is the so-called *daisy chain,* which connects consecutive chronological N images to form the shortest temporal baseline [42–44]. In this network, the reference and secondary images are varied in each interferogram. This network can also minimize the temporal phase wrapping issue and temporal decorrelation owing to its shortest temporal baseline. The simple stacking approach of these interferograms yields mean velocity results [42].

The network in Figure 2.8 (b) is a *single reference network* that forms interferograms with a single reference image. Because all interferograms share a single reference image, the atmospheric phase of the reference signal is also shared in all interferograms. Therefore, simple averaging of interferograms allows estimation of the atmospheric phase of the reference image [20, 45].

Both networks generate $N{-}1$ interferograms, which is the minimum number required to guarantee connectivity [46]. On the other hand, the network shown in Figure 2.8 (c) provides a complete connection where each is connected with all others, which is called a redundant network. This network generates the maximum number of interferograms $n(n{-}1)/2$ for reliable estimation but wastes processing time for redundant pairs.

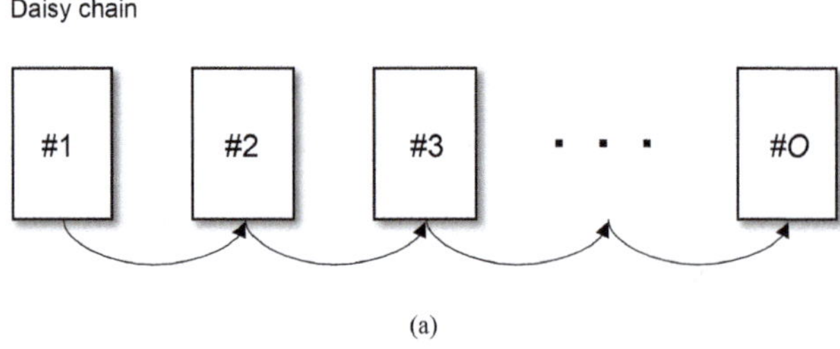

FIGURE 2.8 Three GB-SAR image network types with varying connectivity and number of links. (a) Daisy chain connecting consecutive chronological N images to be the shortest temporal baseline. (b) Single reference network forming interferograms with the single reference image. (c) The redundant network that forms the redundant pair of interferograms to be $M > N{-}1$.

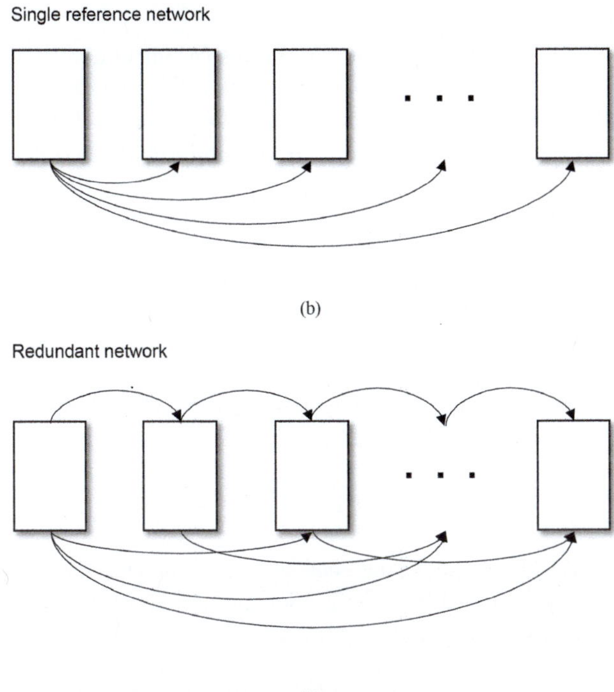

FIGURE 2.8 (Continued)

2.4.2.3 Coherence Image Generation

Coherence is an important parameter that is commonly used to evaluate the quality of an interferogram. Coherence images [11] can be generated by

$$\left|\gamma\left(x,y\right)\right|=\left|\frac{E\left\{\sigma_{s}\left(x,y\right)\sigma_{r}^{*}\left(x,y\right)\right\}}{\sqrt{E\left\{\left|\sigma_{s}\left(x,y\right)\right|^{2}\right\}E\left\{\left|\sigma_{r}\left(x,y\right)\right|^{2}\right\}}}\right| \tag{2.67}$$

where $E\{\cdot\}$ denotes the expectation operation. According to the definition in (2.67), the range of coherence is from 0 to 1. When $\gamma = 1$, perfect coherence is obtained, whereas coherence is totally lost when $\gamma = 0$ [42]. GB-SAR can only obtain an accurate displacement estimation for the pixel with a high coherence value (i.e., with a low phase noise ϕ_{noise}). In practice, the expectation operation in (2.67) can be conducted by the 2D moving average process [43], resulting in

$$\left|\tilde{\gamma}\left(x,y\right)\right|=\left|\frac{\sum_{m=1}^{M}\sum_{n=1}^{N}\sigma_{s}\left(x_{m},y_{n}\right)\sigma_{r}^{*}\left(x_{m},y_{n}\right)}{\sqrt{\sum_{m=1}^{M}\sum_{n=1}^{N}\left|\sigma_{s}\left(x_{m},y_{n}\right)\right|^{2}\sum_{m=1}^{M}\sum_{n=1}^{N}\left|\sigma_{r}\left(x_{m},y_{n}\right)\right|^{2}}}\right| \tag{2.68}$$

where $M \times N$ is the size of the moving window.

2.4.2.4 Pixel Selection

As shown in (2.63), phase noise is one of the displacement-unrelated components of the absolute interferometric phase. If a pixel has a low phase noise ϕ_{noise}, it can be exploited for the following displacement measurement process. The goal of this step is to identify all pixels with low phase noise ϕ_{noise}, referred to as coherent scatterers (CS). Normally, two methods can be used for CS selection: the coherence-based method, which is more suitable for CS selection from a single interferogram, and the amplitude-based method, which is more suitable for pixel selection from multiple interferograms. For these two methods, a user-defined threshold should be provided to balance the phase quality and spatial density of the selected pixels.

2.4.2.4.1 Coherence-Based Method

The generation method of the coherence image was introduced previously, as shown in (2.68). The coherence is influenced by various factors [47] and is given by

$$\gamma(x, y) = \gamma^{thermal}(x, y)\gamma^{spatial}(x, y)\gamma^{temporal}(x, y) \tag{2.69}$$

where $\gamma^{thermal}(x, y)$ is the thermal coherence, $\gamma^{spatial}(x, y)$ denotes the spatial coherence and equals 1 if a zero spatial baseline is adopted, and $\gamma^{temporal}(x, y)$ is the temporal coherence. With an increase in the temporal baseline, decorrelation occurs, which will reduce the temporal coherence. In such a case, it is difficult to achieve accurate displacement estimation results for the pixels selected by the coherence-based method. Therefore, it is usually used for pixel selection from multiple interferograms with a short temporal baseline, or more strictly, from a single interferogram.

With the series of SAR images, temporal mean absolute coherence can be used as a criterion for CS selection. Because the coherence is derived by multi-looking operations, the image resolution is usually degraded. Nonetheless, this operation is suitable for natural distributed areas without a dominant scatterer inside the resolution cell. Note that the coherence might be overestimated when the averaging window size is small, that is, the number of averaging samples is not sufficient [48, 49]. Hence, a trade-off between image resolution and coherence estimation accuracy must always be considered. Note also that the fixed window shape, known as the *boxcar filter,* often leads to bias estimation over the heterogeneous area because of speckle, mixing of different scatterers (i.e., different distribution), and phase disturbances [48].

Figure 2.9 demonstrates the temporal variations (each 6-hour) of the interferometric coherence measured on November 6, 2018, over the post-landslide mountainous area of Minami–Aso, Kumamoto, Japan. The presented interferometric coherence images were estimated using a 5 × 3 boxcar filter between GB-SAR images with a 15-minute time interval. The corresponding field photo taken on October 19, 2018, is presented in Figure 2.10. From the results, the bare surface that is formed by landslides shows higher coherence, whereas the areas with vegetation and trees result in lower coherence. In fact, operators who conducted the earthwork routinely used the excavators as part of slope restoration on that day, where a white dotted circle indicates its location in Figures 2.9(c) and 2.10. They constructed the paths by filling the soil during the daytime (from 8:00 to 17:00). This earthwork causes a change in land surface morphology, apparently producing rapid phase fluctuation

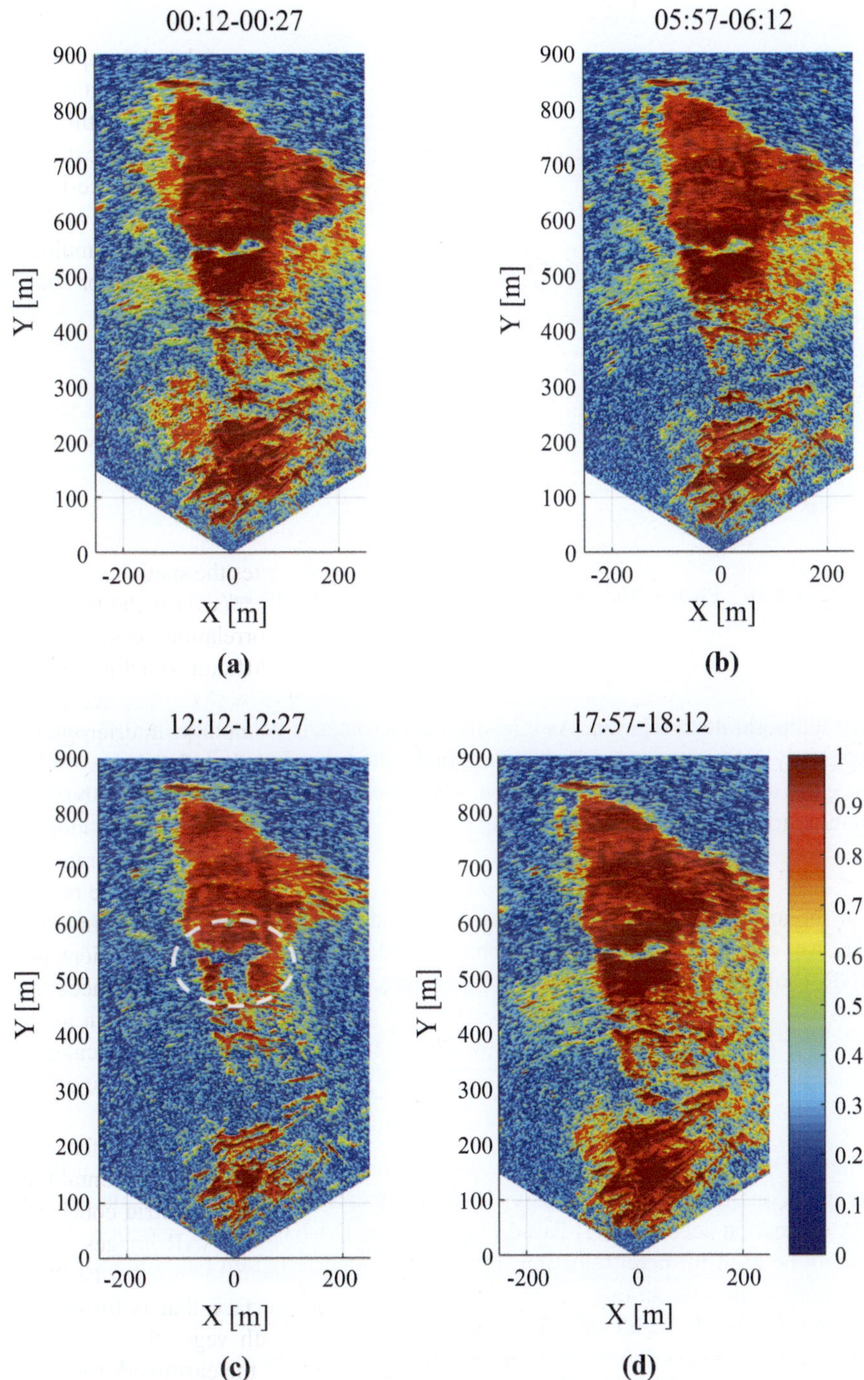

FIGURE 2.9 Interferometric coherence images measured on November 6, 2018, over the post-landslide mountainous area of Minami–Aso, Kumamoto, Japan. Each image is formed by GB-SAR images acquired at (a) 00:12 and 00:27, (b) 05:57 and 06:12, (c) 12:12 and 12:27, and (d) 17:57 and 18:12, with 15-minute time intervals.

FIGURE 2.10 Photo of the observed slope taken on October 19. Earthwork location is indicated by a white dotted circle over the observed slope.

and temporal decorrelation. As a result, interferometric coherence on that part significantly decreases, and only the result in the daytime (Figure 2.9(c)) reveals coherence lost around the indicated location. Therefore, CSs should be updated in a timely manner to avoid decorrelated signals.

2.4.2.4.2 Amplitude-Based Method

For the amplitude-based pixel selection method, the dispersion of amplitude (DA) [11, 20], which is equivalent to the phase standard deviation of a pixel with a high SNR, is always used. Given O GB-SAR images, DA is defined as

$$DA(x,y) = \frac{SD\left[\left|\sigma_1(x,y)\right|,\left|\sigma_2(x,y)\right|,...,\left|\sigma_o(x,y)\right|\right]}{AM\left[\left|\sigma_1(x,y)\right|,\left|\sigma_2(x,y)\right|,...,\left|\sigma_o(x,y)\right|\right]} \tag{2.70}$$

where $SD[\cdot]$ denotes the standard deviation operation and $AM[\cdot]$ denotes the arithmetic mean operation. Because many GB-SAR images are required to achieve a high DA estimation accuracy [11] based on (2.70), the DA-based pixel selection method cannot be used for pixel selection from a single interferogram but is more suitable for multiple interferograms.

This amplitude-based selection enables the full-resolution approach. Therefore, this approach is suited to finding a point-wise scatterer over an urban area that shows a relatively high SNR. For low-SNR terrain, DA generally becomes higher than that in the urban area.

Furthermore, it is worth noting that directly using the amplitude information for pixel selection, that is, selecting pixels with high amplitudes from one GB-SAR

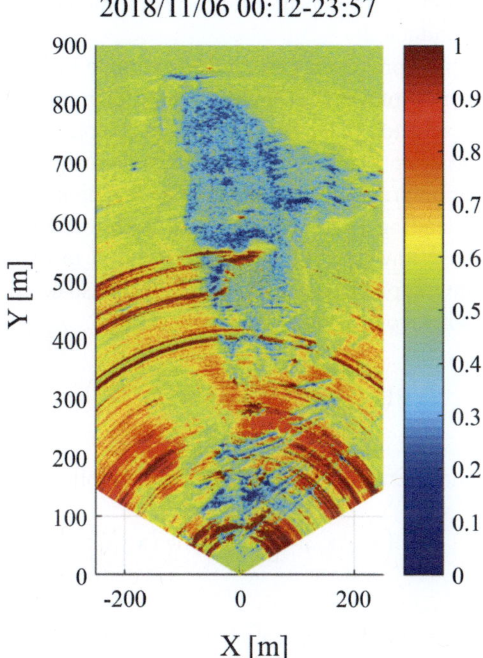

FIGURE 2.11 DA derived from 96 GB-SAR images measured on November 6, 2018, in the post-landslide mountainous area of Minami–Aso, Kumamoto, Japan.

image or with high averaged amplitudes from multiple GB-SAR images, is also a reasonable approach in some applications.

Figure 2.11 presents DA derived by 96 GB-SAR images measured from 00:12 to 23:57 November 6, 2018, at the post-landslide mountainous area, Minami–Aso, Kumamoto, Japan. Low DA values appear over the bare surface zone corresponding to areas with high interferometric coherence, as shown in Figure 2.9. In addition, the middle part of the slope around 500 m of slant range shows relatively higher DA than other bare surface parts due to the previously mentioned earthwork.

2.4.3 Noise Filtering and Phase Unwrapping

After selecting pixels from the GB-SAR interferogram, the phase unwrapping process is applied to achieve the real displacement estimation. However, although the selected pixels are characterized by low phase noise, the presence of phase noise still influences the phase unwrapping process. Therefore, before phase unwrapping and other processing stages, phase noise suppression or filtering should be executed to improve the quality of the interferogram. Many phase noise filtering methods have been proposed and validated in the past decades [50], such as multi-look filters, box-car filters, pivoting mean/median filters, and Goldstein filters. A key consideration for GB-SAR interferogram filtering is to balance noise suppression performance and phase detail preservation performance. After filtering the phase, the phase unwrapping process can be conducted.

The phase unwrapping process is required to estimate the absolute interferometric phase from the interval of $[-\pi, \pi]$; that is, the unknown integral k in (2.62) should be estimated for each selected pixel. To find the unique phase unwrapping solution, the phase continuity assumption, known as the *Itoh condition,* is used [51]. This condition ensures that the unwrapped phase difference (gradient) between two adjacent pixels is less than π, leading to no ambiguity in the unwrapped phase difference. Assuming the 1D phase unwrapping problem described in Figure 2.12(a) with the Itoh condition, the unwrapped phase at $\mathbf{x}_n$ can be reconstructed by integrating the wrapped phase gradients from the reference point $\mathbf{x}_0$, as

$$\phi\left(\mathbf{x}_n\right) = \phi\left(\mathbf{x}_0\right) + \sum_{i=1}^{n}\Delta\psi\left(\mathbf{x}_{i-1},\mathbf{x}_i\right), \tag{2.71}$$

where the wrapped phase gradient can be computed as follows:

$$\Delta\psi\left(\mathbf{x}_{i-1},\mathbf{x}_i\right) = \left[\psi\left(\mathbf{x}_i\right) - \psi\left(\mathbf{x}_{i-1}\right)\right]_{-\pi,\pi} \tag{2.72}$$

where $[\cdot]_{-\pi,\pi}$ indicates the modulo-operator, that is, angle $\left(e^{i\left(\psi(\mathbf{x}_i)-\psi(\mathbf{x}_{i-1})\right)}\right)$. This solution can be easily extended to the 2D case, as shown in Figure 2.12(b), where the solution is path independent when the Itoh condition is held [52]. In this case, the sum of the phase gradient along the loop inside the triangle shown in Figure 2.12(b) is zero, which is an irrotational field. However, phase noise, abrupt changes in atmospheric phase, and deformation may lead to inconsistencies in the Itoh condition and irrotational field. When the loop integration is a non-zero value, that is, $\pm 2\pi$, it is called *a residue,* indicating inconsistencies [53]. The existence of residue makes the phase unwrapping problem difficult. This is the main motivation for various phases unwrapping algorithms in the literature for reliable 2D or 3D unwrapping.

The phase unwrapping algorithms developed so far can be categorized into two groups: (1) path-following and (2) path-independent algorithms [54].

The former algorithm aims to find a reliable integration path. Goldstein's branch-cut algorithm connects nearby plus and minus residues with cuts such that the total length of all branch cuts is minimized [53]. In this way, the unwrapping integration path that passes through any residues can be prevented. Path-following algorithms based on the quality map were also often employed, the so-called *quality-guided method* [55, 56]. This method attempts to guide the integration path through higher-quality pixels. The quality can be evaluated using many types of observation values, such as interferometric coherence and phase derivative variance [52].

On the other hand, the latter algorithm does not depend on the integration path [57]. The path-independent methods find a global solution that minimizes the difference between the wrapped gradient and the unwrapped gradient, as

$$\arg\min\sum_{i=1}^{n}\left|\Delta\phi\left(\mathbf{x}_{i-1},\mathbf{x}_i\right) - \Delta\psi\left(\mathbf{x}_{i-1},\mathbf{x}_i\right)\right|^{p}, \tag{2.73}$$

where p is the degree of the norm. The *minimum cost flow* (MCF) phase unwrapping algorithm developed by Costantini [58–60] deals with the L_1 norm minimization problem, as demonstrated subsequently.

In the first step of MCF, Delaunay triangulation is applied to CSs to generate a non-overlapped spatial triangle network consisting of arcs and nodes [60], as shown in Figure 2.12(b). The MCF assumes that the wrapped and unwrapped gradients differ by integer multiples of 2π:

$$2\pi k_{\Delta_{i-1,i}} = \Delta\phi\left(\mathbf{x}_{i-1}, \mathbf{x}_i\right) - \Delta\psi\left(\mathbf{x}_{i-1}, \mathbf{x}_i\right), \tag{2.74}$$

where $k_{\Delta_{i-1,i}}$ is the integer defined on the arc between $\mathbf{x}_i$ and $\mathbf{x}_{i-1}$. The MCF performs the minimization of the weighted L_1 norm, given as

$$\arg\min \sum_{i=1}^{n} P_{\Delta i, i-1} \left| k_{\Delta i, i-1} \right|, \tag{2.75}$$

where $P_{\Delta_{i-1,i}}$ indicates the weights. To be a path-independent solution, the MCF sets the constraint on the loop sum of the unwrapped phase. For the top triangle example in Figure 2.12(b), the loop sum becomes

$$k_{\Delta 0,1} + k_{\Delta 2,1} + k_{\Delta 0,2} = \frac{1}{2\pi}\left(\Delta\psi\left(\mathbf{x}_0, \mathbf{x}_1\right) - \Delta\psi\left(\mathbf{x}_2, \mathbf{x}_1\right) - \Delta\psi\left(\mathbf{x}_0, \mathbf{x}_2\right)\right) \tag{2.76}$$

Therefore, the estimation of ambiguity in all arcs by the global solution of (2.75) leads to path-independent phase unwrapping.

The original MCF is a 2D phase unwrapping algorithm. This algorithm is further extended to the unwrapping problem for a multi-temporal DInSAR data set, which performs MCF in both temporal and spatial domains [61].

For GB-SAR interferograms, the selected CSs may be discretely and non-uniformly distributed; thus, the MCF-based or least squares-based unwrapping method is commonly used [60]. In addition, in real applications, some artificial targets can be installed to increase the phase unwrapping accuracy. Furthermore, for GB-SAR continuous displacement measurement, 1D phase unwrapping along the time direction can also be conducted [62]. Hence, 3D phase unwrapping (range, azimuth, and time) can provide more accurate phase unwrapping results.

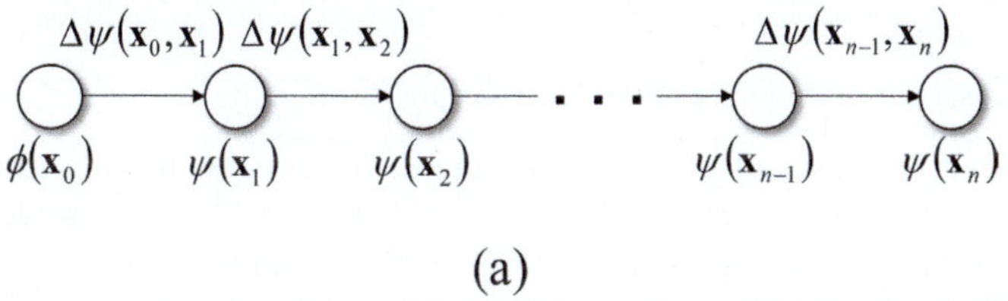

(a)

FIGURE 2.12 (a) Illustration of conditions for 1D phase unwrapping and (b) 2D phase unwrapping.

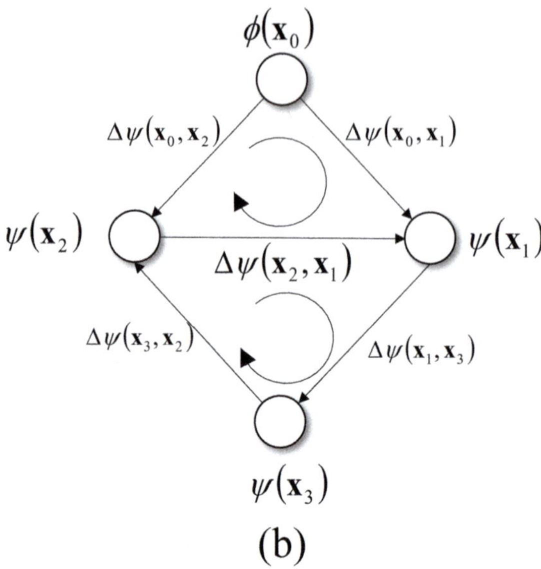

FIGURE 2.12   (Continued)

2.4.4 ATMOSPHERIC PHASE COMPENSATION

The propagation of electromagnetic signals through the atmosphere is influenced by changes in atmospheric properties, such as humidity, temperature, and pressure. Therefore, in the absolute interferometric phase, the atmospheric phase component, which is often called "atmospheric phase screen" (APS), may be the largest phase error source. In this processing stage, the APS should be estimated and compensated, which is critical to the accuracy of the final displacement estimation. In the last decades, different APS compensation methods have been proposed [47, 63–67], such as meteorological observation-based, spatial interpolation-based, and stable target-based methods. The atmospheric phase is normally assumed to be 1D (range) or 2D (range-azimuth/range-height) smoothly varied. Then, an approximated mathematical model, such as the linear model in [63] or the multiple regression model in [67], is adopted to estimate the atmospheric phase and then subtract it from the unwrapped phase obtained in the former stage. Detail of the APS and its compensation will be described in Chapter 4.

2.4.5 DISPLACEMENT ESTIMATION AND GEOCODING

In fact, after the previously mentioned stages, the processing procedure can be stopped with the calculation of displacement based on (2.64). However, because only the displacement along the radar LOS direction can be measured by GB-SAR and the GB-SAR images are obtained by projecting the 3D observation scene onto a 2D imaging plane, geocoding is an additional processing step that needs to be performed, which is important for result interpretation. For example, it can be used for accurate positioning of an area with a large displacement. In general, geocoding involves a transformation from the 2D image space to the 3D object space (for instance, to a local Cartesian coordinate) [39, 47]. Therefore, a 3D reference system (such as a DEM of the observation scene) is required, which can be obtained in advance by, for example, the laser scanning technique.

2.5 POLARIMETRY

2.5.1 SCATTERING MATRIX

Radar polarimetry [12–14] is a powerful tool to analyze the scattering mechanism of radar targets and is wieldy used for target classification and characterization. The scattering matrix is commonly used for this purpose. The uniform plane that has the wave vector $\hat{k}$ parallel to the z-axis, which means that the plane wave is propagating along the z-axis, as shown in Figure 2.13, has the electric field E included in the x-y plane and is given by:

$$E = \left(\hat{x}E_x + \hat{y}E_y \right) e^{-jkz} \tag{2.77}$$

where $k = \omega\sqrt{\varepsilon_0\mu_0}$ ($k = |\hat{k}|$) is the wave number, and E_x and E_y are the x and y components of the electric field, which contains the phase term. Now, we define the incident wave to a target as follows:

$$E^t = \begin{bmatrix} E^t_x \\ E^t_y \end{bmatrix} \tag{2.78}$$

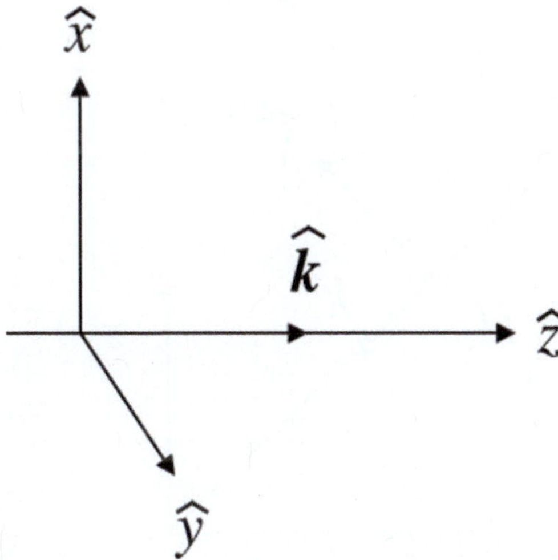

FIGURE 2.13   The Cartesian coordinate and the wave vector.

and then express the scattered wave from the object as follows:

$$\mathbf{E}^s = \begin{bmatrix} E_x^s \\ E_y^s \end{bmatrix} \frac{e^{-jkR}}{R} = \begin{bmatrix} S_{xx} & S_{xy} \\ S_{yx} & S_{yy} \end{bmatrix} \begin{bmatrix} E_x^t \\ E_y^t \end{bmatrix} \frac{e^{-jkR}}{R} = \mathbf{S}\mathbf{E}^t \frac{e^{-jkR}}{R} \tag{2.79}$$

where R is the distance from the target to the observation point and S is the scattering matrix given by

$$\mathbf{S} \equiv \begin{bmatrix} S_{xx} & S_{xy} \\ S_{yx} & S_{yy} \end{bmatrix} \tag{2.80}$$

Note that in (2.79), the amplitude and phase shift due to propagation are excluded by e^{-jkR}/R, and the scattering matrix contains only the amplitude and phase change caused by scattering from the object. In radar polarimetry, only the relative amplitude ratio and phase difference between the elements of the scattering matrix contain polarimetric information.

2.5.2 Coordinate System

In microwave remote sensing, the relative location of the radar system to the ground surface is important. If we define the wave propagation direction $\hat{k}$ on the horizontal infinite plane, which is the ground surface, the incident plane can be uniquely defined as a plane that includes the wave vector and the zenith vector to the ground surface. The horizontal vector $\hat{H}$ is defined as that perpendicular to the incident plane, and the vertical vector $\hat{V}$ is defined by

$$\hat{V} \times \hat{H} = \hat{k} \tag{2.81}$$

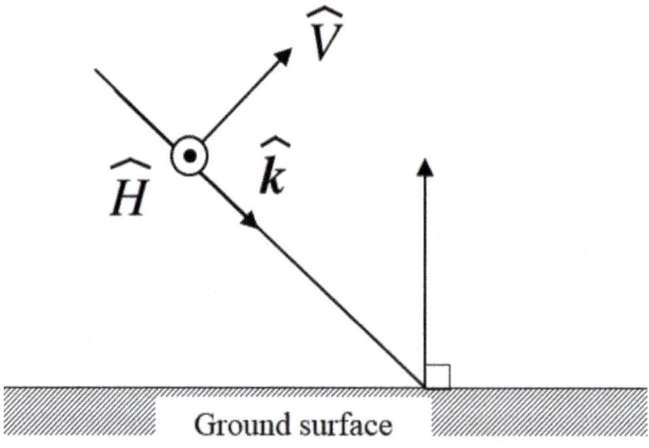

FIGURE 2.14   Incident plane and definition of $\hat{V}$ and $\hat{H}$ vector.

as shown in Figure 2.14. Using this rectangular coordinate, the scattering matrix S is given as follows:

$$S \equiv \begin{bmatrix} S_{VV} & S_{VH} \\ S_{HV} & S_{HH} \end{bmatrix} \tag{2.82}$$

The definition in (2.81) is commonly used in radar polarimetry.

REFERENCES

[1] M. Cheney and B. Borden, Fundamentals of radar imaging, Philadelphia, PA, USA: SIAM Press, 2009.

[2] D. Wehner, High resolution radar, Norwood, MA, USA: Artech House, 1987.

[3] A. J. Fenn, D. H. Temme, W. P. Delaney, et al., "The development of phased-array radar technology," Lincoln Lab. J., vol. 12, no. 2, pp. 321–340, 2000.

[4] W. D. Wirth, Radar techniques using array antennas, London, UK: IEE, 2001.

[5] B. D. Steinberg and H. M. Subbaram. Microwave imaging techniques, New York, NY, USA: Wiley, 1991.

[6] M. Soumekh, Synthetic aperture radar signal processing with MATLAB algorithms, New York, NY, USA: Wiley, 1999.

[7] C. Özdemir, Inverse synthetic aperture radar imaging with MATLAB algorithms, New York, NY, USA: Wiley, 2012.

[8] V. C. Chen and M. Martorella, Inverse synthetic aperture radar imaging: Principles, algorithms and applications, Raleigh, NC, USA: SciTech, 2014.

[9] A. Moreira, P. Prats-Iraola, M. Younis, et al., "A tutorial on synthetic aperture radar," IEEE Geosci. Remote Sensing Mag., vol. 1, no. 1, pp. 6–43, Mar. 2013.

[10] P. A. Rosen, S. Hensley, I. R. Joughin, et al., "Synthetic aperture radar interferometry," Proc. IEEE, vol. 88, no. 3, pp. 333–382, Mar. 2000.

[11] R. Bamler and P. Harti, "Synthetic aperture radar interferometry," Inverse Probl., vol. 14, no. 4, pp. R1–R54, Aug. 1998.

[12] J. Lee and E. Pottier, Polarimetric imaging: From basics to applications, Boca Raton, FL, USA: CRC Press, 2009.

[13] S. R. Cloude and K. P. Papathanassiou, "Polarimetric SAR interferometry," IEEE Trans. Geosci. Remote Sens., vol. 36, no. 5, pp. 1551–1565, Sep. 1998.

[14] L. Pipia, X. Fabregas, A. Aguasca, et al., "Polarimetric differential SAR interferometry: First results with ground-based measurements," IEEE Geosci. Remote Sens. Lett., vol. 6, no. 1, pp. 167–171, Dec. 2008.

[15] C. Elachi, T. Bicknell, R. L. Jordan, et al., "Spaceborne synthetic-aperture imaging radars: Applications techniques and technology," Proc. IEEE, vol. 70, no. 10, pp. 1174–1208, Oct. 1982.

[16] G. Fornaro, "Trajectory deviations in airborne SAR: Analysis and compensation," IEEE Trans. Aerosp. Electron. Syst., vol. 35, no. 3, pp. 997–1009, Jul. 1999.

[17] W. Wang, Q. Peng, and J. Cai, "Waveform-diversity-based millimeter-wave UAV SAR remote sensing," IEEE Trans. Geosci. Remote Sens., vol. 47, no. 3, pp. 691–700, Mar. 2009.

[18] O. Frey, C. L. Werner, U. Wegmuller, et al., "A car-borne SAR and InSAR experiment," Proc. IEEE Int. Geosci. Remote Sens. Symp., pp. 93–96, Jul. 2013.

[19] O. Monserrat, M. Crosetto, and G. Luzi, "A review of ground-based SAR interferometry for deformation measurement," ISPRS J. Photogramm. Remote Sens., vol. 93, pp. 40–48, Jul. 2014.

[20] A. Ferretti, C. Prati, and F. Rocca, "Permanent scatterers in SAR interferometry," IEEE Trans. Geosci. Remote Sens., vol. 39, no. 1, pp. 8–20, Jan. 2001.

[21] A. Ferretti, A. Fumagalli, F. Novali, et al., "A new algorithm for processing interferometric data-stacks: SqueeSAR," IEEE Trans. Geosci. Remote Sens., vol. 49, no. 9, pp. 3460–3470, May 2011.

[22] D. Leva, G. Nico, D. Tarchi, et al., "Temporal analysis of a landslide by means of a ground-based SAR interferometer," IEEE Trans. Geosci. Remote Sens., vol. 41, no. 4, pp. 745–752, Jun. 2003.

[23] R. Iglesias, A. Aguasca, X. Fabregas, et al., "Ground-based polarimetric SAR interferometry for the monitoring of terrain displacement phenomena-Part I: Theoretical description," IEEE J. Sel. Top. Appl. Earth Obs., vol. 8, no. 3, pp. 994–1007, Mar. 2015.

[24] Y. Wang, W. Hong, Y. Zhang, et al., "Ground-based differential interferometry SAR: A review," IEEE Geosci. Remote Sens. Mag., vol. 8, no. 1, pp. 43–70, Mar. 2020.

[25] A. Meta, P. Hoogeboom, and L. Ligthart, "Signal processing for FMCW SAR," IEEE Trans. Geosci. Remote Sens., vol. 45, no. 11, pp. 3519–3532, Nov. 2007.

[26] R. Wang, O. Loffeld, H. Nies, et al., "Focus FMCW SAR data using the wavenumber domain algorithm," IEEE Trans. Geosci. Remote Sens., vol. 48, no. 4, pp. 2109–2118, Apr. 2010.

[27] S. Rödelsperger, A. Coccia, D. Vicente, et al., "The novel Fast GB-SAR sensor: Deformation monitoring for dike failure prediction," in Proceedings Asia-Pacific Conference on Synthetic Aperture Radar, Tsukuba, Japan, pp. 420–423, 2013.

[28] J. FortunyGuasch, "A fast and accurate far-field pseudopolar format radar imaging algorithm," IEEE Trans. Geosci. Remote Sens., vol. 47, no. 4, pp. 1187–1196, Apr. 2009.

[29] K. Han, Y. Wang, X. Chang, et al., "Generalized pseudopolar format algorithm for radar imaging with highly suboptimal aperture length," Sci. China Inf. Sci., vol. 58, no. 4, Apr. 2015.

[30] M. Pieraccini and L. Miccinesi, "Ground-based radar interferometry: A bibliographic review," Remote Sens., vol. 11, no. 9, p. 1029, Apr. 2019.

[31] W. Feng, L. Yi, and M. Sato, "Near range radar imaging based on block sparsity and cross-correlation fusion algorithm," IEEE J. Sel. Topics Appl. Earth Observ. Remote Sens., vol. 11, no. 6, pp. 2079–2089, 2018.

[32] R. Wang, Y. K. Deng, O. Loffeld, et al., "Processing the Azimuth-Variant Bistatic SAR data by using monostatic imaging algorithms based on 2-D principle of stationary phase," IEEE Trans. Geosci. Remote Sens., vol. 49, no. 10, pp. 3504–3520, Oct. 2011.

[33] D. Garcia, L. Le Tarnec, S. Muth, et al., "Stolt's f-k migration for plane wave ultrasound imaging," IEEE Trans. Ultrason. Ferroelectr. Freq. Control, vol. 60, no. 9, pp. 1853–1867, Sep. 2013.

[34] S. R. Degraaf, "SAR imaging via modern 2-D spectral estimation methods," Proc. SPIE, vol. 2230, pp. 36–47, Apr. 1994.

[35] J. Li and P. Stoica, "An adaptive filtering approach to spectral estimation and SAR imaging," IEEE Trans. Sig. Process., vol. 44, pp. 1469–1484, Jun. 1996.

[36] C. Jakowatz, D. Wahl, P. Eichel, et al. Spotlight-mode synthetic aperture radar: A signal processing approach, Norwell: Kluwer Academic Publishers, 1996.

[37] G. Nico, D. Leva, G. Antonello, et al., "Ground-based SAR interferometry for terrain mapping: Theory and sensitivity analysis," IEEE Trans. Geosci. Remote Sens., vol. 42, pp. 1344–1350, Jun. 2004.

[38] G. Nico, D. Leva, J. Fortuny-Guasch, et al., "Generation of digital terrain models with a ground-based SAR system," IEEE Trans. Geosci. Remote Sens., vol. 43, no. 1, pp. 45–49, Jan. 2005.

[39] O. Monserrat, "Deformation measurement and monitoring with Ground-Based SAR," Ph.D. dissertation, Universitat Politècnica de Catalunya, Barcelona, Spain, 2012.

[40] M. Crosetto, O. Monserrat, G. Luzi, et al., "Discontinuous GB-SAR deformation monitoring," ISPRS J. Photogramm. Remote Sens., vol. 93, pp. 136–141, Jul. 2014.

[41] R. Scheiber and A. Moreira, "Coregistration of interferometric SAR images using spectral diversity," IEEE Trans. Geosci. Remote Sens., vol. 38, pp. 2179–2191, Sept. 2000.

[42] T. Strozzi, U. Wegmüller, L. Tosi, et al., "Land subsidence monitoring with differential SAR interferometry," Photogramm. Eng. Remote Sens., vol. 67, no. 11, pp. 1261–1270, 2001.

[43] S. H. Hong, S. Wdowinski, S. W. Kim, et al., "Multi-temporal monitoring of wetland water levels in the Florida Everglades using Interferometric Synthetic Aperture Radar (InSAR)," Remote Sens. Environ., vol. 114, no. 11, pp. 2436–2447, 2010.

[44] M. C. Garthwaite, V. L. Miller, S. Saunders, et al., "A simplified approach to operational InSAR monitoring of volcano deformation in low-and middle-income countries: Case study of Rabaul Caldera, Papua New Guinea," Front. Earth Sci., vol. 6, p. 240, 2019.

[45] E. Tymofyeyeva and Y. Fialko, "Mitigation of atmospheric phase delays in InSAR data, with application to the eastern California shear zone," J. Geophys. Res. Solid Earth, vol. 120, pp. 5952–5963, 2015.

[46] D. Perissin and T. Wang, "Repeat-pass SAR interferometry with partially coherent targets," IEEE Trans. Geosci. Remote Sens., vol. 50, no. 1, pp. 271–280, 2012.

[47] S. Rödelsperger, "Real-time processing of Ground Based Synthetic Aperture Radar (GB-SAR) measurements," Ph.D. dissertation, Technische Universität Darmstadt, Darmstadt, Germany, 2011.

[48] M. Jiang, X. Ding, and Z. Li, "Hybrid approach for unbiased coherence estimation for multitemporal InSAR," IEEE Trans. Geosci. Remote Sens., vol. 52, no. 5, pp. 2459–2473, 2014.

[49] S. Lee, A. R. Miller, and S. A. Mango, "Intensity and phase statistics of multilook polarimetric and interferometric SAR imagery," IEEE Trans. Geosci. Remote Sens., vol. 32, no. 5, pp. 1017–1028, 1994.

[50] X. Lin, F. Li, D. Meng, et al., "Nonlocal SAR interferometric phase filtering through higher order singular value decomposition," IEEE Geosci. Remote Sens. Lett., vol. 12, no. 4, pp. 806–810, Apr. 2015.

[51] K. Itoh, "Analysis of the phase unwrapping algorithm," Appl. Opt., vol. 21, no. 14, p. 2470, 1982.

[52] H. Yu, Y. Lan, Z. Yuan, et al., "Phase unwrapping in InSAR : A review," IEEE Geosci. Remote Sens. Mag., vol. 7, no. 1, pp. 40–58, 2019.

[53] R. M. Goldstein, H. A. Zebker, and C. L. Werner, "Satellite radar interferometry: Two-dimensional phase unwrapping," Radio Sci., vol. 23, no. 4, pp. 713–720, 1988.

[54] B. Osmanoğlu, F. Sunar, S. Wdowinski, et al., "Time series analysis of InSAR data: Methods and trends," ISPRS J. Photogramm. Remote Sens., vol. 115, pp. 90–102, 2016.

[55] T. J. Flynn, "Consistent 2-D phase unwrapping guided by a quality map," in International Geoscience and Remote Sensing Symposium (IGARSS), 1996.

[56] M. Zhao, L. Huang, Q. Zhang, et al., "Quality-guided phase unwrapping technique: Comparison of quality maps and guiding strategies," Appl. Opt., vol. 50, no. 33, pp. 6214–6224, 2011.

[57] C. Esch, J. Köhler, K. Gutjahr, et al., "On the analysis of the phase unwrapping process in a D-InSAR stack with special focus on the estimation of a motion model," Remote Sens., vol. 11, no. 19, p. 2295, 2019.

[58] M. Costantini, "A novel phase unwrapping method based on network programming," IEEE Trans. Geosci. Remote Sens., vol. 36, no. 3, pp. 813–821, 1998.

[59] M. Costantini, A. Farina, and F. Zirilli, "A fast phase unwrapping algorithm for SAR interferometry," IEEE Trans. Geosci. Remote Sens., vol. 37, no. 1, pp. 452–460, 1999.

[60] M. Costantini and P. A. Rosen, "Generalized phase unwrapping approach for sparse data," in International Geoscience and Remote Sensing Symposium (IGARSS), 1999.

[61] A. Pepe and R. Lanari, "On the extension of the minimum cost flow algorithm for phase unwrapping of multitemporal differential SAR interferograms," IEEE Trans. Geosci. Remote Sens., vol. 44, no. 9, pp. 2374–2383, 2006.

[62] B. Osmanoglu, T. H. Dixon, and S. Wdowinski, "Three-dimensional phase unwrapping for satellite radar interferometry: DEM generation," IEEE Trans. Geosci. Remote Sens., vol. 52, no. 2, pp. 1059–1075, Feb. 2014.

[63] G. Luzi, "Ground-based radar interferometry for landslides monitoring: Atmospheric and instrumental decorrelation sources on experimental data," IEEE Trans. Geosci. Remote Sens., vol. 42, no. 11, pp. 2454–2466, Nov. 2004.

[64] L. Noferini, M. Pieraccini, D. Mecatti, et al., "Permanent scatterers analysis for atmospheric correction in ground-based SAR interferometry," IEEE Trans. Geosci. Remote Sens., vol. 43, no. 7, pp. 1459–1471, Jul. 2005.

[65] L. Pipia, X. Fbregas, A. Aguasca, et al., "Atmospheric artifact compensation in ground-based DInSAR applications," IEEE Geosci. Remote Sens. Lett., vol. 5, no. 1, pp. 88–92, Jan. 2008.

[66] L. Iannini and A. M. Guarnieri, "Atmospheric phase screen in ground-based radar: Statistics and compensation," IEEE Geosci. Remote Sens. Lett., vol. 8, no. 3, pp. 537–541, May 2011.

[67] R. Iglesias, X. Fabregas, A. Aguasca, et al., "Atmospheric phase screen compensation in ground-based SAR with a multiple regression model over mountainous regions," IEEE Trans. Geosci. Remote Sens., vol. 52, no. 5, pp. 2436–2449, May 2014.

3 Installation and Operation of GB-SAR

3.1 SITE MODELING

3.1.1 3D GIS Model Design for GB-SAR Illumination

The three dimensional (3D) GIS model provides much information about the ground surface when visual inspection is not necessarily sufficient for user-specified analysis. It can be used to render significant information about the spatial dissemination of electromagnetic signals emitted by any active sensor system. This information can be used to optimize the sensor performance to achieve successful GB-SAR field experiments. We introduce new software tools that can be used to find the best location for the GB-SAR installation and the optimum hardware configuration to receive the maximum reflection from the area of interest. It will significantly improve the received data quality and the operating efficiency of the equipment.

3.1.2 Environmental Modeling in the Spatial Domain

With the rapid development of computer technology, space technology, and modern information infrastructure, the application of 3D GIS is increasingly widespread, covering battlefield simulations, digital cities, resource management, environmental assessment, regional planning, flood modeling, public facility management, telecommunications, mobile communications, and many other aspects. For environmental monitoring systems, terrain is an important object of visual expression in geographic information science based on very complex natural scenery. In addition, 3D terrain visualization is an extremely important research field in regions of the Earth, which can reflect complex topography. Figure 3.1(a) depicts an experiment conducted at the Kawauchi campus of Tohoku University, Japan. The main objective of this experiment is to create a 3D spatial model in a suburban environment to understand the power distribution of the radar sensor. The radar was mounted on a portable tripod and pointed towards the GPR laboratory.

The environment in front of the radar system was mapped into a geocoded Google image in the preliminary stage. The geocoded image was used to extract the ground truth data such as building locations; tree locations; and fixed obstacles such as fences, road signs, and vehicles observed during the experiment. This information was separated into stranded vector data formats of point, line, and polygon according

DOI: 10.1201/9781003316312-3

to the physical geometries accepted by the ESRI ArcGIS software platform, and depicted in Figure 3.1(b). It will help:

- To understand the illumination area and area that is not illuminated by the source due to its installation location,
- To estimate the angle of incidence in a specific illumination area,
- To estimate the incidence power in a specific illumination area,
- To determine the relative location of the AOI,
- To interpret the information more accurately and in a user-friendly manner.

(a)

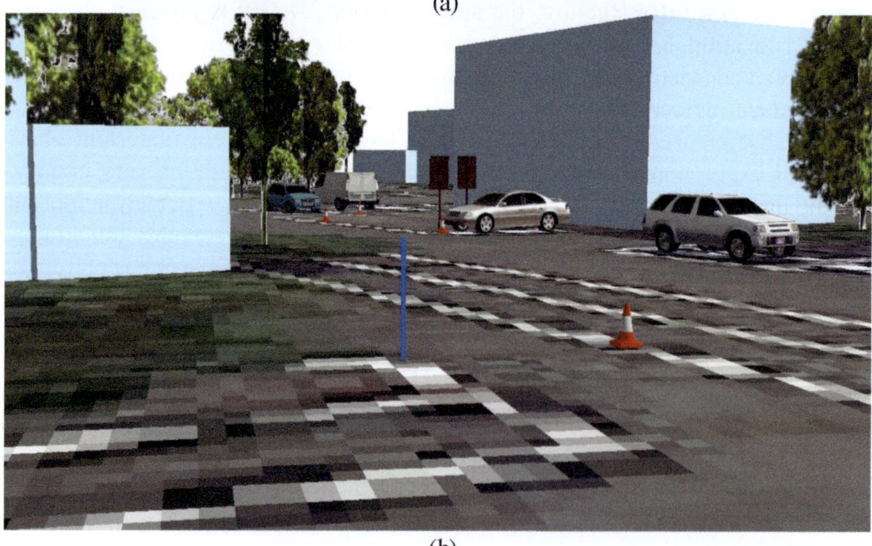

(b)

FIGURE 3.1 (a) Image taken near the GPR room. (b) 3D model interpretation of the corresponding area.

With the rapid development of information technology and major needs, software needs to be developed to support the trends of being dynamic, fast, user friendly, and able to be used by wider society. Therefore, it is necessary to develop an integrated application for the data processing of special 3D visualization resource information systems. With compatible software that is widely used today, the spatial database can be used more effectively and efficiently. Therefore, we have proposed a 3D GIS model-assisted survey to find the best location for GB-SAR installation and estimate the best GB-SAR configuration before site installation. The new software tool was designed to estimate the optimum radar illumination. The developed technique was successfully used for feasibility surveys in large-scale landslide locations in Minami–Aso, Kumamoto. The simulation results show the expected illumination from different installation locations in the designed 3D model [1].

3.1.3 3D Model Design

The Kumamoto earthquake was the largest catastrophic earthquake recorded in recent years in Kyushu Island, Japan. In April 2016, the highest number of foreshocks and aftershocks were recorded near the Futagawa and Hinagu fault zones, which expanded northeastwards and southwestwards along Kyushu Island. An M6.5 earthquake occurred at 21:26 JST on April 14, and an M7.3 earthquake occurred at 01:25 JST on April 16 [2]. The two earthquakes killed at least 50 people and injured approximately 3,000 others. Severe damage occurred in Kumamoto and Oita prefectures, with numerous structures collapsing and catching fire. More than 44,000 people were evacuated from their homes because of the disaster [3].

The number of earthquake-triggered landslides has significantly increased, and the largest one was reported near National Road 57. It was named after the 200-m-long Aso-shashi bridge that formerly spanned the 80-m-deep gorge of the Kurokawa River [3]. In addition, the main debris flow of the landslide shut off the main railway and the road connecting the two prefectures.

Immediate recovery projects were launched to remove the debris flow after the landslide. They were started as part of the disaster recovery plan. The deposited soil layer of the main debris flow near the mountain foot was removed by manned vehicles, and the unstable soil layer close to the mountain peak was removed by unmanned vehicles controlled by a nearby remote station. This total soil removal process was

FIGURE 3.2 The Kumamoto earthquake (April, 2016) epicenter and the location of the landslide are shown in the elevation map. The image taken during the first field observation in Minami–Aso is shown on the right.

completed, and the road construction project was launched in January 2017. Due to the instability of the top soil layer on the hillside and the number of recorded earthquakes in the fault zone, stability information of the open soil layer is significantly important for workers involved in daily road construction work. Against this background, the GB-SAR remote sensing technique is considered the best solution for retrieving stability information on the total landslide area in order to give early warnings of landslides to workers from the remote monitoring station for disaster prevention.

3.1.4 LiDAR Survey

LiDAR has become an established method for collecting very dense and accurate elevation data across landscapes and project sites. This active remote sensing technique is similar to radar but uses laser light pulses instead of radio waves. LiDAR is typically "flown" or collected from planes where it can rapidly collect points over large areas. LiDAR, as a remote sensing technique, has several advantages. The foremost are high accuracy, high point density, large coverage area, and the ability for users to resample areas quickly and efficiently. This creates the ability to map distinct changes at a very high resolution and cover large areas uniformly and accurately [4, 5].

After the earthquake triggered a landslide in the western tip of the Aso caldera, it was highlighted as the largest landslide recorded in the post-disaster period. On the other hand, it stopped the main transportation roads that connect Kumamoto city and Oita city and distorted the main access bridge that crosses the Kurokawa River. The LiDAR survey conducted in May 2016 was processed into 30-cm pixel resolution with a 32-bit pixel depth raster layer, as shown in Figure 3.3. It was

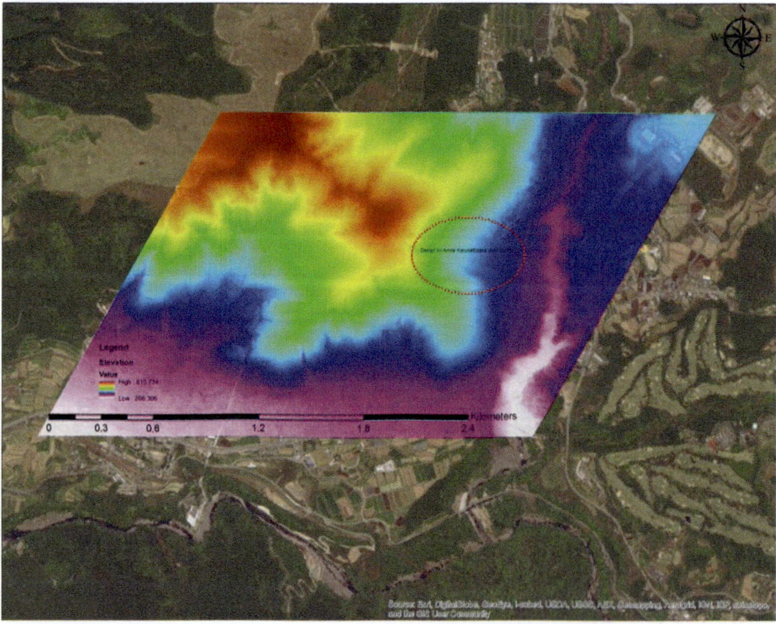

FIGURE 3.3 LiDAR information of the Minami–Aso area. It was geocoded and overlaid with the base map to find the relative location.

TABLE 3.1

LiDAR Information

Parameters	Value
Data format	Laser file format
Survey month	May 2016
Survey area	3 km × 2 km
Number of bands	1
Pixel depth	32 bit
Elevation range	266.306–815.734 m
Top extent	3641054.86312
Left extent	683532.209701
Right extent	686819.797314
Bottom extent	3639496.40515

geocoded, projected into the JGD_2001_54N standard projected coordinate system, and overlaid in the satellite image base layer to find relative locations. The total area of the processed LiDAR information (Table 3.1) covered around 3 km × 2 km in ground area, including the western tip of the Aso caldera in the middle, National Road 57 in the west–east and south–north directions near the bottom and right corner, and the Kurokawa River from the north–south direction in the right-side corners of Figure 3.3.

After geocoding, the elevation profile of the processed LiDAR data was estimated to be 265 m at the Kurokawa River, and the highest value was recorded as 815 m at the peak of the western tip of the mountain cliff. By the color ramp of the elevation profile, rapid elevation change can be observed within a short distance in the horizontal plane. This has a higher potential and is a favorable factor for triggering landslides. The overall results of the LiDAR survey show the number of landslides triggered after the earthquake. The area covered by the circle recorded the largest landslide and debris flow that collapsed the Aso Ohaishi Bridge after the landslide. Because the area was restricted for access, the LiDAR survey gave good results with highly accurate ground truth information. Geocoded LiDAR information can be used to visualize the existing environment more effectively than any other remote sensing and pinpoint measurement. This information can be further processed to obtain a 3D model. It will render more realistic information, which is important to estimate the spatial dissemination of EM radiation prior to deploying any ground-based remote sensing technique such as GB-SAR.

3.1.5 Digital Elevation Model

The designed digital elevation model (DEM) model in Figure 3.4 was geocoded using the local coordinated system. Both horizontal and vertical coordinate systems were used to preserve the accuracy of the LiDAR information in the horizontal and vertical planes. This information is summarized in Table 3.2.

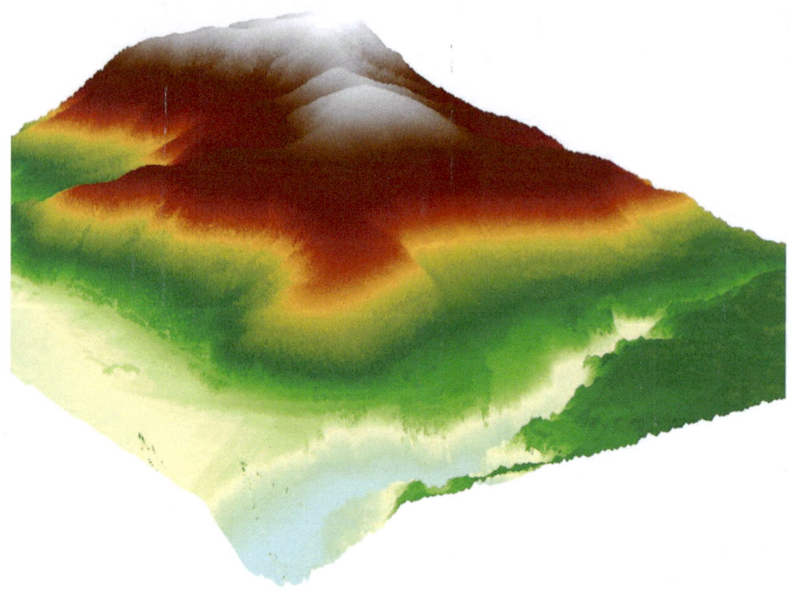

FIGURE 3.4 DEM generated with 30 cm × 30 cm pixel resolution.

TABLE 3.2
DEM Information

Parameters	Value
Spatial reference	JGD_2000_UTM_Zone_52N
Survey month	May 2016
Liner unit	Meters (1.000000)
Angular unit	Degrees (0.01745329251994)
False easting	50000
False northing	0
Central meridian	129
Scale factor	0.9996
Datum	D_JGD_2000

3.2 ESTIMATION OF THE ILLUMINATION AREA

3.2.1 Antenna Radiation Pattern

An antenna radiation pattern describes the far-field directional properties of the antenna. By virtue of reciprocity, a receiving antenna has the same directional antenna pattern as the pattern it exhibits when operated in transmission mode. The strength of the radiated field or the power density of the antenna can be specified by the zenith angle θ and azimuth angle ϕ. For an antenna with a single main lobe, the pattern solid angle Ω_p describes the ebullient width of the main lobe of the antenna pattern [6]. It can be described as the integral of the normalized radiation intensity $F(\theta,\phi)$ over a sphere as follows:

$$\Omega_p = \iint F(\theta,\phi)\,d\Omega \tag{3.1}$$

This pattern solid angle characterizes the directional properties of the 3D radiation pattern. The only practical way to increase the directivity of an antenna waveguide is to flare its end into a horn. Figures 3.5(a) and (b) show the rectangular horn antenna used in GB-SAR and Figure 3.5(c) shows the sensor, and its dimensions. Using these geometrical parameters, the H- and E-plane radiation patterns can be estimated.

The antenna radiation pattern in each plane gives the directivity properties of the GB-SAR, as depicted in Figures 3.6(a) and (b). The half-power beam width of

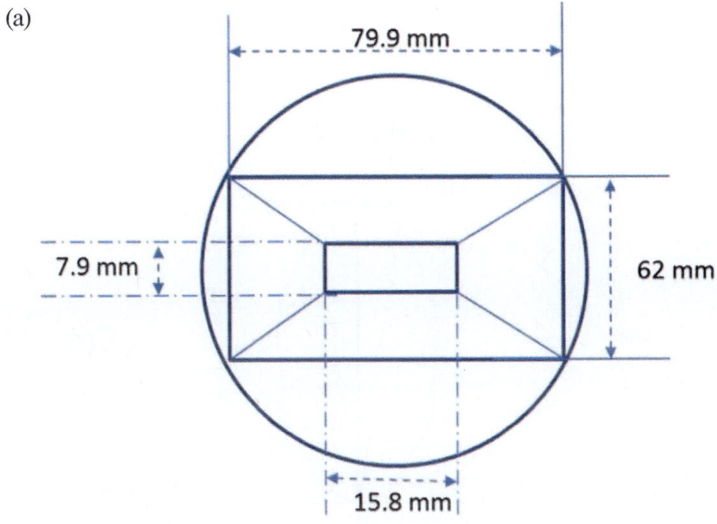

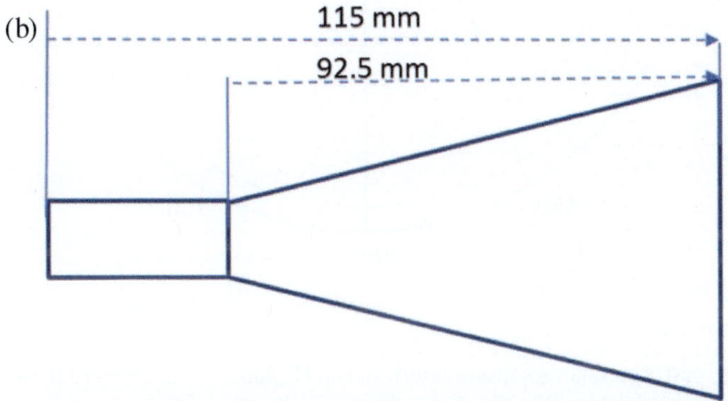

FIGURE 3.5 (a) and (b) Dimensions of the rectangular horn antenna used in (c) GB-SAR.

FIGURE 3.5 (Continued)

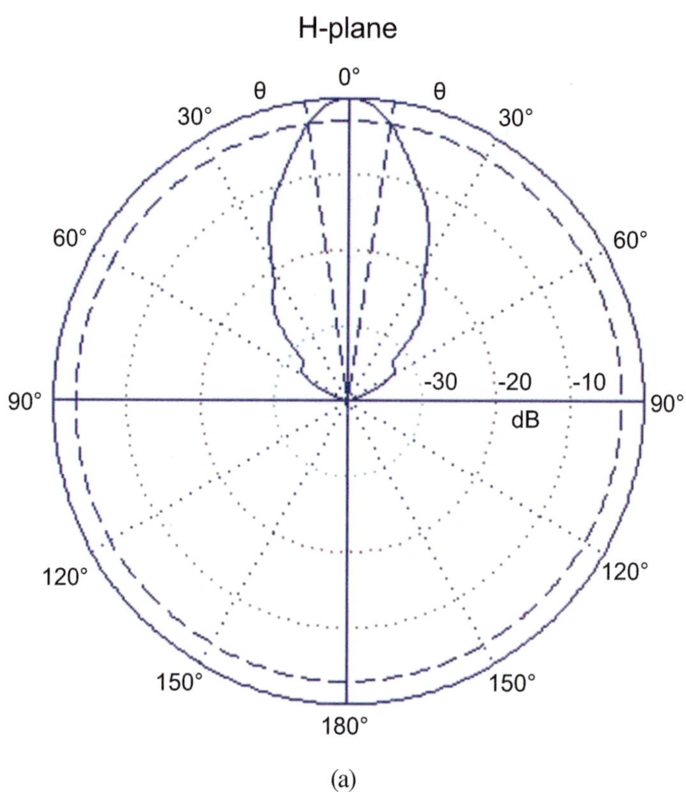

(a)

FIGURE 3.6 (a) Antenna radiation pattern in the H plane, (b) antenna radiation pattern in the E plane, and (c) comparison of radiation power in the H and E planes.

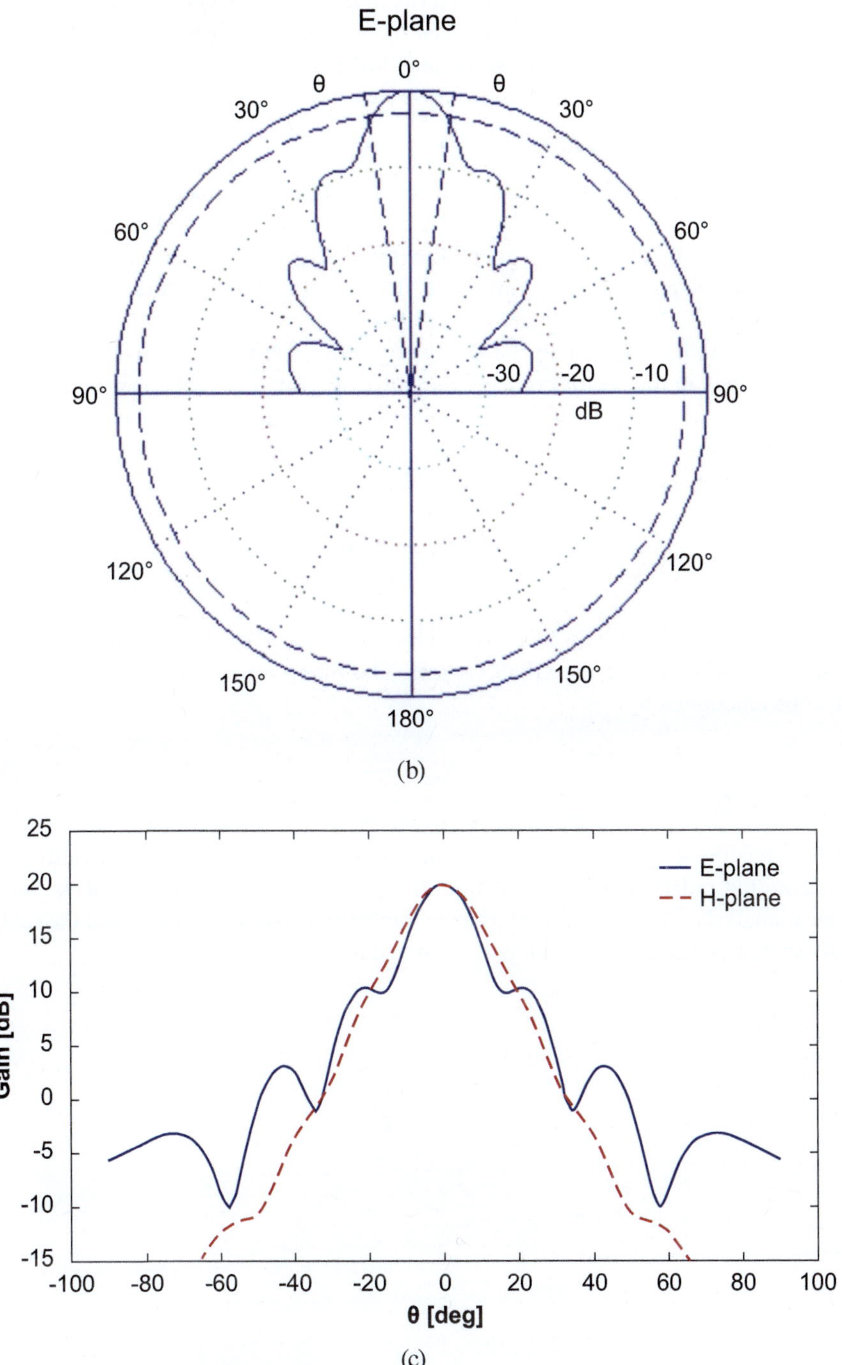

FIGURE 3.6   (Continued)

the H-plane and E-plane radiation patterns in the GB-SAR system is defined as the angular width of the main lobe between the two angles at which the magnitude of $F(\theta,\phi)$ is equal to half of its peak value (–3dB point).

3.2.2 Vector Grid Design

We developed a new software tool to extract spatial information from the terrestrial model. It was developed on the Python platform with the Arcpy and Numpy libraries [7]. The conical polygonal area given by the estimated radiation pattern was further decomposed according to the range and azimuth resolution of the SAR image.

The azimuth resolution of the SAR image corresponds to the distance between the two lines in the sector circle of the polygon grid. The range resolution of the SAR image corresponds to the space between two circular arcs in the polygon grid (Figure 3.7). Ignoring the angular dependence, the azimuth resolution of the GB-SAR image can be calculated as follows:

$$R_{AZ} = \frac{\lambda_C}{2L} R \tag{3.2}$$

where λ_C is the wavelength of the center frequency of the transmitted signal, and L is the length of the GB-SAR rail. The range resolution R_{RN} of the system can be calculated by

$$R_{RN} = \frac{c}{2B} \tag{3.3}$$

where c is the speed of light in free space, and B is the frequency bandwidth of the transmitted radar signal. The location of a ground patch can be critical for subsequent analysis. This can be used as a favorable information to understand the spatial dispersion of radar signal from particular ground location.

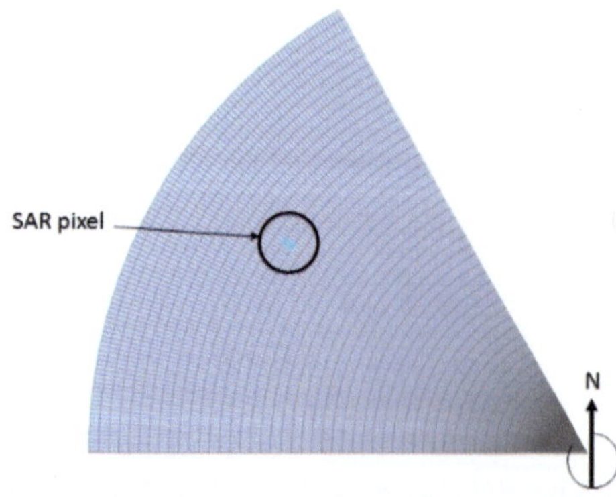

FIGURE 3.7   SAR polygon grid tool.

3.2.3 ESTIMATION OF ILLUMINATION PIXELS

3.2.3.1 Estimation of Pixel Orientation

A gridded surface is assumed to be mathematically continuous. It is possible to derive mathematical derivatives at any location. In practice, since the raster surface has been discretized, the derivatives along two rectangular planes in Figure 3.8(a) and (b) can be used to approximate the pixel orientation in both the vertical plane (plane perpendicular to the earth surface) and the horizontal plane (the plane parallel to the earth surface) [8]. In this analysis, the pixel orientation in the vertical plane was estimated by gradient, the maximum rate of change of altitude, using (3.4). The pixel orientation in the horizontal plane was estimated by (3.5):

$$\tan S = \left[\left(\frac{\delta z}{\delta x} \right)^2 + \left(\frac{\delta z}{\delta y} \right)^2 \right]^{\frac{1}{2}} \tag{3.4}$$

$$\tan A = \left[\left(\frac{\delta z}{\delta y} \right) + \left(\frac{\delta z}{\delta x} \right) \right] \tag{3.5}$$

The corresponding finite difference estimator using all outer points of the kernel in Figure 3.8(c) in the window is given as follows.

For the east–west and south–north gradients of the raster layer

$$\left[\frac{\delta z}{\delta x} \right] = \left[\left(z_{i+1,j+1} + 2z_{i+1,j} + z_{i+1,j-1} \right) - \left(z_{i-1,j+1} + 2z_{i-1,j} + z_{i-1,j-1} \right) \right] / 8\delta x \tag{3.6}$$

$$\left[\frac{\delta z}{\delta y} \right] = \left[\left(z_{i+1,j+1} + 2z_{i+1,j} + z_{i+1,j-1} \right) - \left(z_{i-1,j+1} + 2z_{i-1,j} + z_{i-1,j-1} \right) \right] / 8\delta x \tag{3.7}$$

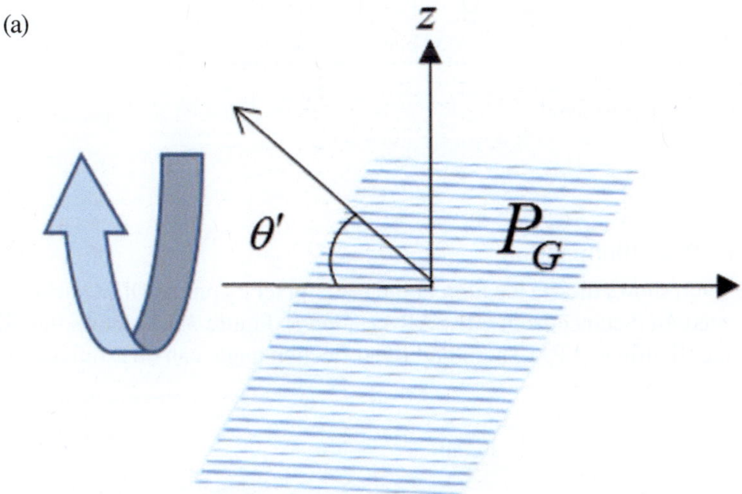

FIGURE 3.8 Raster pixel orientation in (a) vertical plane, (b) horizontal plane, and the 3 × 3 kernel used to estimate the east–west and south–north gradients in (c).

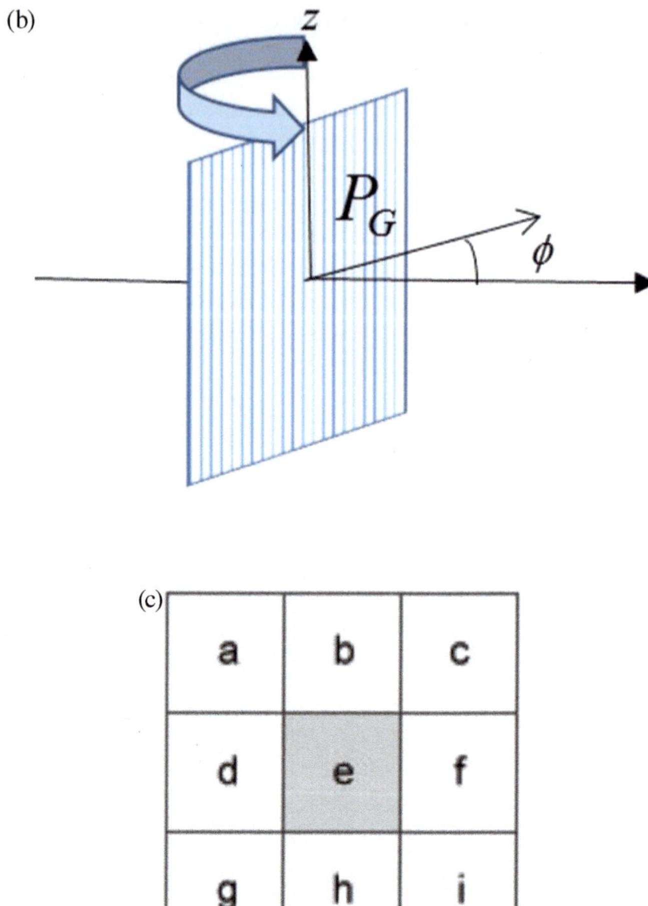

FIGURE 3.8   (Continued)

3.2.3.2   Pixel Illumination

Figure 3.9(a) shows the 2D location of the raster pixel P_G on the DEM surface, which was located Δd distance from GB-SAR location R. Figure 3.9(b) shows the 3D location of the illuminated $P_{G'}$. Therefore, the reflection angle can be estimated by [3]:

$$\alpha = \arctan\left[\frac{\Delta z}{\sqrt{(\Delta x)^2 + (\Delta y)^2}}\right] \tag{3.8}$$

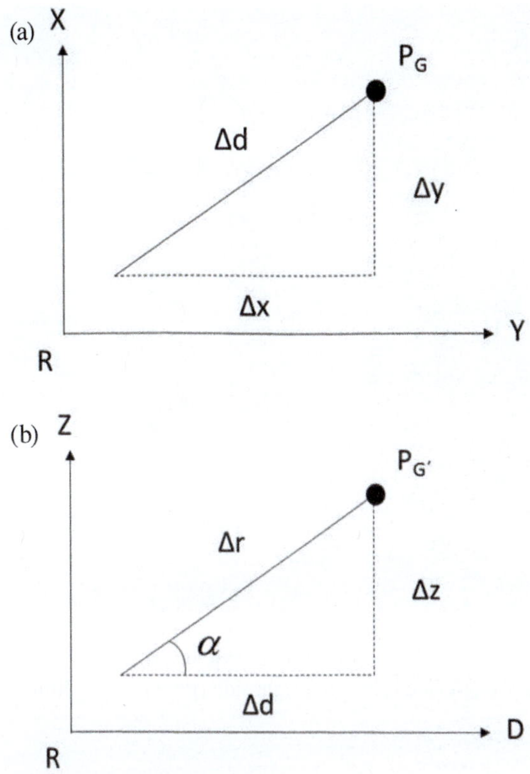

FIGURE 3.9　(a) 2D location of the ground patch; (b) 3D location of the ground patch.

3.3　CASE STUDY OF MINAMI–ASO

3.3.1　SITE DESCRIPTION

The thematic map in Figure 3.10 shows the 1 km × 1 km area around the main landslide area of Minami–Aso. Spatial dissemination of the radiated signal is an indispensable factor for environmental monitoring systems. It renders significant information about equipment usability before field installation. A 2D map shows the possible location for the GB-SAR installation, which was investigated after the field survey. After the catastrophic landslide, the geography of the surrounding area was enormously changed, and many places became inaccessible. Meanwhile, infrastructure such as transportation and electricity were shut off, and equipment maintenance facilities were not available in each place. Against this background, preliminary estimation of equipment applicability becomes mandatory to estimate the system deployment parameters and appropriate configuration. The road reconstruction project was actively carried out at the bottom of the mountain, and GB-SAR illumination was not necessarily needed in all views of the open soil layer. This requires a more sophisticated method to estimate GB-SAR illumination

FIGURE 3.10 The proposed location for GB-SAR installation is shown in Loc1 to Loc6, which is located on the left side (blue) and right side (green) of the Kurokawa River.

before site deployment to minimize the calibration time for fast deployment in a disaster situation.

After the earthquake, the main debris was floved with a dense forest canopy. The open soil layer of the mountain creep was divided into two layers according to the disaster recovery work. The soil layer near the crown of the landslide and the surrounding area (area A) was prioritized. The remaining unstable soil layer of this area was removed by unmanned vehicles operated by a remote station. This process was systematically performed until December 2016. Starting from January 2017, road reconstruction projects were launched, and workers started to work at the bottom of the mountain. Due to the instability of the topsoil layer on the hillside and the number of recorded earthquakes in the fault zone, stability information of the open soil layer and the newly created geological boundaries in areas B and C became significant for the workers involved in daily road construction work. The proposed locations for GB-SAR installation are shown in Figure 3.11.

3.3.2 Selection of the GB-SAR Site

The expected illumination for Loc1 to Loc6 was estimated using the developed 3D model. It gives the maximum range, optimum height of the system platform, azimuth angle in the horizontal plane, and view angle in the vertical plane. The simulation results with possible installation parameters are summarized in Table 3.3. The

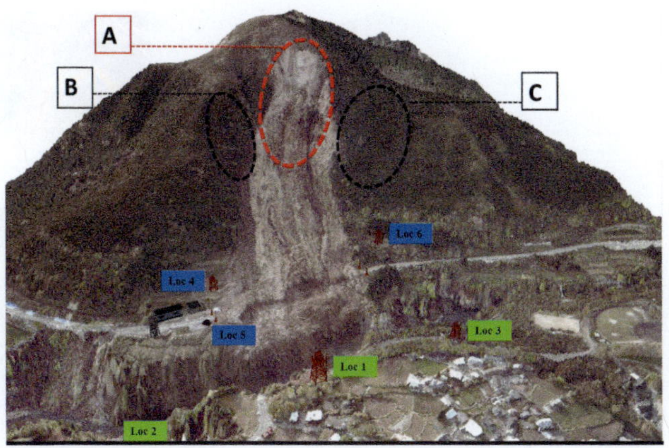

FIGURE 3.11 3D simulation model of the Minami–Aso area.

TABLE 3.3

Antenna Parameters

System Parameters	Value
Horn antenna model	FGBS-ANT-HRN15
Gain	15 dBi
Frequency range	17.05–17.35 GHz
E-plane beam width	32.2°
H-plane beam width	31.9°
Waveguide shape	WG19(WR51, R180)

estimated illumination information was geocoded by the local coordinate system and projected onto a 3D model. Figure 3.12(a) shows the expected illumination from Loc6 (in left riverbank), and Figures 3.12(b) and (c) show the expected illumination from Loc1 and Loc3.

3.3.3 Optimization of the GB-SAR Configuration

In Figure 3.13, blue indicates the estimated illumination for location 5. Zones A, B, and C show higher illumination than the surrounding area.

Location 5 was selected for system installation by considering optimum illumination from the expected zones. In advance, this location was highlighted due to the accessibility of maintenance work and supply of electricity and other infrastructure facilities, which were highly important for real-time monitoring. The GB-SAR system was installed, and data acquisition began in mid-January 2017. The linear rail was mounted in 1.5-m rigid concrete blocks to achieve the estimated height for a stable platform. The system was aligned to 315° LOS with 23° views (tilt) angle, as shown in Figure 3.14.

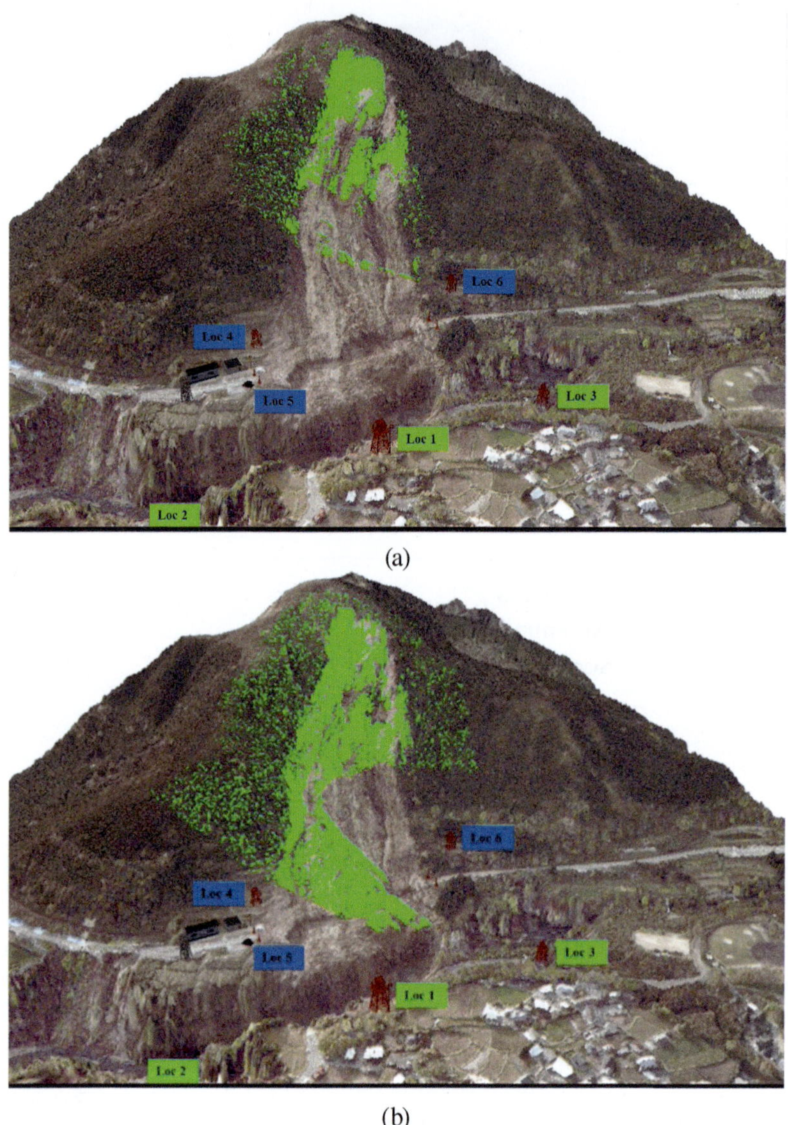

FIGURE 3.12   Estimated illumination by simulation. (a) Location 6 in left side of the river and (b) Location 3 and (c) Location 1 on the right side of the river.

3.3.4 FIELD VALIDATION

The estimated illumination and the acquired SAR-processed power image (20–30 dBm) were projected in the 3D model. Figure 3.15 shows a better correlation between the simulation results and the experimental results. In the simulation results, the GB-SAR illumination shows a discontinuity in illumination area, which is approximately 350 m from the GB-SAR sensor. This can be seen in the real illumination observed in the SAR-processed

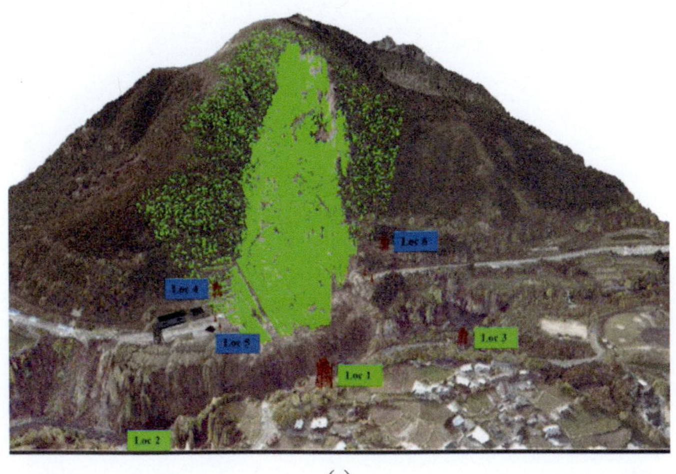

(c)

FIGURE 3.12 (Continued)

TABLE 3.4

Estimated Parameters for GB-SAR Installation

Installation Location	Latitude	Longitude	Radar Range (m)	Base Height (m)	Azimuth Angle (°)	View Angle (°)
Loc1	32.883945	130.990160°	1000	8	290	15
Loc2	32.880030	130.989195°	1000	10	300	15
Loc3	32.886185	130.989912°	1000	10	265	15
Loc4	32.883408	130.986888°	900	7	310	30
Loc5	32.882468	130.987070°	900	2	315	23
Loc6	32.886487	130.987352°	900	7	265	15

image. This could be due to the elevation difference in the mountain cliff along the radar LOS. In addition to the open soil layer of the main landslide area, the reflection becomes more random due to the tree canopy. Therefore, the backscatter power decreases. In the 3D model designed by LiDAR data can use to assess this ground truth information. Thus, we can see very clear illumination differences in the simulation results of the soil layer and the tree canopy. Furthermore, three separate linear wedge shape structures of P, Q, and R can be seen in the simulation results of the main landslide area. These structures become more identical and visible to the processed GB-SAR images.

3.4 SYSTEM INSTALLATION

During the post-disaster period, recovery projects were quickly launched for the Minami–Aso road reconstruction. Furthermore, the establishment of landslide monitoring systems and landslide early warning systems is briefly discussed.

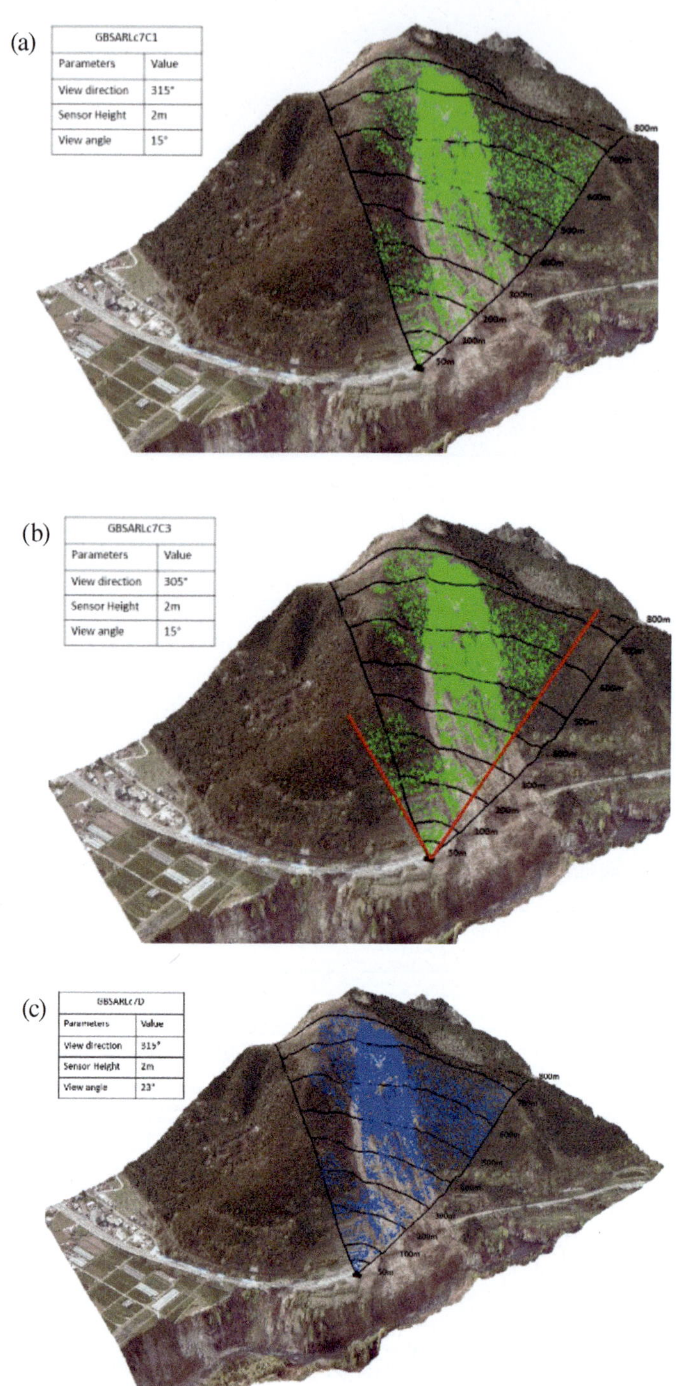

FIGURE 3.13 Estimated illumination for configurations (a), (b), (c), and (d) based on simulation results.

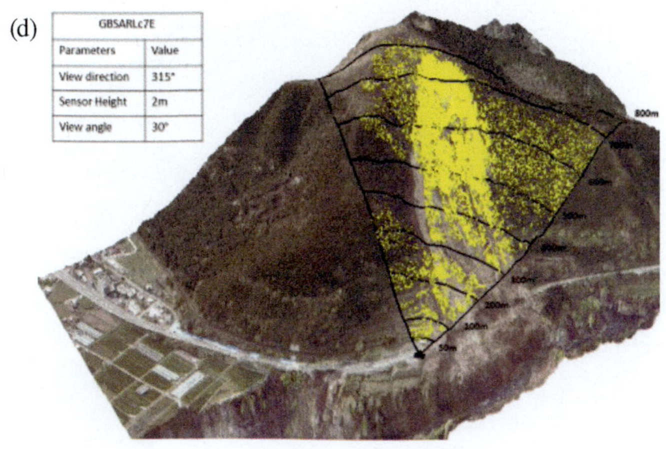

FIGURE 3.13 (Continued)

TABLE 3.5
Summary of the GB-SAR Configuration

Configuration	Base Height (m)	Azimuth Angle (°)	View Angle (°)	Illumination Area (%)
(a)	2	290	15	76.1
(b)	2	305	15	73.1
(c)	2	315	23	77.4
(d)	1.5	315	30	70.4

3.4.1 Hardware Installation

The GB-SAR system shown in Figure 3.16 was installed, and data acquisition began in mid-January 2017. The linear rail was mounted on 1.5-m rigid concrete blocks to achieve the estimated height for a stable platform.

The system was aligned to 315° LOS with 23° views (tilt) angle (Figure 3.17). System calibration was performed, and sudden atmospheric changes showed a high influence on the interferometric phase, which needs to be compensated for in order to estimate the ground displacement accurately.

Differential SAR interferometry works under a coherent condition where the received waveforms correlate with the compared SAR pair. One of the benefits of the GB-SAR sensor is the opportunity to gather zero-baseline repeated scans for differential measurements. Therefore, terms usually affecting differential coherence, such as image co-registration approximation, baseline construction uncertainties, and digital elevation model removal residual errors, become negligible [9].

However, GB-SAR works on the Ku band and has high sensitivity to small-scale changes, which tends to decorrelate within the time frame and complicates

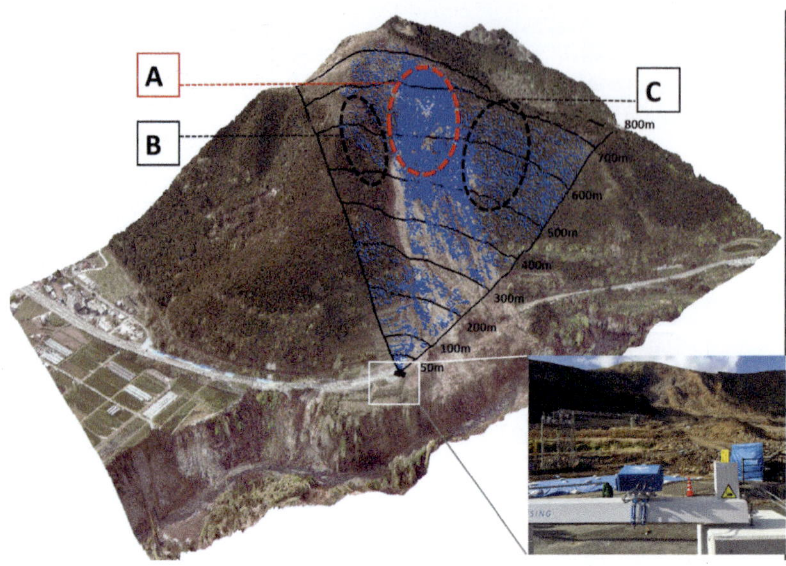

FIGURE 3.14 Estimated GB-SAR illumination for location 5. Blue indicates the estimated illumination, Zones A, B, and C show higher illumination than the surrounding area. The GB-SAR system was installed at this location, and continuous monitoring was started in January 2017.

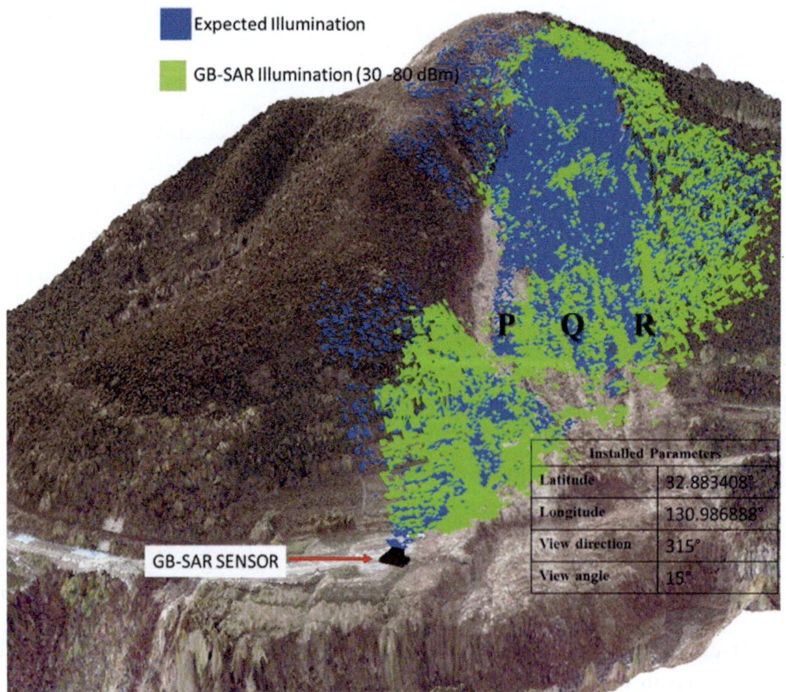

Installed Parameters	
Latitude	32.883408°
Longitude	130.986888°
View direction	315°
View angle	15°

FIGURE 3.15 Comparison of estimated illumination in blue and real illumination in green.

FIGURE 3.16   GB-SAR system deployed in Minami–Aso.

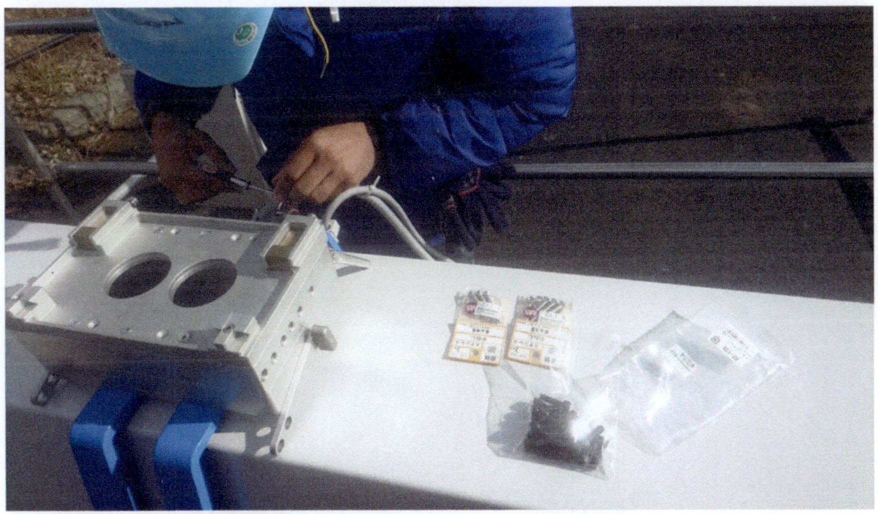

FIGURE 3.17   GB-SAR system configuration

quantitative long-term analysis. The problem of phase wrapping is one of the main issues observed in long-range SAR interferograms. Typically, phase unwrapping is performed on data, regularly sampled and stored in two-dimensional matrices [10]. Unwrapping the phase without any ambiguity requires the absolute phase gradient along adjacent samples, which is less than π phase change. Here we use a multi-interferogram framework to identify highly coherent targets along radar LOS to overcome most of the difficulties related to phase unwrapping and to discern the different signals that occur with the interferometric phase in order to calculate the displacement represented in Figure 3.18.

3.4.2 SOFTWARE PARAMETER SETTINGS

3.4.2.1 Interferometric Phase Quality Estimation

The complex coherence of interferograms describes the correlation between consecutive data acquisitions and is a vital measurement of where the phase is exploitable. It should be emphasized that coherence serves as a quality measurement for both acquisitions. The complex coherence between two images is defined as [11]:

$$\gamma_c = \frac{E\{M \cdot S^*\}}{\sqrt{E\{M \cdot M^*\} \cdot E\{S \cdot S^*\}}}$$

(3.9)

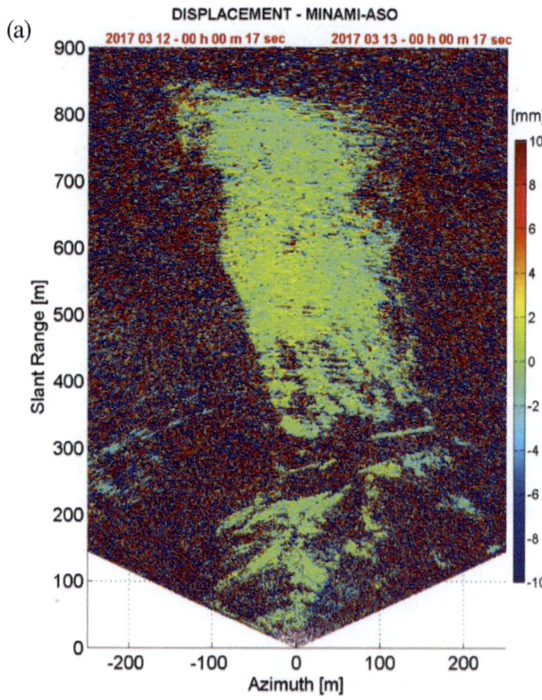

FIGURE 3.18 Interferograms of cumulative displacement of (a) 1-day observation, (b) 5-day observation, (c) 10-day observation, and (d) 15-day observation.

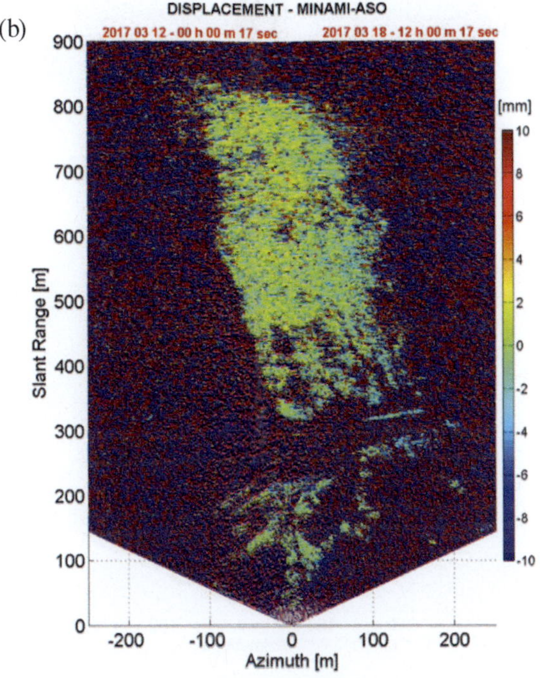

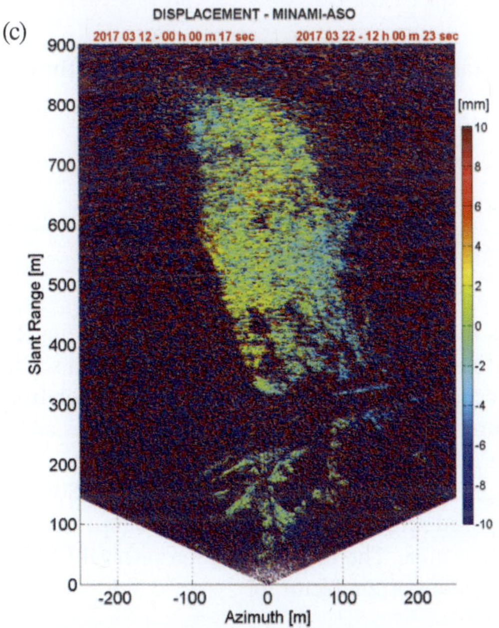

FIGURE 3.18 (Continued)

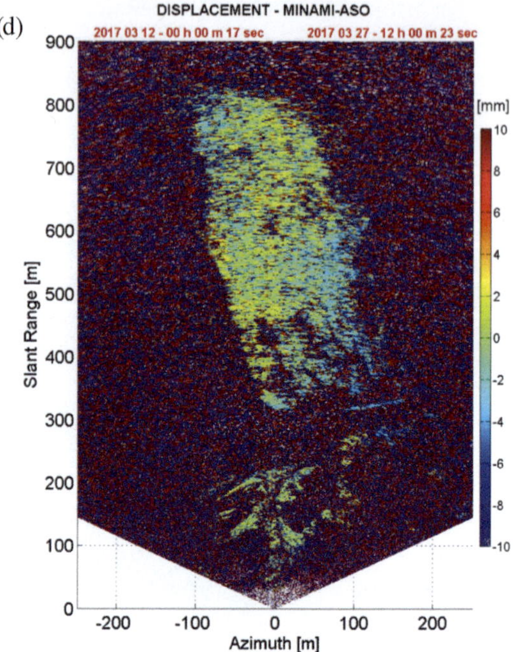

FIGURE 3.18   (Continued)

where $E\{\cdot\}$ is the statistical expectation. Coherence is defined by $\hat{\gamma}$, and it is estimated as [12]:

$$\hat{\gamma} = \frac{\dfrac{1}{n}\sum_{i=1}^{n} M_i \cdot S_i^*}{\sqrt{\dfrac{1}{n}\sum_{i=0}^{n} M_i M_i^* \dfrac{1}{n}\sum_{i=0}^{n} S_i S_i^*}} \tag{3.10}$$

Coherence resides in the range [0,1], with high values being coherent. Coherence decorrelates with time because of changes in the observed scene. The material and the shape of the scene strongly affect coherence. Vegetation has low coherence due to its entropic nature, whereas solid materials such as rocks and structures maintain high coherence for a longer time.

In this campaign, the observation area ranged up to 800 m and contained the open soil layer exposed after the landslide and the dense tree canopy spread beside the main observation area. The data acquisition interval of the remotely operated GB-SAR sensor is an important factor for

- Temporal phase compensation
- Deciding the data transfer rate

concerning displacement estimation in real-time operation. In advance, this can be used to preserve the coherence of the interferometric phase to minimize the temporal decorrelation. On the other hand, the problem of phase wrapping due to wave propagation in the range direction can be technically minimized by selecting the optimum data acquisition interval. It will improve the quality of the interferograms and provide reliable information for displacement monitoring.

3.4.2.2 Experimental Results

The location near the GB-SAR monitoring station is a notable factor for the water vapor content of the atmosphere in low-elevation areas. The wind speed and direction have a high influence on the sudden change in humidity measurements. This can be identified as a desirable factor that affects reflectivity due to highly dynamic weather conditions [13, 14].

Due to the compatibility of the latest GB-SAR system, the data acquisition interval can be increased up to 40 s. During the preliminary site observation, fast data acquisition was not a necessity for real time, more than the appropriate data rate. Therefore, the data acquisition interval was tested for 24 h (both daytime and nighttime) up to a 1-min interval, and complex coherency was estimated as a qualitative and quantitative parameter for the interferometric phase, as depicted in Figure 3.19. Table 3.6 shows the estimated complex coherence coefficient for each data acquisition interval in the sample data set acquired from March 12 to 27. This statistical estimation shows that $\hat{\gamma}_{mean}$ has high fluctuation for long data acquisition intervals. Apart from the number of data sets, the estimated $\hat{\gamma}_{mean}$ for daytime has a lower value than the estimated $\hat{\gamma}_{mean}$ during the nighttime. This emphasizes that the observed interferometric phase during daytime is more inconsistent than that during the nighttime in the sample data set. Furthermore, the estimated interferometric phase distortion due to environmental changes becomes more significant during the daytime than during the nighttime.

3.4.2.3 Optimum Data Acquisition Interval

The estimated $\hat{\gamma}_{mean}$ variation for data acquisition intervals of 1 day, 12 hours, 6 hours, and up to 1 min is summarized in Figure 3.20. A higher coherency value (close to 1) was noticed for a small data acquisition interval, and the coherency became weaker (close 0.6) in larger data acquisition intervals. This shows that the temporal decorrelation was significantly changed in a 24-h to 30-min time interval. The temporal decorrelation of the interferometric phase showed high fluctuation in this region, which advisedly affected the reliability of the phase estimation. The estimated $\hat{\gamma}_{mean}$ in 15-min to 1-min data acquisition intervals has high coherency and remained stable for 15-, 10-, 5-, and 1-min data acquisition intervals. Therefore, the 15-min data acquisition interval was fixed as an optimum data acquisition interval.

3.5 LONG-TERM OBSERVATION

The flow chart in Figure 3.21 shows the complete data processing steps and a prototype of the developed landslide monitoring and early warning system. The

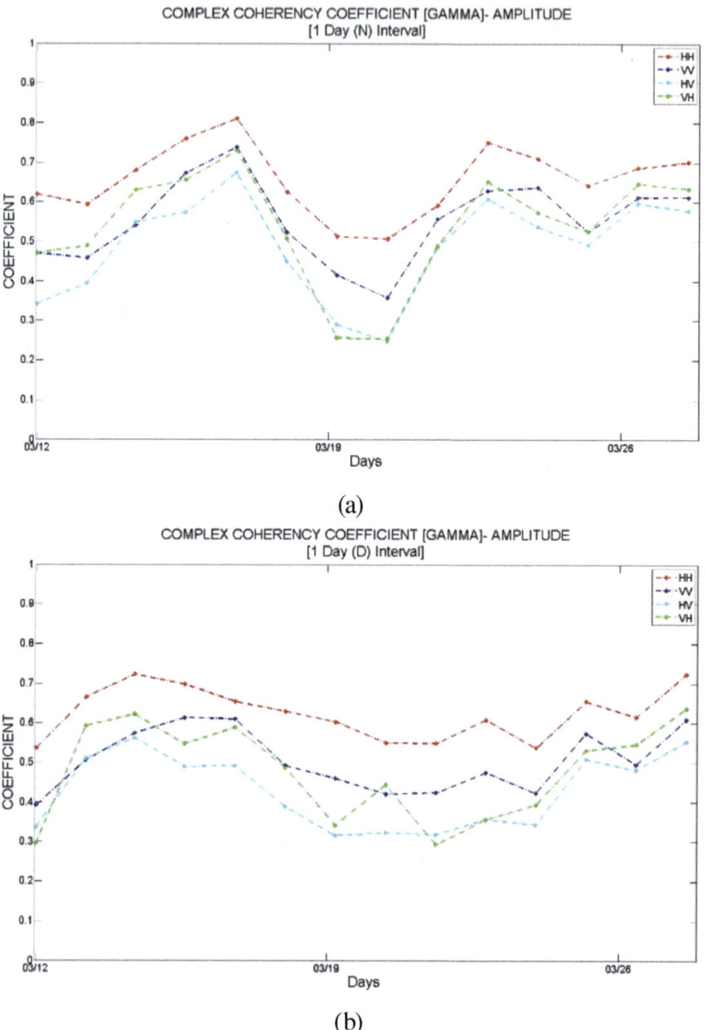

FIGURE 3.19   Estimated complex coherence for (a) 1-day data acquisition interval (nighttime), (b) 1-day data acquisition interval (daytime) day observation, (c) 12-h data acquisition interval, (d) 6-hour data acquisition interval, (e) 30 min data acquisition interval, (f) 15-min data acquisition interval.

pre-processing of the acquired data was done on a PC kept at the monitoring site, and after preliminary processing was completed, it was transferred to Tohoku University through the internet. This process was performed automatically. In this experimental setup, we investigated many data transfer mechanisms. Finally, we used the most cost-efficient and secure Google server system for data upload. We used a fixed Wi-Fi router with an appropriate internet connection for data transfer.

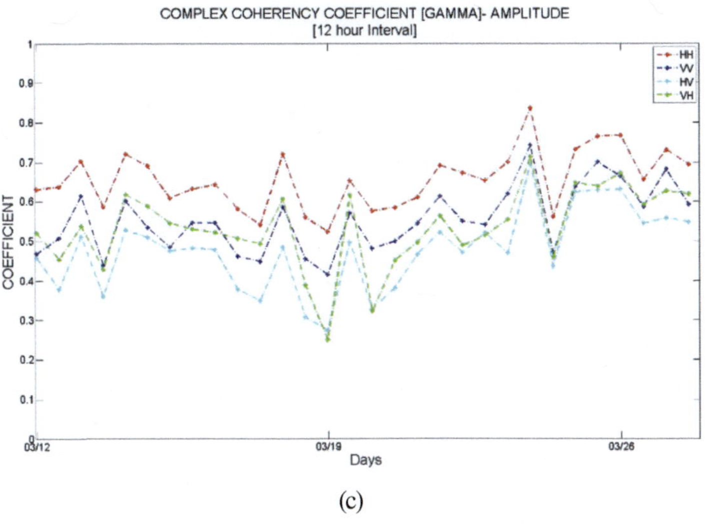

(c)

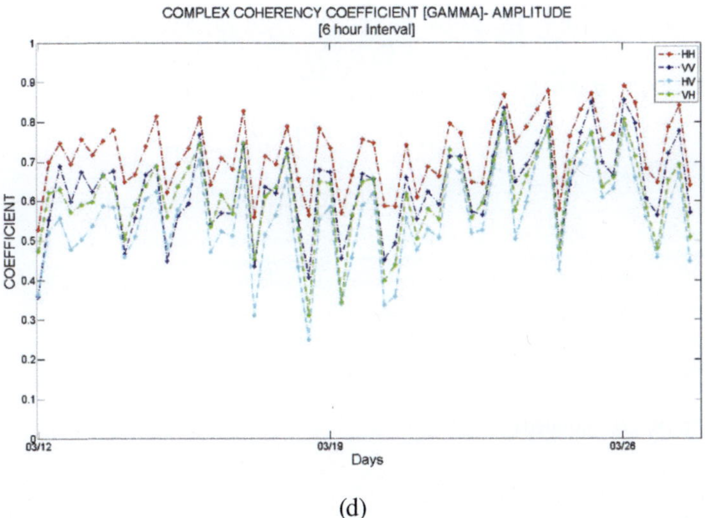

(d)

FIGURE 3.19   (Continued)

The operating PC was permanently connected to the internet and synchronized with the Google server. Thus, the newly acquired data were automatically updated in the cloud database. The same cloud server was shared by both the Minami–Aso PC and the monitoring PC dedicated to Tohoku University for real-time observation. It minimized the data download time and automated the complete data transfer process.

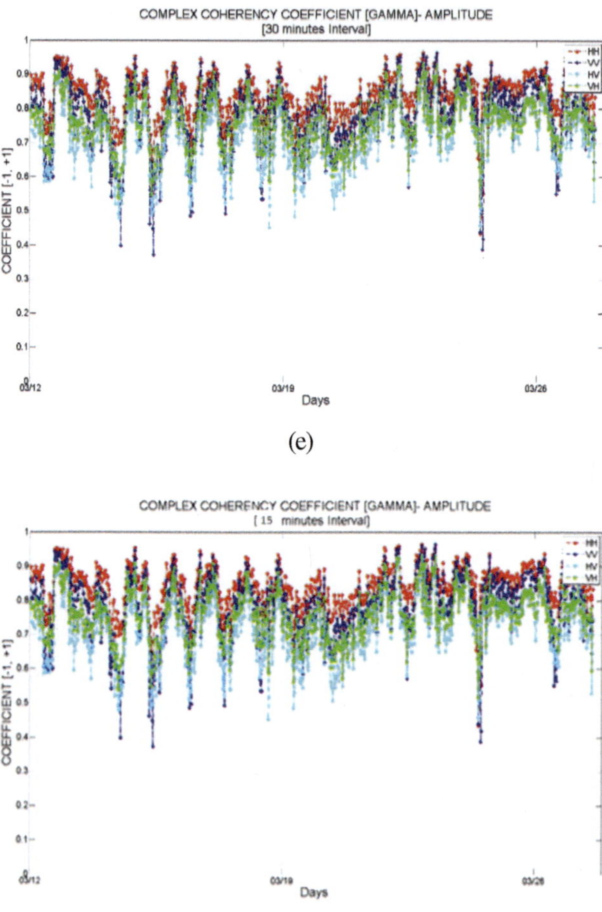

(e)

(f)

FIGURE 3.19   (Continued)

3.5.1 Automatic Mapping

After completing the secondary signal processing in the monitoring PC at Tohoku University, the estimated displacement was retrieved in two different ways. The first one was the 2D map projected onto the 3D DEM surface, which was updated by every data acquisition. Figure 3.22 depicts the interface of the real-time 3D displacement map in the monitoring PC. The displacement of the estimated CS point locations was updated with every 15 min of data acquisition completed by Minami–Aso GB-SAR. The color scale shows the displacement of each CS point in the two alternative directions. Blue represents movement toward the radar, and red shows movement in the opposite direction. After the road construction projects started, most of the CS points near the radar were distorted by ongoing project work, which was neglected by known movements of construction machinery. The displacement

TABLE 3.6

Estimated Complex Coherency Coefficients

Date	Number of Data Sets	$\hat{\gamma}_{mean}$
1 day (nighttime)	16	0.6635
1 day (daytime)	16	0.6021
12 h	32	0.6540
6 h	63	0.7209
3 h	127	0.7663
1 h	312	0.8102
30 min	616	0.8390
15 min	1231	0.8555
10 min	2251	0.8607
5 min	4501	0.8735
1 min	9000	0.8980

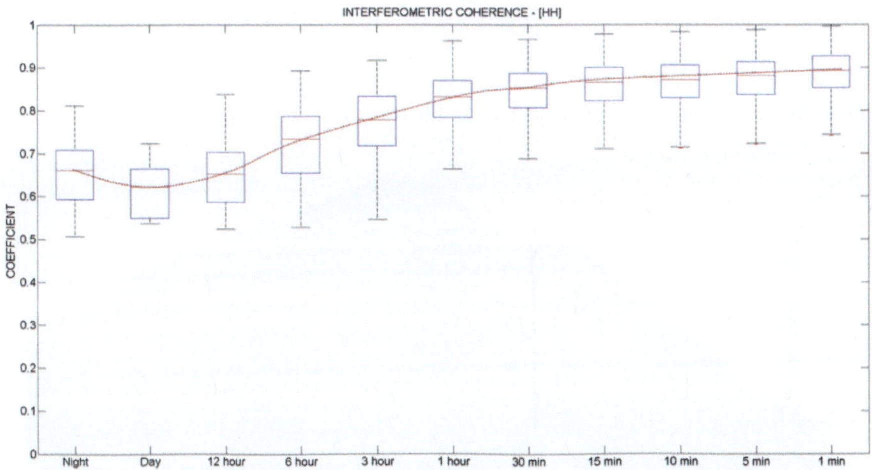

FIGURE 3.20 The interferometric phase quality estimation. The interferometric coherence coefficient change by temporal baseline 24 h to a 1-min interval is summarized in the box plots.

inside the forest canopy was considerable for the stability of the untouched soil layer, which had less coherency due to the frequent movement of tree leaves. Therefore, 10 corner reflectors were deployed and monitored remotely. Due to the fast-moving tree canopy, three reflectors lost their coherency, which was not shown by any CS estimation method. However, all seven of the other reflectors remained stable and continuously monitored.

The real-time displacement of the corner reflectors was retrieved in the secondary interface, as shown in Figure 3.23. It shows real-time monitoring results of

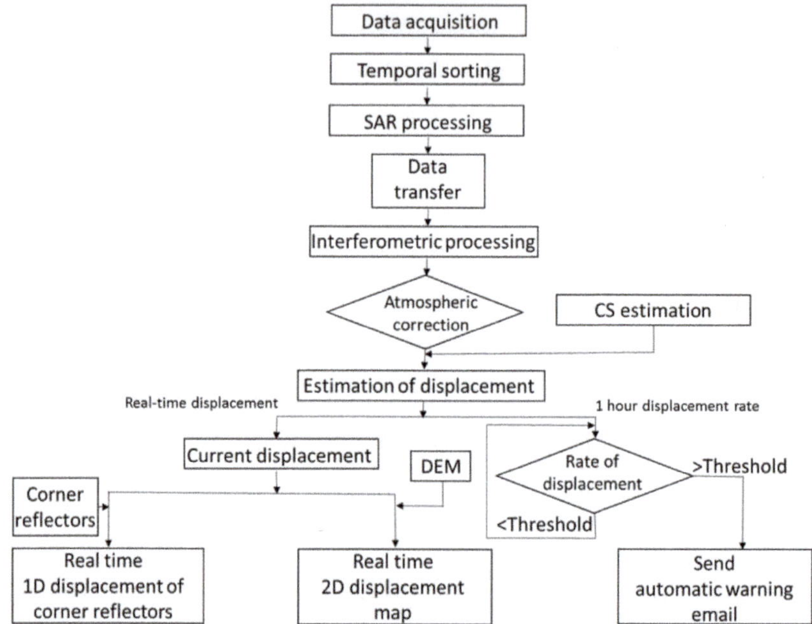

FIGURE 3.21   Data processing diagram of the real-time landslide monitoring system.

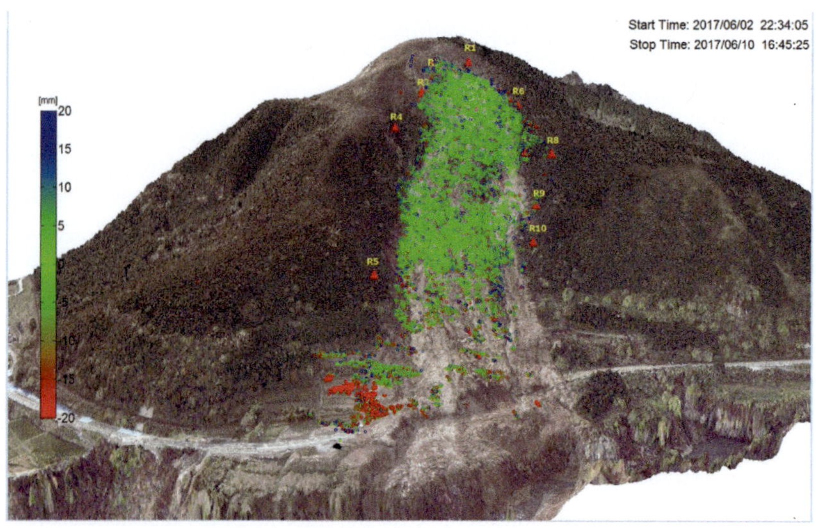

FIGURE 3.22   Real-time 2D displacement map.

7 trihedral corner reflections out of 10. After the start of the rainy season in mid-June, the R1 and R3 reflectors, which were located inside the forest, showed considerable movement from June 15 to 21, as shown in Figure 3.24.

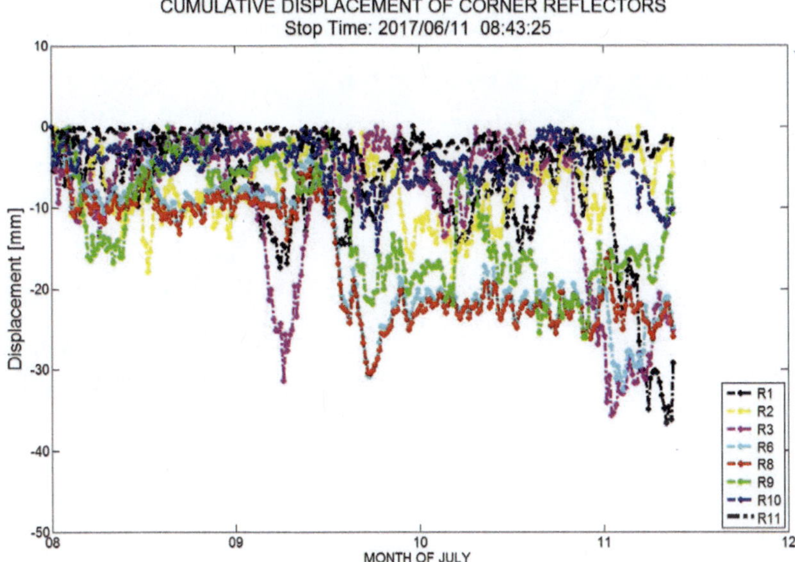

FIGURE 3.23 Displacement of the corner reflector before the rainy season (+10 to –50–mm scale).

3.5.2 SURFACE DISPLACEMENT OBSERVATION

As mentioned in Section 3.1, the main objective of implementing a GB-SAR monitoring station in Kumamoto was to identify the locations that showed abnormal displacement in the landslide area and investigate high possible areas to trigger a sudden slope failure in the near future. In the Minami–Aso GB-SAR monitoring station, the total coverage area was spread around 0.45 km^2. Because this area is wider and includes a steep slope and a dense forest canopy, it is impossible to capture the locations that have a high potential to trigger a landslide within the total observation area by conventional pinpoint measuring techniques. GB-SAR has 4.5m / rad × 0.5m ground patch resolution that covers the total observation area by 2000 × 404 pixels within a minimum of 10 s time lag. This makes it a great advantage to monitor, detect, and inform about the locations that show abnormal displacement. After finding such a location, more precise and accurate ground analysis of a particular location can be performed using non-remote sensing techniques such as extension meters and inclinometers to find the reason for the displacement in the geological background.

3.5.2.1 Estimation of Displacement Using GB-SAR

The theoretical explanation of displacement monitoring by GB-SAR differential interferograms is shown in Figure 3.25. The radar is mounted in the fixed position

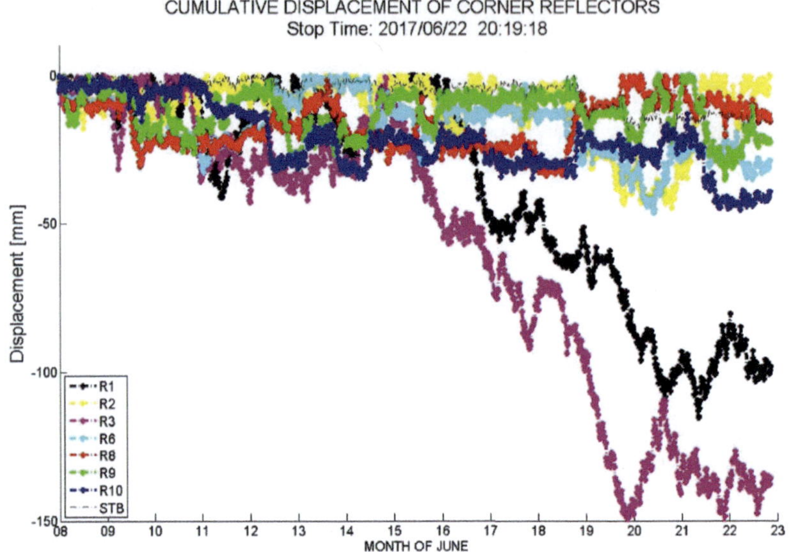

FIGURE 3.24 Displacement of the corner reflector during the rainy season.

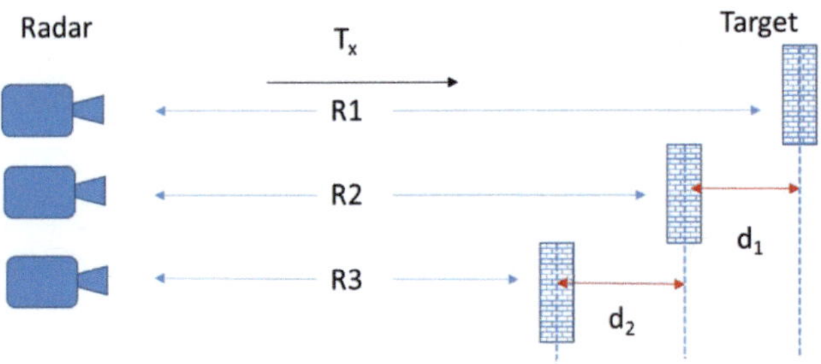

FIGURE 3.25 Schematic diagram of target monitoring by GB-SAR.

on the left side of the figure, and the target is on the right side. ϕ_1 is the calculated phase in T_{R1} time, ϕ_2 is the calculated phase in T_{R2} time, and ϕ_3 is the calculated phase in T_{R3} time. If we assume that the target shows random displacement in the radar LOS direction, the estimated distance between the radar and the target is R1, R2, and R3. Therefore, the displacement of the target in the real world can be estimated as $d_1, d_2 \dots \dots$. In this derivation, we assumed that the target that moves towards the radar is a negative value (blue color scale in 2D maps), the target that moves far from the radar is a positive value (red color scale in 2D maps), and the target monitored as stable in time uses the green color scale. By considering the phase of the received EM wave toward the radar, ϕ_1, ϕ_1, and ϕ_3 can be written as follows:

$$\phi_1 = \frac{2\pi}{\lambda}\left(2R_1\right), \; \phi_2 = \frac{2\pi}{\lambda}\left(2R_2\right), \; \phi_3 = \frac{2\pi}{\lambda}\left(2R_3\right) \tag{3.11}$$

If we assume that the phase difference does not exceed $\pm\,\pi$ (no phase wrapping), the true displacement during consecutive acquisitions can be estimated as follows:

$$\phi_{21} = -\frac{4\pi}{\lambda}\left(R_2 - R_1\right) = -\frac{4\pi}{\lambda}d_1, \; \phi_{32} = -\frac{4\pi}{\lambda}\left(R_3 - R_2\right) = -\frac{4\pi}{\lambda}d_2$$

$$\phi_{31} = -\frac{4\pi}{\lambda}\left(R_3 - R_1\right) = -\frac{4\pi}{\lambda}\left(d_1 + d_2\right) \tag{3.12}$$

$$\phi_{n\ldots1} = -\frac{4\pi}{\lambda}\left(R_n - R_{n-1}\right) = \frac{4\pi}{\lambda}\left|\sum_{r=1}^{n-1}d_r\right| \tag{3.13}$$

The total displacement of each CS point in the 2D InSAR image during the time intervals T_{d_1} and $T_{d_{n-1}}$ can be estimated by $\phi_{n\ldots1}$. However, in natural environmental monitoring, the phase change of each and every pixel will not be considered as a reliable indication of the target movement due to the RCS, the number of noise pixels, and the pixel affected by the atmospheric phase screen. The atmospheric phase screen can be removed by the method described in Chapter 4, and the noise pixel locations can be removed by considering the relevant CS estimation method. These models can be used to minimize observed phase distortions in InSAR pixels, but they cannot guarantee 100% phase compensation. Therefore, to enhance the reliability of the GB-SAR remote sensing technique, the acquired data were compared with non-remote sensing techniques for validation.

3.5.2.2   Estimation of 2D Displacement: Spring Season

GB-SAR monitoring of Minami–Aso was started in January, at the end of the winter. At the beginning of the measurement in January to mid-June, the total monitoring area did not show any critical displacement for the GB-SAR observation. During this time period, a few millimeter displacements were detected only in some temporary wall structures (Figures 3.26(a) and (b)) at the bottom of the mountain, and other places remained stable.

Points P2 and P8, P9, and P11 were selected from the surface soil layer of the landslide area. These points showed high coherency in the interferometric phase. In the 15-day sample data set, P2 and P10 showed similar displacements of less than 4 mm. However, point P11 on the wall structure was bounded near the temporary roads that were constructed to give access to the area in order to remove the deposited debris after the landslide. This process was performed using heavy vehicles. Therefore, a small displacement was recorded from the wall during the day at point P11, depicted in Figure 3.27.

In mid-June, the rainy season started in the southern part of Japan, and it continuously affected all islands from June to August. Minami–Aso in Kumamoto prefecture, located in south Japan, reportedly started rain after June 15. Continuous GB-SAR measurements were carried out from the beginning of the year, including the rainy season, both before and after. After the earthquake triggered the landslide

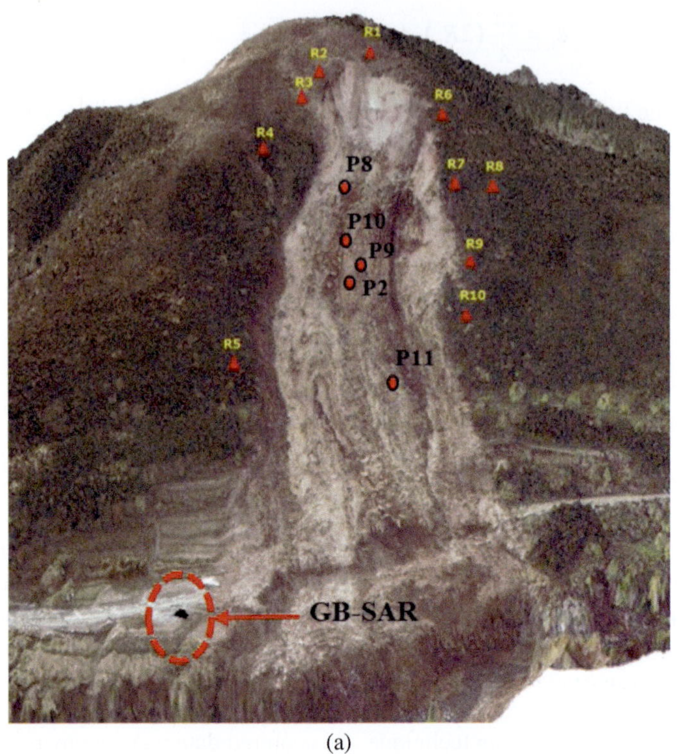

(a)

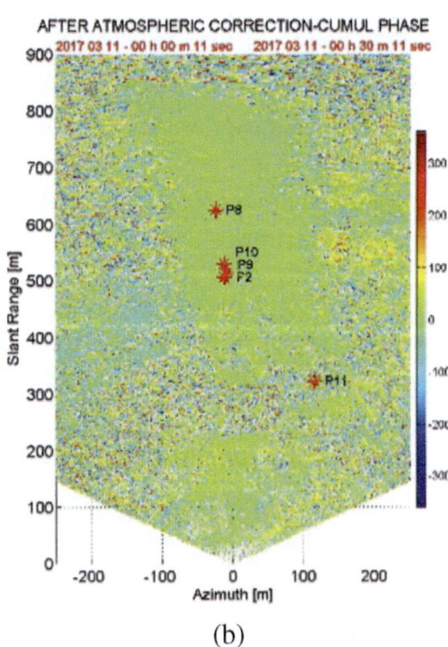

(b)

FIGURE 3.26 Schematic of target monitoring in GB-SAR.

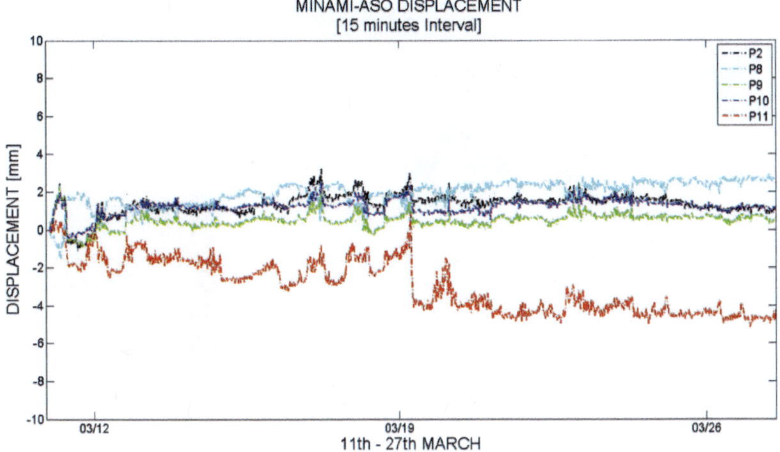

FIGURE 3.27 Recorded displacement of the coherence points during the 15-day period.

in Kumamoto, the unstable soil layer was removed after the previous rainy season in 2016. The stability of the crown of the landslide, in addition to the forest canopy during the rainy season, was not confirmed by any kind of terrestrial field measurement after completion of the soil removal process. Therefore, the GB-SAR monitoring results during the rainy season become an important factor to confirm the stability of the entire area for future construction plans. On the other hand, daily construction work was continuously performed throughout the entire period, including extreme weather conditions. Access to the installed conventional displacement monitoring system was restricted during the rainy period due to the site conditions. Therefore, GB-SAR is the only method that gives reliable remote sensing information from the entire construction site as a real-time displacement monitoring system.

Since the displacement inside the forest canopy was monitored by estimating the movement of the trihedral corner reflected within the GB-SAR illumination area, the reflected signal (Figure 3.26) from the trihedral corner reflector was observed before and during the rainy season.

3.5.2.3 Estimation of Displacement before the Rainy Season

The estimated movement of the corner reflectors is depicted in Figure 3.28(a), and the estimated displacement of the coherent scatters is shown in a green layer in the 3D model of the Minami–Aso area in Figure 3.28(b). Before start the rainy season in the middle of the year, the estimated movement of the corner reflectors of R1 to R10 does not show any continuous movement. They show some sudden fluctuation in the displacement axis, which is possibly due to the sudden change in the surrounding area of the corner reflector by the movement of the tree canopy due to the wind during the GB-SAR monitoring period.

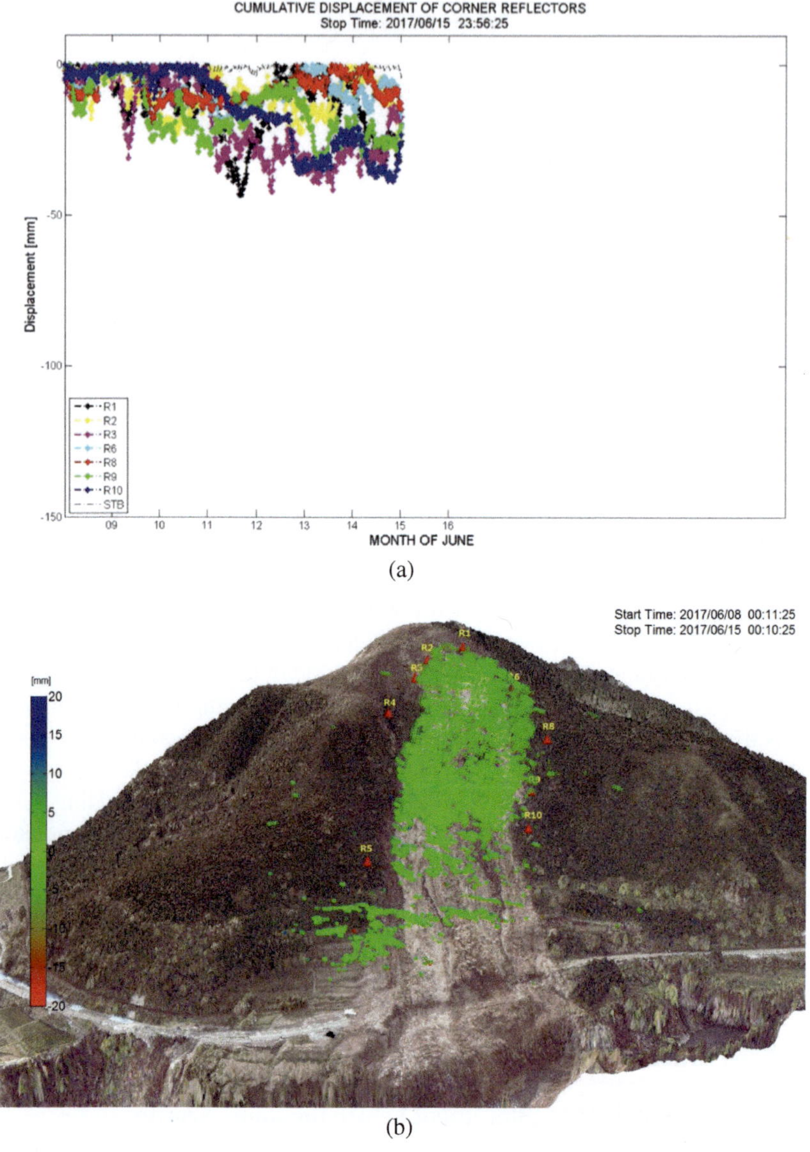

FIGURE 3.28 Recorded displacement from June 8 to 15, 2017, of (a) corner reflectors and (b) coherent scatters.

However, 2D observations of the overall displacement of the coherent scatters in the 3D model (Figure 3.28(b)) confirm that no significant movement in particular parts or regions of the geolugical boundaries was recorded before the start of the rainy season.

3.5.2.4   Estimation of Displacement during the Rainy Season

The rainy season started in June, and the weather conditions at the site changed significantly during the second week of the month. Figure 3.28(a) shows June 15 as a starting period of the movement in the corner reflector R1 to R10 (except R4, R5, and R7, which permanently lose their coherency due to random movements of the surrounding tree clutter which was not considered for displacement estimation) and STB coherent point pick from the open soil layer. The red dotted line in Figures 3.28(a) to 3.33(a) indicates the corresponding monitoring period of GB-SAR, and Figures 3.28(b) to 3.33(b) show the cumulative displacement of the coherent points in the 3D model.

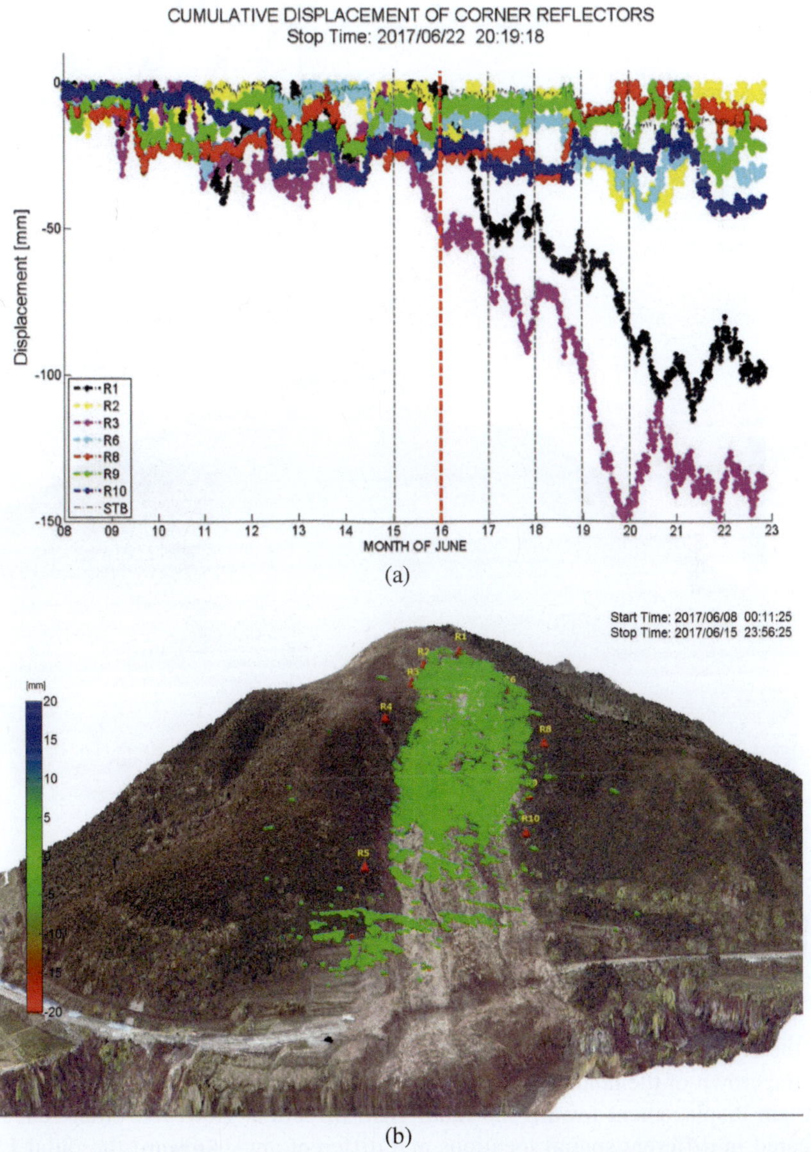

FIGURE 3.29 Recorded displacement from June 8 to 16, 2017: (a) corner reflectors, (b) coherent scatters.

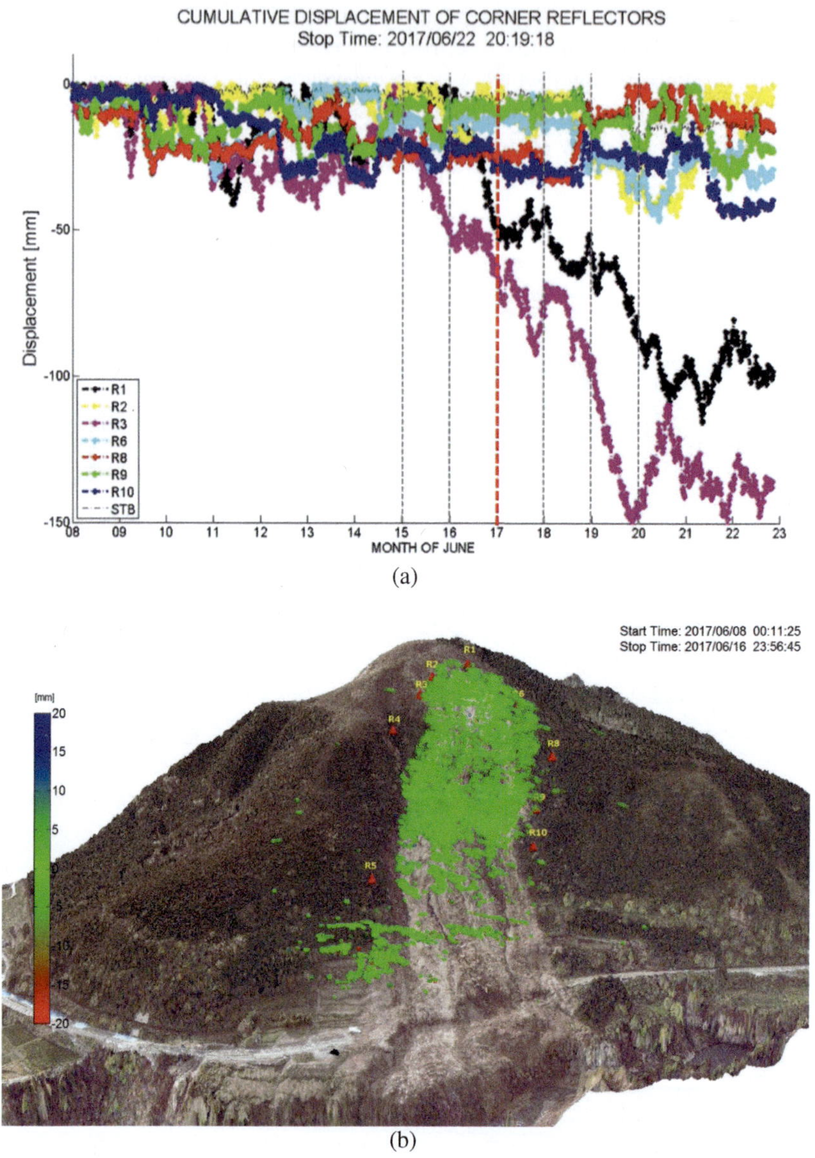

FIGURE 3.30 Recorded displacement from June 8 to 17, 2017: (a) corner reflectors (b) coherent scatters.

After June 15, the GB-SAR monitoring result indicated considerable movement in the corner reflectors. On 15th, reflector R3, which is located in the left corner of the crown of the landslide, started to move, and at the end of the day, it showed 3.2 cm displacement toward the radar LOS direction. However, other reflectors located in different spatial locations and different angles toward the radar LOS, including the nearby reflectors R1, R4, and the coherent points, did not show

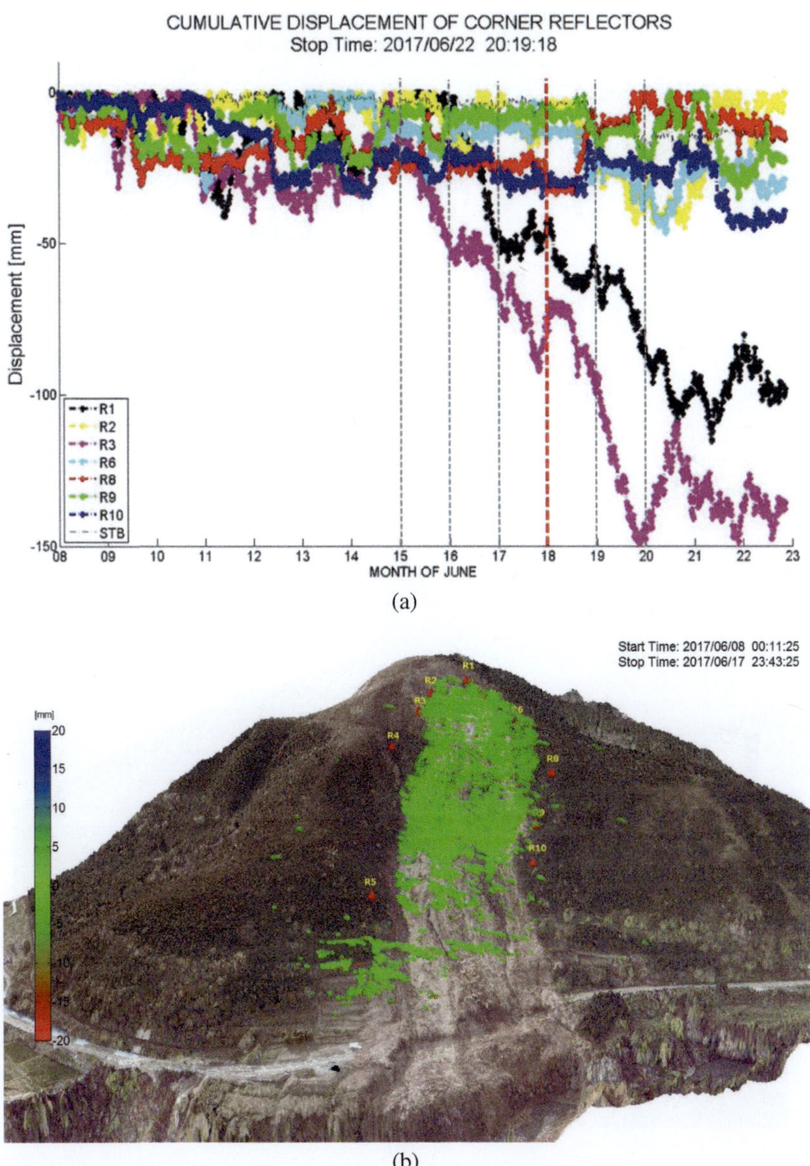

FIGURE 3.31 Recorded displacement during June 8–18, 2017: (a) corner reflectors and (b) coherent scatters.

any correlation or considerable indication of displacement during this period. It indicates that the directional movement of reflector R3 was 308° and showed 0.0003 m/min movement. But in the next 24-h period (June 16–17), reflector R3's moving tendency was dramatically reduced. The other reflectors, R2 and R6 to R10, recorded similar results. This indicates that no continuous movement can be observed in particular locations. Figure 3.32(a) shows that at the beginning of the

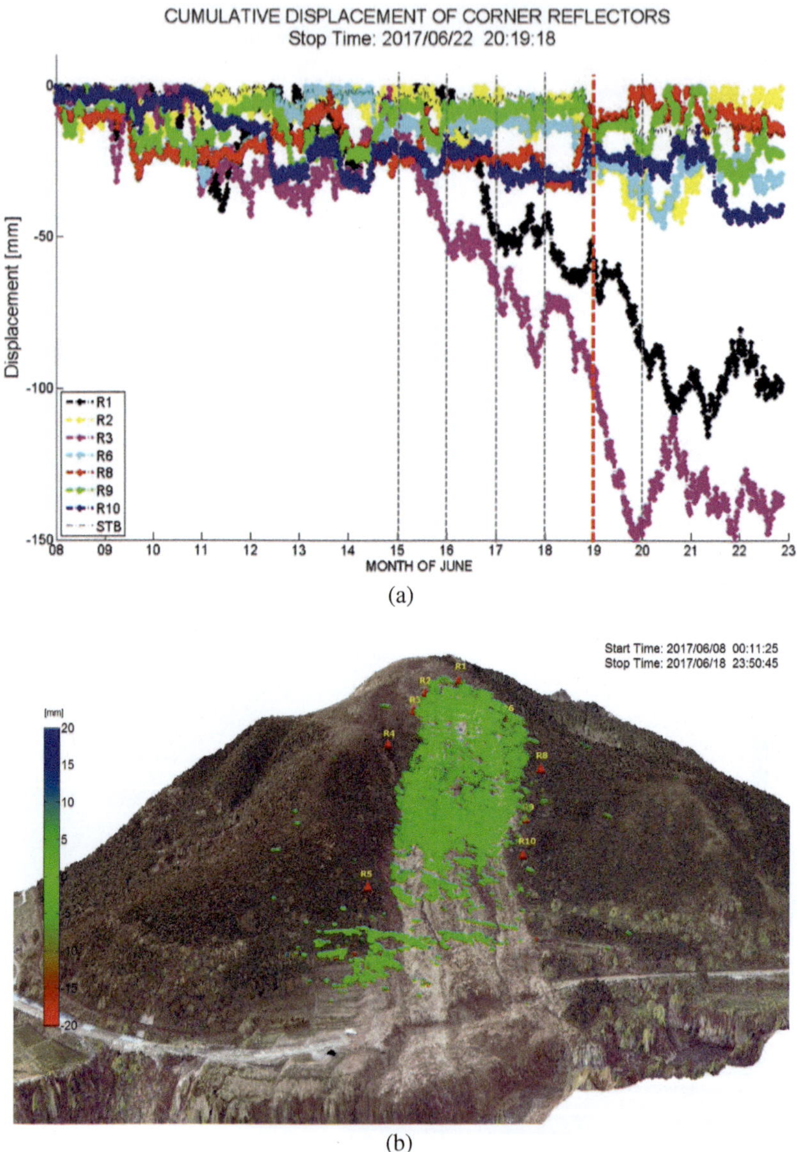

FIGURE 3.32   Recorded displacement during June 8–19, 2017: (a) corner reflectors and (b) coherent scatters.

18th, reflectors R2 to R10 did not indicate any continuous movement. Moreover, rain was not recorded on this day, and the weather conditions of the Minami–Aso area became stable. Figure 3.33(b) depicts the cumulative displacement of coherent scatters from June 8 to 19, 2017. This confirms that no significant displacement in a particular part or region in the post-landslide area was recorded during the period.

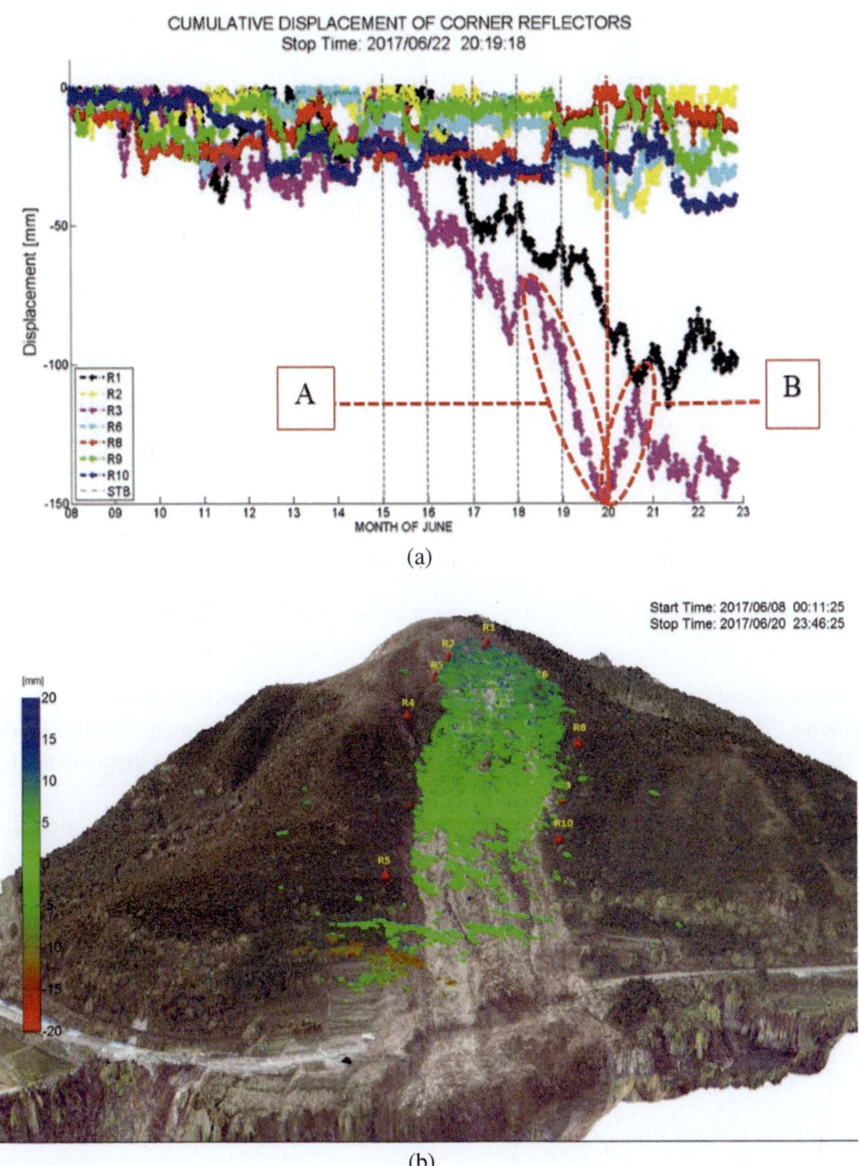

(a)

(b)

FIGURE 3.33 Recorded displacement during June 8–20, 2017: (a) corner reflectors and (b) coherent scatters.

The coherent scatter depicted in Figure 3.33(b) indicates considerable displacement in the crown of the landslide at the end of the day on June 20. The blue color scale indicates that this area moved toward the GB-SAR sensor direction. A similar indication can be observed in the corner reflectors (Figure 3.33(a)) inside the forest area. However, the movement of the corner reflectors and the estimated

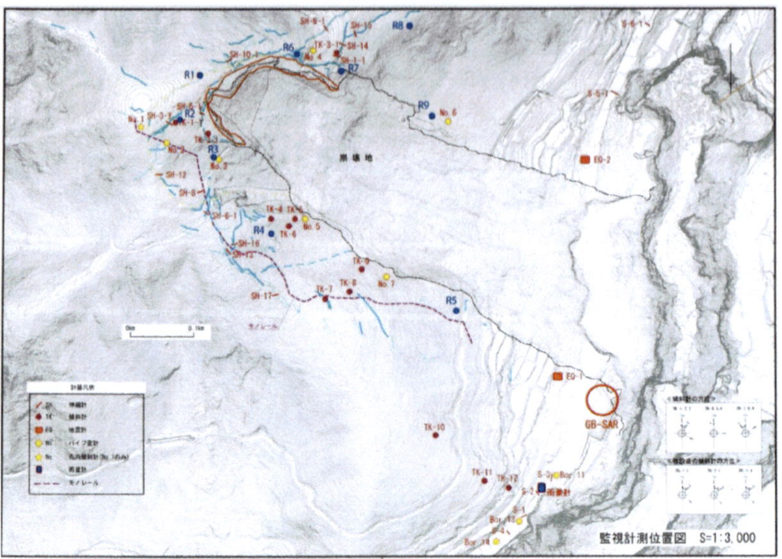

FIGURE 3.34	Locations of the displacement measuring instruments.

displacement of the coherent scatter show a diverse scale for displacement measurements. Therefore, we compare these results with field measurements to distinguish the factors that affected the movement of the corner reflectors and displacement of coherent scatters.

The terrain map of the main landslide and surrounding area is depicted in Figure 3.34. Due to the instability of the top region of the mountain and the newly created fractures (blue lines) zones, some surface movement measuring techniques were implemented in the most critical locations. To observe lateral movement in fracture zones, extensometer measurements were examined. The installed extensometer and inclinometer area are depicted in orange and red dots in Figure 3.34. The GB-SAR measurements show considerable displacement in the crown of the landslide during the rainy season during July 15 to 21, 2017 [15].

### 3.5.2.5	Ground Displacement Measurement—Discussion

In this campaign, GB-SAR was operated as a real-time remote landslide early warning system that monitors large areas in a short time. The main objective of implementing a GB-SAR monitoring station in Kumamoto is to identify the locations that show abnormal displacement in the post-landslide area and further investigate highly possible areas that could trigger a sudden slope failure in order to prevent harmful situations for the people who actively involved with road reconstruction work at the bottom of the mountain. This situation is quite similar to GB-SAR deployment in the open-pit mining industry worldwide. There, the measurement accuracy of each and every DInSAR image pixel and its movement were not considered critical issues since the overall 2D displacement image can

TABLE 3.7

Extensometer Readings from 06/07/2017 to 06/21/2017

Reading #	Extensometer	Displacement (mm)	Nearby CS/Corner Reflector
1	SH-3-1	0.3	CR 2
2	SH-9-1	0.6	CR 6
3	SH-15	0.4	CS 1
4	SH-16	0.6	CR 3
5	SH-17	0.4	CS 2

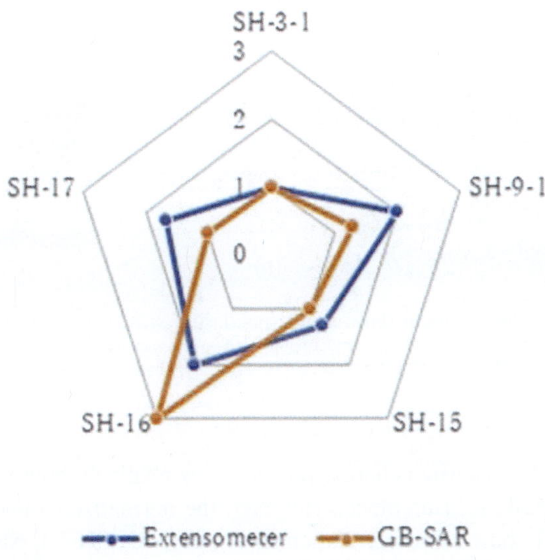

FIGURE 3.35 Comparison of the extensometer meter measurements and displacement of coherent scatters estimated by GB-SAR.

distinguish the area that is stable and an area that has a high potential to trigger slope failure.

However, in the case of post-landslide monitoring sites such as Minami–Aso, the probability of trigging another landslide is minimal. Therefore, the accuracy of the estimated displacement is also important to estimate the long-term stability of critical areas such as fracture zones. Table 3.7 summarizes the displacement data measured by the extensometer [16]. Figure 3.35 depicts the comparison of the extensometer measurements and estimated movement of the nearby coherent point. The extensometer readings indicate the lateral movement between fracture locations bound between lower scales.

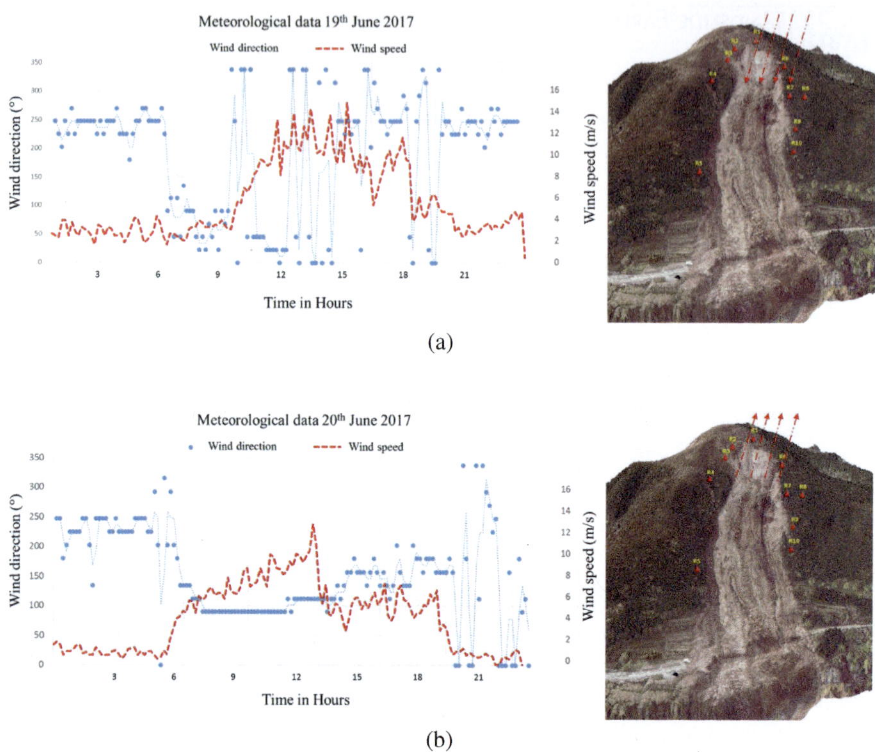

FIGURE 3.36 Meteorological data, wind speed, and wind speed during June (a) 19 and (b) 20, 2017.

The displacement of nearby coherent points shows relatively high values when compared with GB-SAR measurements. However, the normalized values of both pieces of the equipment show some correlation in Figure 3.35; SH-3-1, SH-9-1, and SH-15 for scaling factor 30. This could happen because GB-SAR measures the displacement in the radar LOS direction, but the field measurement has a slight deviation, which can be estimated by terrain slope information in the future. However, extensometer meter SH-16 and corner reflector point R3 show clear differences since R3 shows movement in alternative directions in Figure 3.33(a), marked A and B. This was not shown in the interferometric results in Figure 3.33(b) and only affects the corner reflectors. Theoretically, this type of phase accumulation could occur (Figure 3.25) if the corner reflector undergoes vibration by an external force, such as a strong wind. The meteorological data at a nearby weather station recorded that on the 19th, the fast wind flow (Figure 3.36(a) during 09:00–2100) moved toward the radar in the northwest direction, and on June 20, fast wind (Figure 3.36(b) during 06:00–19:00) moved outwards from the radar in the southeast direction. It adds an additional phase error that does not relate to ground displacement. This type of error can be overcome by averaging the multiple CR phase components prior to estimating the displacement in the future.

3.5.3 Landslide Early Warning System

3.5.3.1 Automatic Cluster Movement Tracking System

In a practical manner, automatic early warning becomes a necessity in order to send a message to people involved in road construction work. In this automatic early warning system, the upper part of the radar illumination area in the landslide location was divided into 12 clusters (Figure 3.37), and the mean displacement of each cluster was dynamically calculated.

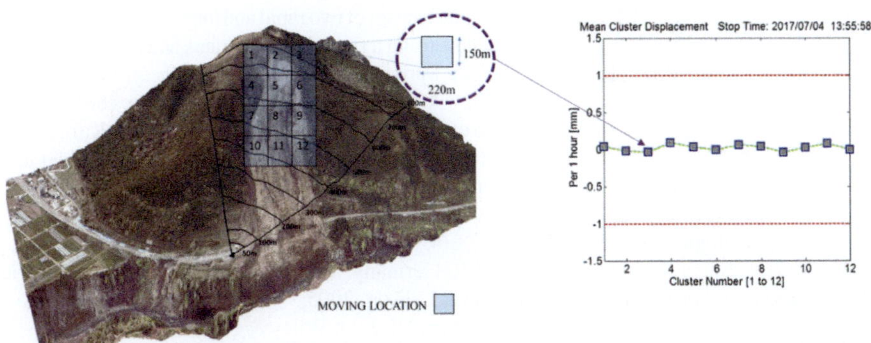

FIGURE 3.37   Automatic cluster displacement tracking system.

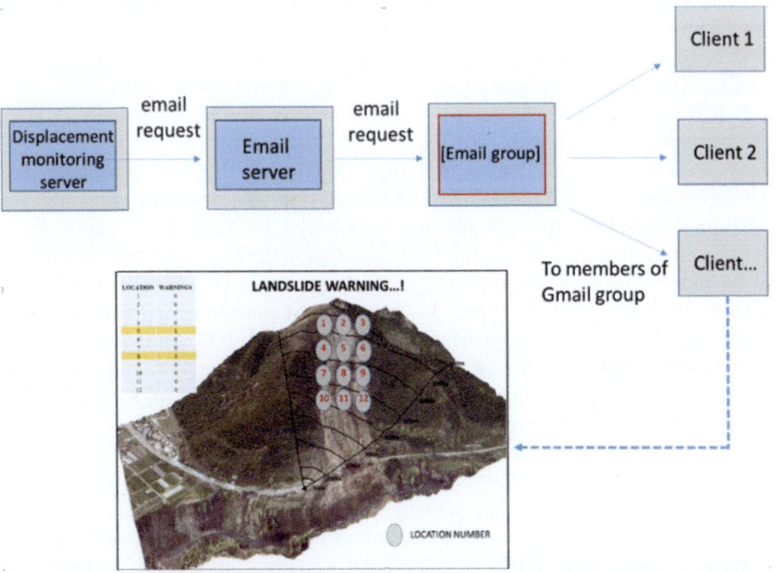

FIGURE 3.38   Automatic warning system.

3.5.3.2 Automatic Email System

The automatic email system was configured to activate, if the cluster displacement exceeds the threshold level of 1 mm per hour. Then, the monitoring PC generates an automatic email and sends it to the responsible parties to investigate the area by immediate site inspection [17].

REFERENCES

[1] A. Karunathilake and M. Sato, "3D model assisted survey to design and estimate ground based SAR illumination," in 42nd Remote Sensing Symposium, Chiba. 2017, pp. 35–38.

[2] K. Dang, K. Sasa, H. Fukuoka, et al., "Mechanism of two rapid and long-runout landslides in the 16 April 2016 Kumamoto earthquake using a ring-shear apparatus and computer simulation (LS-RAPID)," Recent Landslides, Springer-Verlag Berlin Heidelberg, August 2016. Online resource, Wikipedia. https://en.wikipedia.org/wiki/2016_Kumamoto_earthquakes

[3] Y. Yagi, R. Okuwaki, B. Enescu, et al.," Rupture process of the 2016 Kumamoto earthquake in relation to the thermal structure around Aso volcano," Earth Planet and Space, Springer, vol. 68, no. 118, pp 1–6, Jul. 2016.

[4] H. K. Heidemann, LiDAR base specification, chapter 4 of section B, Virginia, USA: U.S. Geological Survey Standards, 2014.

[5] U. Wandinger, Introduction to LiDAR, Germany: TeLeibniz Institute for Tropospheric Research, Permoserstraße 15, D-04318 Leipzig, 2005.

[6] C. A. Balanis, Antenna theory, Hoboken, NJ: John Wiley & Sons, 2005.

[7] P. A. Zandbergen, PYTHON scripting for ArcGIS, Redland: Esri Press, 2013.

[8] P. A. Burrough, M. F. Goodchild, R. A. McDonnell, et al., Principles of geographical information systems, Oxford, NY: Oxford University press, Great Clarendon Street, Oxford ox2 6DP, 2000.

[9] G. Luzi, M. Pieraccini, D. Mecatti, et al., "Ground-based radar interferometry for landslides monitoring: Atmospheric and instrumental decorrelation source on experimental data," IEEE Trans. Geosci. Remote Sens., vol. 42, no. 11, pp 2454–2466, Nov. 2004.

[10] G. Nico and J. Fortuny, "Using the matrix pencil method to solve phase unwrapping," IEEE Trans. Signal Process., vol. 51, no. 3, pp. 886–888, Mar. 2003.

[11] R. F. Hanssen, Radar interferometry: Data interpretation and error analysis, Dordrecht, The Netherlands: Kluwer, 2001.

[12] M. S. Seymour and I. G. Cumming, "Maximum likehood estimation for SAR interferometry," in Proceeding IGARSS-Surface Atmospheric Remote Senseing: Technologies, Data Analysis Interpretation, vol. 4, pp. 2272–2275, Aug. 8–12, 1994.

[13] E. K. Smith Jr. and S. Weintraub, "The constant in the equation for atmospheric reflective index at Radio frequency," Proc. I. R. E, pp 1035–1037, October 1953.

[14] O. A. Alduchov and R. E. Eskridge, "Improved Magnus form approximation of saturation vapor pressure," J. Appl. Meteol., vol. 35, pp. 601–609, April 1996.

[15] A. Karunathilake, L. Zou, K. Kikuta, et al., "Implementation and configuration of GB-SAR for landslide monitoring: Case study in Minami-Aso, Kumamoto," Explor. Geophys., vol. 50, no. 2, pp. 210–220, Apr. 2019.

[16] Online Data Achieve, Japan Meteorological Agency (JMA). http://www.jma.go.jp/en/warn/f_4343300.htm

[17] A. Karunathilake, "An implementation and configuration of ground based synthetic aperture radar for environmental monitoring: Chapter 6," Doctoral dissertation, Tohoku University, 2017. https://tohoku.repo.nii.ac.jp/record/124499/files/170925-KARUNATHILAKE-245-1.pdf

4 Atmospheric Phase Delay

4.1 INTRODUCTION

The velocity of an emitted electromagnetic (EM) wave undergoes modulation as it traverses the tropospheric medium, which is influenced by the atmospheric refractive index. The temporal variation in the distribution of the refractive index introduces an additional differential phase in InSAR, commonly referred tó as APS. While APS is generally regarded as a nuisance in deformation monitoring, certain applications have repurposed this term to map atmospheric water vapor distribution [1–3]. Initial observations of phase disturbances due to atmospheric delay were documented over Mount Etna and Landers, California [4, 5]. Further analysis of the APS was performed using data acquired by the Spaceborne Imaging Radar-C and X-band SAR (SIR-C/X-SAR) sensor over the Mojave Desert in California and the island of Hawaii, with the aim of evaluating its impact on the InSAR performance [6–8].

Many attempts to estimate atmospheric water vapor have been made with external auxiliary data sets in the context of spaceborne SAR applications. One approach involves using global positioning system (GPS) measurement data [9–11]. Measurement of atmosphere-induced path delay by GPS is exploited for APS correction. Another strategy employs integrated water vapor data from multispectral sensors such as the medium resolution imaging spectrometer (MERIS) and the moderate resolution imaging spectroradiometer (MODIS) [12–14]. In addition, a different approach uses numerical weather prediction (NWP) models, such as the fifth-generation Penn State/NCAR mesoscale model (MM5) and the weather research and forecasting (WRF) model or global meteorological reanalysis data to predict 3D fields of the refractive index even in cloudy regions [15–18].

On the other hand, the use of multi-temporal SAR acquisitions has been proposed as an alternative method for APS removal, eliminating the need for external data sets [8]. The technique involves averaging multiple interferograms, referred to as the stacking SAR interferogram technique, which mitigates APS by assuming the temporal uncorrelatedness of the phenomenon [19–21]. Advanced time-series InSAR algorithms also offer a potential solution by isolating APS through spatial low-pass and temporal high-pass filtering, assuming that APS exhibits spatially lower and temporally higher frequency signals compared with deformation [22–26].

Given that observations occur within the troposphere, the interferometric phase of GB-SAR signals is significantly affected by atmospheric phase delay. APS emerges as the most pertinent disturbance in ground-based radar measurements, potentially eliminating the spatial signature of actual land deformation. This becomes particularly critical in near-real-time monitoring scenarios, such as hazard anticipation for steep mountainous slopes, where strong APS effects can trigger false alarms in early warning systems. Consequently, compensation for APS is crucial to provide reliable displacement values in such applications.

DOI: 10.1201/9781003316312-4

Because of differences in observation geometry, spatial scale, and data acquisition intervals between GB-SAR and spaceborne SAR systems, the approach for APS correction also diverges. Notably, external data sets such as GPS, multispectral data, and NWM are no longer applicable in the context of GB-SAR. This limitation arises primarily from the resolution of these external products, which is comparable to the spatial extent of GB-SAR observations. Additionally, the observation geometry of these external data sets differs from that of GB-SAR [27].

Furthermore, the assumption of the temporal uncorrelatedness of APS, commonly employed in spaceborne SAR applications, becomes uncertain in the case of near-real-time GB-SAR monitoring. This uncertainty stems from the fact that the acquisition interval for GB-SAR is typically much shorter than that for spaceborne SAR [27]. Consequently, direct adaptation of conventional spatio-temporal spectrum filtering methods is not feasible in this particular case.

4.2 THEORETICAL BACKGROUND

4.2.1 REFRACTIVITY CHANGE EFFECT IN THE PROPAGATION PHASE

GB-SAR observations are typically conducted within the troposphere. In this context, the two-way propagation phase of the electromagnetic wave, backscattered from the slant range distance r_s, can be expressed as follows:

$$\varphi = \frac{2\pi}{\lambda} 2r_s + \varphi_{atm} + \varphi_{scat} + \varphi_{noise} + 2\pi k. \tag{4.1}$$

where φ_{atm}, φ_{scat}, and φ_{noise} represent the atmospheric phase, the scatterer's intrinsic phase, and the noise term, respectively. As shown in (4.1), the backscattered EM wave propagating through the troposphere is influenced by various phase factors, in addition to the propagation phase term $\frac{2\pi}{\lambda} 2r_s$. The atmospheric phase φ_{atm} can be expressed in terms of the refractive index $n_r(r_s,t)$, a spatiotemporal function of temperature T (in Kelvin), pressure P (in millibars), and the partial pressure of water vapor e at time t, as:

$$\varphi_{atm}(t) = \frac{4\pi}{\lambda} \int_l n_r(r_s,t)\,dr_s. \tag{4.2}$$

where λ accounts for wavelength.

When the signal propagates through the atmosphere, $n_r(r_s,t)$ varies only slightly more than the refractive index of vacuum (i.e., $n_r = 1$) [1]. Therefore, n_r is generally expressed as

$$n_r = 1 + 10^{-6} N_r, \tag{4.3}$$

where N_r is referred to as a refractivity index and acts as a scale-up index. For radio frequency, N_r is empirically approximated as follows [28]:

$$N_r = \frac{77.6 \cdot P}{T} + \frac{3.37 \cdot 10^5 \cdot e}{T^2} = N_{hyd} + N_{wet} \tag{4.4}$$

where e is defined as

$$e = \frac{6.107 \cdot H}{100} \exp\left(\frac{12.27(T-273)}{T-35.85}\right), \tag{4.5}$$

where H % is the relative humidity. The two components in (4.4), N_{hyd} and N_{wet}, are called the hydrostatic (dry) term and the wet term of the refractivity index. Temporal variations in the refractivity index result in the APS, which is predominantly influenced by the relative humidity H. According to (4.4), at low temperatures, N_{hyd} is a dominant term, and variations in H do not significantly impact the refractivity index [29]. In contrast, H is the primary contributor to refractivity index variations as temperature increases [30].

The APS can be derived by computing the difference in atmospheric phase with respect to the temporal separation Δt between times t_1 and t_2.

$$\phi_{APS}(\Delta t) = \varphi_{atm}(t_1) - \varphi_{atm}(t_2) = 10^{-6}\frac{4\pi}{\lambda}\left(\int_l N_r(r_s,t_1)dr_s - \int_l N_r(r_s,t_2)dr_s\right). \tag{4.6}$$

In accordance with (4.6), it is evident that modeling N_r along the propagation path provides the solution for ϕ_{APS}. This deterministic approach involves approximating the spatial distribution of N_r to estimate ϕ_{APS}, demonstrated in Section 4.3.1.

4.2.2 STOCHASTIC MODELING

Atmospheric turbulence is generated by solar heating of the Earth's surface by the sun, resulting in a winding process characterized by irregular motion in terms of both direction and velocity. The large-scale air motion is eventually fragmented into smaller-scale motions in a self-similar cascade, leading to the formation of randomly sized and distributed pockets of air known as turbulent eddies [31]. This cascading process continues until the energy is dissipated from the large spatial scales to the inner spatial scale [32]. These turbulent wind vortices act as carriers of atmospheric constituents, such as water vapor [1]. Consequently, a highly heterogeneous and random refractive distribution emerges, producing turbulent APS effects in radar interferometry.

Under such refractivity distributions, the wavefront of the EM wave no longer remains planar after propagation. The Kolmogorov turbulent theory provides a description of the structure-function of the spatial variation of the refractivity index, exhibiting a power-law distribution with spatial frequency within both inner and outer spatial scales [32].

Precise analysis of the structural characteristics employing statistical measures such as covariance and variogram proves to be suitable for modeling the stochastic contribution of APS based on the Kolmogorov turbulence theory. To apply such statistical measures, stationarity must be defined in accordance with the random function (RF) property.

When the probability density function remains invariant under the arbitrary translation of points by a vector, the RF has strict stationarity. Recognizing that strict stationarity is a stringent requirement, alternative stochastic measures are defined under more relaxed stationarity conditions, specifically second-order stationarity and intrinsic stationarity. Second-order stationarity assumes the stationarity of the first and second moments (i.e., mean and variance) of a random function. Under this stationarity, the mean m and covariance C exist at an arbitrary point x with a separation vector h [33], given as follows:

$$E\left[\phi(x)\right] = m,$$

$$E\left[\phi(x)\phi(x+h)\right] = C(h). \tag{4.7}$$

where $\phi(x)$ represents the interferometric phase, which is the random variable realized from the RF. Consequently, the mean is constant, and the covariance solely depends on the separation vector h. If the covariance depends only on the length and not on the direction, it is termed isotropy. The covariance function is defined as an even function as well as a positive definite function, and it is bounded by its value at the origin [34].

Intrinsic stationarity is a more relaxed hypothesis than second-order stationarity and assumes the second-order stationarity of the increment $\Delta\phi(x+h,x) = \left[\phi(x) - \phi(x+h)\right]$, characterized by

$$E\left[\phi(x+h) - \phi(x)\right] = \Delta m,$$

$$\mathrm{var}\left[\phi(x+h) - \phi(x)\right] = 2\gamma(h). \tag{4.8}$$

where var[·] indicates the variance operator, Δm is the mean of the increment, referred to as drift, and $\gamma(h)$ is the variance of increment, also known as a variogram. Thus, the variogram is defined as

$$\gamma(h) = \frac{1}{2}\mathrm{var}\left[\phi(x+h) - \phi(x)\right], \tag{4.9}$$

which depends only on the separation vector h. The variogram quantifies dissimilarity with respect to separation h and is equivalent to the structure function, which is often referred to in geostatistics.

If the random function has second-order stationarity, it has intrinsic stationarity, and the variogram exists. Consequently, the variogram can be derived from the following covariance:

$$\gamma(h) = C(0) - C(h). \tag{4.10}$$

However, the covariance cannot always be derived from the variogram because it is not necessarily a bounded function (e.g., fractional Brownian motion has an unbounded variogram). If the variogram is bounded by a finite value, the covariance can be obtained.

4.3 COMPENSATION METHODS

The compensation for APS is a critical step in GB-SAR deformation monitoring and is applied to all generated interferograms before estimating deformation. The general correction approach involves modeling the spatial distribution of atmospheric refractivity, which is known as the model-based approach. However, in many cases, ϕ_{APS} cannot be estimated deterministically, especially when spatially local heterogeneity in the refractivity index is induced by turbulent processes in the troposphere, which should be described stochastically [1]. Therefore, ϕ_{APS} can be modeled with both deterministic and stochastic contributions as follows:

$$\phi_{APS} = \phi_{model} + \phi_{res} = X\beta + \phi_{res}. \tag{4.11}$$

where ϕ_{model} is the APS component given by the deterministic approach; ϕ_{res} is the residue treated as stochastic in this section; X is a matrix of regressors, which are functions of the coordinates at corresponding locations; β is the vector of regression coefficients; and ϕ_{res} is described by spatial covariance or variogram, assuming spatial correlation with the geostatistical solution.

This section illustrates the application of both model-based and geostatistical solutions in GB-SAR interferometric monitoring. Two case studies employing a model-based approach are presented, covering scenarios involving soft and steep topography. Additionally, the efficacy of these methods is evaluated using the Ku-band GB-SAR data set.

4.3.1 Model-Based Approach

The model-based approach is aimed at estimating the APS component, which has a low spatial frequency. An approximation of the refractivity index distribution in (4.6) is attempted in a deterministic manner. Therefore, the effectiveness of the APS correction is highly dependent on the assumption of the spatial distribution of N_r. In this chapter, two types of approximated N_r models are demonstrated in the following subsections.

4.3.1.1 Soft Topography Scenarios

Under spatially homogeneous refractivity distribution scenarios, such as soft topography areas, indicating the position independence of N_r, the unwrapped APS in (4.6) can be simplified as follows:

$$\phi_{linear} = 10^{-6}\frac{4\pi}{\lambda}\left(\int_l N_r(r_s,t_1)dr_s - \int_l N_r(r_s,t_2)dr_s\right) = 10^{-6}\frac{4\pi\Delta N_r}{\lambda}r_s = \beta_1 r_s, \tag{4.12}$$

where $\Delta N_r = N_r(t_1) - N_r(t_2)$, and β_1 is the unknown coefficient. Consequently, the APS under homogeneous reflectivity conditions can be expressed as a linear relationship with respect to the position in the range [35, 36]. The unknown coefficient β_1 can be estimated by linear regression over coherent pixels. An estimated linear phase-ramp is then subtracted from the measured interferograms to achieve compensation. Only stable pixels with less decorrelation, referred to as CS, are considered for the estimation. CSs are typically selected by interferometric coherence or DA criteria, demonstrated in Chapter 2.

The Ku-band GB-SAR images obtained by Tohoku University were analyzed to illustrate the linear behavior described in (4.12). The study focused on a site situated on Mt. Kurikoma in Kurihara, Miyagi, Japan, where a landslide formed a cliff approximately 150 m in height, as depicted in Figure 4.1. Figure 4.2(a) displays the differential phase image, showing only stable pixels, derived from GB-SAR images acquired at 02:00 and 03:34 on June 8, 2012. Additionally, Figures 4.2(b) and (c) represent the estimated APS and the compensated differential phase, respectively.

In Figure 4.3(a), the measured differential phase (indicated by black dots) and the estimated APS (indicated by red circles) based on the approximate model in (4.12) are projected on the slant-range axis. Figure 4.3(b) illustrates the compensated differential phase.

The measured differential phase depicted in Figure 4.3(a) demonstrates an almost linear behavior with respect to the slant range, indicating a homogeneous refractivity distribution. Also, the estimated APS shown in Figure 4.2(b) aptly captures the overall trend of the APS. As a result, the compensated differential phase results, as illustrated in Figures 4.2(c) and 4.3(b), indicate a significant improvement due to the compensation process.

FIGURE 4.1　Observed post-landslide cliff located on Mt. Kurikoma, Kurihara, Miyagi, Japan.

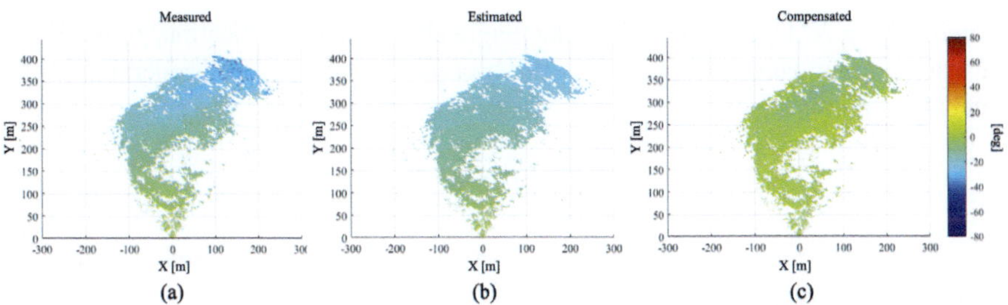

FIGURE 4.2 APS compensation result, displaying only CSs. (a) Measured differential phase image formed by GB-SAR images acquired at 02:00 and 03:34 June 8, 2012; (b) estimated APS; (c) compensated differential phase.

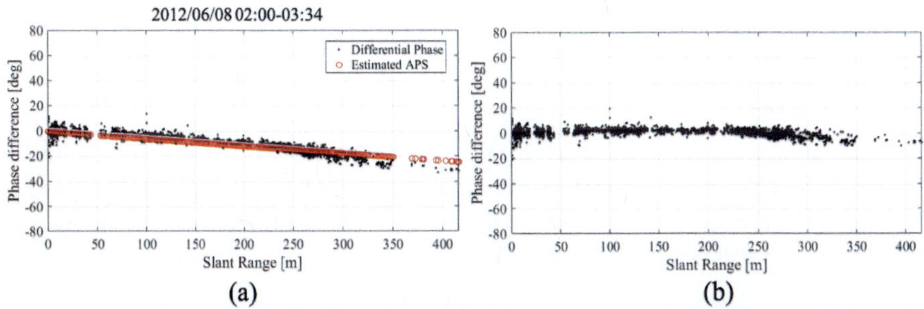

FIGURE 4.3 (a) Differential phase (indicated by black dots) and estimated APS (indicated by red circles) at CSs projected on the slant-range axis. (b) Compensated differential phase projected on the slant-range axis.

4.3.1.2 Steep Topography Scenarios

The assumption of a homogeneous refractivity distribution in (4.12) is no longer applicable to a steep topographic field. In such cases, the refractivity index N_r is presumed to be related to the topographic height z through an exponential profile [37]:

$$N_r(z) = N^s e^{-\alpha z}, \tag{4.13}$$

where N^s is the refractivity index at sea level, z is the height above sea level, and α represents the decay parameter. This expression is further approximated by the first two terms of the Taylor series as follows [29]:

$$N_r(z) = N^s - N^s \alpha z. \tag{4.14}$$

By substituting (4.14) into (4.6), the unwrapped stratified APS can be obtained as follows:

$$\phi_{str} = 10^{-6} \frac{4\pi}{\lambda} \left(\Delta N^s r_s - \Delta N^s \alpha \frac{r_s z_d}{2} \right) = \beta_1 \cdot r_s + \beta_2 \cdot z_d \cdot r_s = \beta_1 \cdot x_1 + \beta_2 \cdot x_2, \tag{4.15}$$

where $\Delta N^s = N^s\left(t_1\right) - N^s\left(t_2\right)$, z_d is the height above the radar location, β_1 and β_2 are unknown parameters, and x_1 and x_2 are observation variables. For generality, (4.15) includes a phase offset β_0:

$$\phi_{str} = \beta_0 + \beta_1 \cdot x_1 + \beta_2 \cdot x_2. \tag{4.16}$$

The matrix notation of (4.16) dealing with all k candidate pixels are given as follows:

$$\phi = X\beta + \epsilon, \tag{4.17}$$

where ϕ is a $[k \times 1]$ vector of the measured interferometric phase, β is a $[3 \times 1]$-dimension vector of unknown parameters, and X is a $[k \times 3]$-dimension vector of observation variables. Ordinary least square (OLS) inversion solves for the unknown parameters β as

$$\hat{\beta} = \left(X^T X\right)^{-1} X^T \phi. \tag{4.18}$$

Finally, the estimated unknown parameters are substituted into (4.17), and the stratified APS can be estimated as follows:

$$\hat{\phi}_{str} = X\hat{\beta}. \tag{4.19}$$

It is essential to note that the selected k CSs must be spatially unwrapped before the inversion to make the system in (4.17) linear.

The GB-SAR data acquired over a mountainous slope in Kumamoto, Japan, were analyzed to demonstrate APS behavior in a steep topographic scenario. The deployed GB-SAR system, MetaSensing Fast GB-SAR, is an FMCW radar, operating at the Ku-band (17.2 GHz) center frequency with a 300-MHz frequency bandwidth, as shown in Figure 4.4.

Figure 4.5(a) showcases the differential phase image, showing selected CSs observed over the post-landslide mountainous field in Minami–Aso, Kumamoto, Japan. This interferogram is formed by two GB-SAR images acquired at 01:27 and 01:42 on November 11, 2018. Figure 4.5(a) displays the differential phase change, characterized by low spatial frequency, attributed to the temporal change in stratified refractivity. The measured differential phase at the CSs is then used for APS estimation through (4.18) and (4.19). The measured differential phase and estimated APS at CSs projected on a slant-range axis are presented in Figure 4.6(a), where black dots and red circles indicate the measured differential phase and estimated APS, respectively.

Observing Figure 4.6(a), the differential phase deviates from ~450 m, and this trend is no longer linear. The estimated APS, based on the stratified refractivity model in (4.13), captures the overall trend of the differential phase, as depicted in Figure 4.5(a). Consequently, APS compensation is achieved by subtracting the estimated APS from the measured differential phase, as shown in Figures 4.5(c) and 4.6(b).

FIGURE 4.4 GB-SAR system deployed for monitoring post-landslide mountainous slopes at Mt. Aso, Kumamoto, Japan.

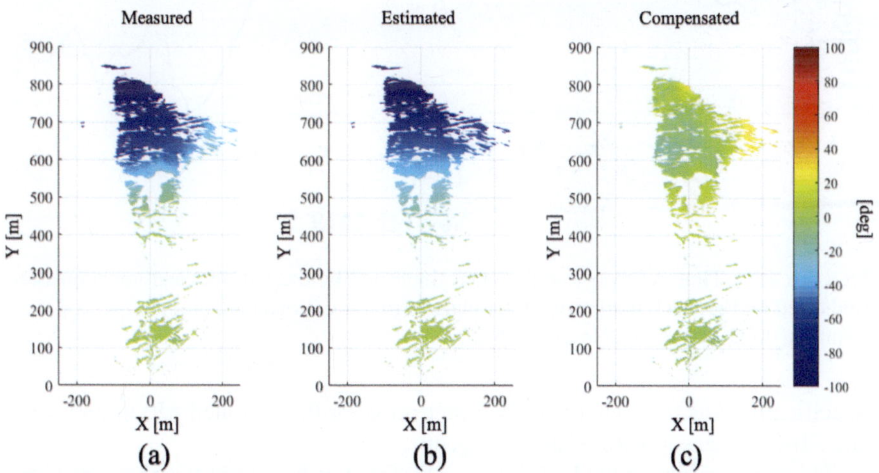

FIGURE 4.5 Results of stratified APS compensation, displaying only CSs. (a) Measured differential phase formed by GB-SAR images acquired at 01:27 and 01:42 on November 6, 2018. (b) Estimated APS. (c) Compensated differential phase.

To demonstrate the validity of the 24-hour time-series results, Figure 4.7 compares the differential phase time series of APS-compensated (red line) and uncompensated (black line) data sets. The time series spans 00:12 to 23:57 on November 6, 2018, computed by the cumulative sum of consecutive phase differences. Figure 4.7

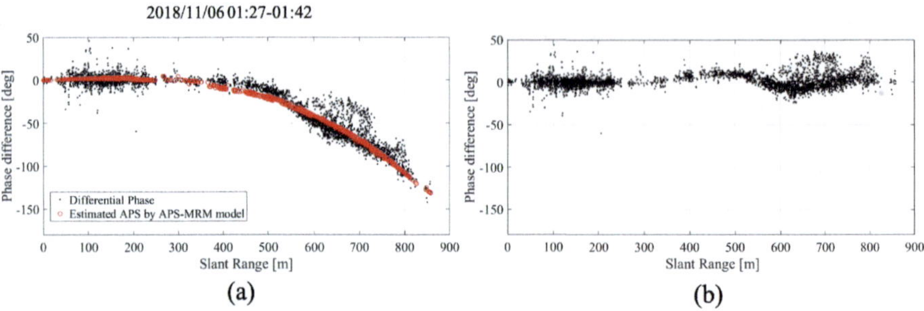

FIGURE 4.6 (a) Differential phase (indicated by black dots) and estimated stratified APS (indicated by red circles) at sparse points out of CPs projected on the slant range axis. (b) Compensated differential phase projected on the slant range axis.

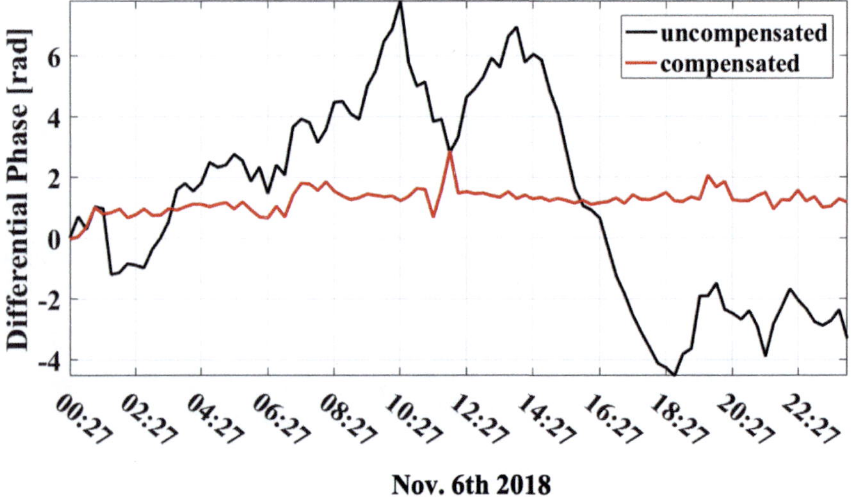

FIGURE 4.7 Time series of compensated (indicated by the red line) and uncompensated (indicated by the black line) differential phases from a CR installed on the observed slope, spanning 00:12 to 23:57 on November 6, 2018.

specifically plots the signal from a corner reflector (CR) located 850 m away from the GB-SAR sensor, with no displacement.

The comparison results highlight a substantial reduction in the phase deviation after applying the model-based compensation method, with the SD decreasing significantly from 3.32 to 0.38 rad. This indicates the effectiveness of the APS compensation approach in enhancing the accuracy of the measured differential phase.

4.3.2 GEOSTATISTICAL APPROACH

As shown in (4.11), the residual APS ϕ_{res}, induced by atmospheric turbulence, often referred to as turbulent APS, cannot be adequately described by the model-based approach because of its inherent local heterogeneity. In this context, the geostatistics interpolator known as Kriging interpolation, which considers the spatial correlation of observations, facilitates the estimation of the ϕ_{res}.

Kriging interpolation involves predicting ϕ_{res} through a weighted average of neighboring observation values, where the assigned weights are determined on the basis of spatial covariance. According to (4.11), ϕ_{res} is assumed to be zero-mean and spatially correlated and exhibit second-order stationarity. In such cases, the so-called simple Kriging (SK) method is deemed appropriate among the various Kriging methods [33, 38]. It is essential to note that both the model-based and geostatistical approaches must be concurrently employed to comprehensively address all potential components of the APS.

In the Kriging method, the estimation ϕ_{res} at a prediction pixel x_0 is obtained as the weighted average of ϕ_{res} at neighboring pixels:

$$\hat{\phi}_{res}(x_0) = \sum_{i=1}^{n} w_i \phi_{res}(x_i), \tag{4.20}$$

where i denotes the index of neighboring CSs (or interpolating CSs), and w_i represents the weights assigned to neighboring CSs. The Kriging approach can achieve an unbiased estimator, expecting a zero-mean estimation error through the expression of (4.20). The weights assigned to each interpolating CS are derived to minimize the estimation variance, resulting in the SK system [33]:

$$C_1 w_{SK} = C_0, \tag{4.21}$$

where

$$C_1 = \begin{bmatrix} C_{11} & \cdots & C_{1n} \\ \vdots & \ddots & \vdots \\ C_{n1} & \cdots & C_{nn} \end{bmatrix}, \tag{4.22}$$

$$w_{SK}^T = \begin{bmatrix} w_1, w_2, \cdots, w_n \end{bmatrix}, \tag{4.23}$$

$$C_0^T = \begin{bmatrix} C_{01}, C_{02}, \cdots, C_{0n} \end{bmatrix}, \tag{4.24}$$

where $C_{\alpha\beta}$ represents spatial covariance, its index corresponds to the pixel number (with $\alpha, \beta = 0, \cdots, n$), and w_{SK} is weight vector of the SK. In this way, the assigned weights are determined based on the spatial autocorrelation structure. In the case of second-order stationarity, where both the expectation and covariance are assumed to be independent of the location, the covariance function is represented by the bounded variogram estimator γ:

$$C(h) = \gamma(\infty) - \gamma(h). \tag{4.25}$$

Here, the variogram can be obtained as follows:

$$\gamma(h) = \frac{1}{2} E\left[\left(\phi_{res}(x) - \phi_{res}(x+h) \right)^2 \right]. \tag{4.26}$$

In practical applications, the empirical variogram is derived from field observations. Subsequently, a parametric variogram, such as an exponential model, is fit to the empirical variogram to obtain a smooth function [33]. The variogram expression in (4.26) assumes spatial isotropy of the observation valuables, implying that the covariance depends only on the spatial lag h and is independent of the direction.

Finally, the estimation of the residual APS at the prediction pixel x_0 is given as an unbiased, minimum variance estimator through the SK:

$$\hat{\phi}_{res}\left(x_0\right)=\left(C_1^{-1}C_0\right)^T z_{res}. \tag{4.27}$$

where $z_{res}^T=\left[\phi_{res}\left(x_1\right),\phi_{res}\left(x_2\right),\cdots,\phi_{res}\left(x_n\right)\right]$.

Specifically, the compensation process for residual APS compensation generally involves three key steps. The first step entails masking the pixels of interest, which typically correspond to the displaced region. Subsequently, the residual APS over the masked area is predicted by extrapolating from CSs in the non-masked area through the Kriging approach. It is crucial to compute the Kriging covariance over the nondisplaced region to prevent the inclusion of displacement signal effects. The third step involves compensation by subtracting the predicted residual APS from the interferograms. The entire procedure is repeated for each interferogram before velocity inversion.

It is important to note that prior knowledge of the displaced region is essential for the Kriging approach. Therefore, this method is particularly practical when information about the displaced region is available in advance, such as in applications like volcano lava dome monitoring, landslide monitoring, and glacier displacement measurements.

To demonstrate the effectiveness of the Kriging-based correction, a Ku-band GB-SAR data set observed in a mountainous field, also used in Section 4.3.1.2, is employed. In the observation field, the atmospheric conditions exhibited a heterogeneous spatial distribution of the refractivity index, as confirmed by on-site meteorological sensors [39]. Consequently, model-based APS compensation alone is not sufficient to eliminate all APS from the interferograms. Because of the basin structure in the observation field, high temporal variability of local winds is expected, leading to significant errors even after stratified APS correction.

Three image subsets were chosen from the observed data, each comprising 25 images measured over a 2-hour period. As a post-processing step, the MCF phase unwrapping algorithm [40] and model-based APS compensation are applied to all interferograms.

Throughout the observation campaign, no major associated deformation was observed. In other words, the observed data set includes only an APS signal, along with potential temporal and noise decorrelation. In Figure 4.8, three displacement velocity images (projected on the digital elevation model (DEM)) estimated from the three image subsets are shown. These selected subsets are labeled subsets I, II, and III, with the acquisition time in Japan Standard Time (JST) (YYYY.MM.DD) at the top of each map in Figure 4.8. It is noteworthy that all three displacement maps reveal significant

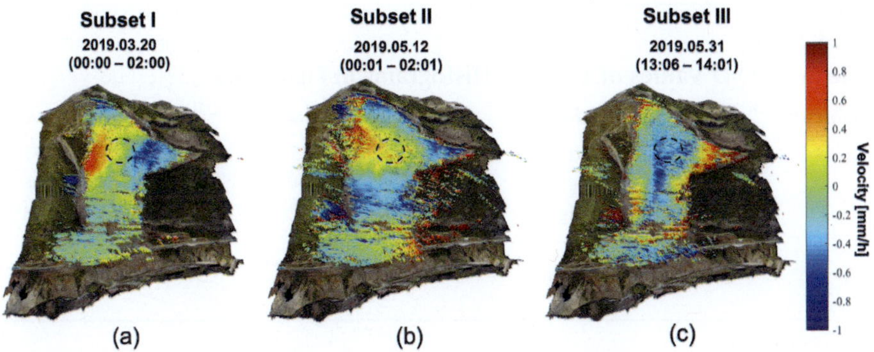

FIGURE 4.8 Velocity images estimated from three image subsets acquired over the post-landslide mountainous field without residual APS compensation. Each figure is generated from images observed over a 2-hour period with a 5-minute time interval. (a) 2019.03.20 00:00–02:00, (b) 2019.05.12 00:01–02:01, and (c) 2019.05.31 13:06–14:01 (YYYY.MM.DD JST).

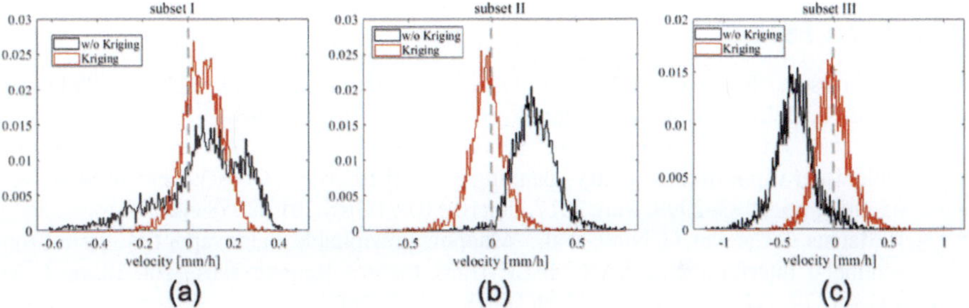

FIGURE 4.9 Histograms of inverted velocity for three subsets. (a) subset I, (b) subset II, and (c) subset III. The black corresponds to the result without the Kriging correction, while the red corresponds to the result with the Kriging correction.

residual APS, indicating the insufficiency of the model-based APS correction. This fact demands further APS compensation to achieve reliable velocity observation.

The evaluation of the compensation performance is conducted on the final velocity inversion results. For this evaluation, the pixels were divided into two samples: one used for interpolating pixels to infer the spatial covariance and the other used for prediction (interpolation) pixels to assess the correction performance. The prediction area is arbitrarily defined and identified by black dashed circles in Figure 4.8. In this evaluation, the residual APS over the prediction area is predicted using Kriging interpolation from the CSs in the nonmasked area. Compensation is then executed by subtracting the predicted residual APS from the interferogram. This process is repeated for all interferograms. Finally, velocity inversion was performed using corrected interferograms over the predicted area.

Figure 4.9 summarizes the histograms of inverted velocity for three subsets with different approaches: (1) a model-based approach without Kriging-based correction and (2) a model-based approach plus Kriging-based correction. The mean and SD values of each histogram are compared and tabulated in Table 4.1. According to the

TABLE 4.1

Mean and SD Values of Velocity Histograms in Figure 4.9

Subset #	Method	Mean [mm/h]	SD [mm/h]
I	w/o Kriging	0.062	0.187
	Kriging	0.064	0.086
II	w/o Kriging	0.252	0.125
	Kriging	−0.031	0.113
III	w/o Kriging	−0.359	0.18
	Kriging	−0.016	0.164

results, the Kriging-based correction demonstrates a noticeable improvement in the APS correction when compared with the results without Kriging, showing lower mean and SD values.

REFERENCES

[1] R. Hanssen, "Radar interferometry: Data interpretation and error analysis," Ph.D. dissertation, Delft University of Technology, The Netherlands, 2001.

[2] P. Mateus, J. Catalao, and G. Nico, "Sentinel-1 interferometric SAR mapping of precipitable water vapor over a country-spanning area," IEEE Trans. Geosci. Remote Sens., vol. 55, no.5, pp. 2993–2999, May 2017. doi:10.1109/TGRS.2017.2658342

[3] P. Mateus, J. Catalao, G. Nico, et al., "Mapping precipitable water vapor time series from sentinel-1 interferometric SAR," IEEE Trans. Geosci. Remote Sens., vol. 58, no.2, pp. 1373–1379, Feb. 2020. doi:10.1109/TGRS.2019.2946077

[4] D. Massonnet, K. Feigl, M. Rossi, et al., "Radar interferometric mapping of deformation in the year after the landers earthquake," Nature, vol. 369, pp. 227–230, May 1994. doi:10.1038/369227a0

[5] D. Massonnet and K. Feigl, "Discrimination of geophysical phenomena in satellite radar interferograms," Geophys. Res. Lett., vol. 22, pp. 1537–1540, June 1995. doi:10.1029/95GL00711

[6] P. Rosen, S. Hensley, H. Zebker, et al., "Surface deformation and coherence measurements of Kilauea Volcano, Hawaii, from SIR-C radar interferometry," J. Geophys. Res. E Planets, vol. 101, pp. 23109–23125, 1996. doi:10.1029/96JE01459

[7] R. Goldstein, "Atmospheric limitations to repeat-track radar interferometry," Geophys. Res. Lett. vol. 22, pp. 2517–2520, Sept. 1995. doi:10.1029/95GL02475

[8] H. Zebker, P. Rosen, and S. Hensley, "Atmospheric effects in interferometric synthetic aperture radar surface deformation and topographic maps," J. Geophys. Res. Solid Earth, vol. 102, pp. 7547–7563, Apr. 1997. doi:10.1029/96jb03804

[9] G. Wadge, P.Webley, I. James, et al., "Atmospheric models, GPS and InSAR measurements of the tropospheric water vapour field over Mount Etna," Geophys. Res. Lett., vol. 29, Oct. 2002. doi:10.1029/2002gl015159

[10] P. Webley, R. Bingley, A. Dodson, et al., "Atmospheric water vapour correction to InSAR surface motion measurements on mountains: Results from a dense GPS network on Mount Etna," Phys. Chem. Earth, vol. 27, no. 4–5, pp. 363–370, 2002. doi:10.1016/S1474-7065(02)00013-X

[11] F. Onn and H. Zebker, "Correction for interferometric synthetic aperture radar atmospheric phase artifacts using time series of zenith wet delay observations from a GPS network," J. Geophys. Res. Solid Earth, vol. 111, no. 19, Sept. 2006. doi:10.1029/2005JB004012

[12] A. Eff-Darwich, J. Pérez, J. Fernández, et al., "Using a mesoscale meteorological model to reduce the effect of tropospheric water vapour from DInSAR data: A case study for the Island of Tenerife, Canary Islands," Pure Appl. Geophys., vol. 169, pp. 1425–1441, 2012. doi:10.1007/s00024-011-0401-4

[13] Z. Li, J. Muller, P. Cross, et al., "Interferometric Synthetic Aperture Radar (InSAR) atmospheric correction: GPS, moderate resolution imaging spectroradiometer (MODIS), and InSAR integration," J. Geophys. Res. Solid Earth, vol. 110, 2005. doi:10.1029/2004JB003446

[14] Z. Li, E. Fielding, P. Cross, et al., "Advanced InSAR atmospheric correction: MERIS/MODIS combination and stacked water vapour models," Int. J. Remote Sens., vol. 110, pp. 1425–1441, 2009. doi:10.1080/01431160802562172

[15] G. Nico, R. Tome, J. Catalao, et al., "On the use of the WRF model to mitigate tropospheric phase delay effects in SAR interferograms," IEEE Trans. Geosci. Remote Sens., vol. 49, no. 12, pp. 4970–4976, Dec. 2011. doi:10.1109/TGRS.2011.2157511

[16] J. Foster, B. Brooks, T. Cherubini, et al., "Mitigating atmospheric noise for InSAR using a high resolution weather model," Geophys. Res. Lett., vol. 33, 2006. doi:10.1029/2006GL026781

[17] R. Jolivet, R. Grandin, C. Lasserre, et al., "Systematic InSAR tropospheric phase delay corrections from global meteorological reanalysis data," Geophys. Res. Lett., vol. 38, 2011. doi:10.1029/2011GL048757

[18] P. Mateus, G. Nico, R. Tome, et al., "Experimental study on the atmospheric delay based on GPS, SAR interferometry, and numerical weather model data," IEEE Trans. Geosci. Remote Sens., vol. 51, no.1, pp. 6–11, Jan. 2013. doi:10.1109/TGRS.2012.2200901

[19] D. Sandwell and E. Price, "Phase gradient approach to stacking interferograms," J. Geophys. Res. Solid Earth, vol. 103, pp. 30183–30204, 1988. doi:10.1029/1998jb900008

[20] D. Sandwell and L. Sichoix, "Topographic phase recovery from stacked ERS interferometry and a low-resolution digital elevation model," J. Geophys. Res. Solid Earth, vol. 105, pp. 28211–28222, 2000. doi:10.1029/2000jb900340

[21] T. Strozzi, U. Wegmüller, L. Tosi, et al., "Land subsidence monitoring with differential SAR interferometry," Photogramm. Eng. Remote Sens., vol. 67, pp. 1261–1270, 2001.

[22] A. Ferretti, C. Prati, and F. Rocca, "Nonlinear subsidence rate estimation using permanent scatterers in differential SAR interferometry," IEEE Trans. Geosci. Remote Sens., vol. 38, vol. 5, pp. 2202–2212, May 2000. doi:10.1109/36.868878

[23] A. Ferretti, C. Prati, and F. Rocca, "Permanent scatterers in SAR interferometry," IEEE Trans. Geosci. Remote Sens., vol. 39, no. 1, pp. 8–20, Jan.2001. doi:10.1109/36.898661

[24] P. Berardino, G. Fornaro, R. Lanari, et al., "A new algorithm for surface deformation monitoring based on small baseline differential SAR interferograms," IEEE Trans. Geosci. Remote Sens., vol. 40, no. 11, pp. 2375–2383, 2002. doi:10.1109/TGRS.2002.803792

[25] O. Mora, J. Mallorqui, and A. Broquetas, "Linear and nonlinear terrain deformation maps from a reduced set of interferometric SAR images," IEEE Trans. Geosci. Remote Sens., vol. 41, no. 10, pp. 2243–2253, Oct. 2003. doi:10.1109/TGRS.2003.814657

[26] P. Blanco-Sánchez, J. Mallorquí, S. Duque, et al., "The Coherent Pixels Technique (CPT): An advanced DInSAR technique for nonlinear deformation monitoring," Pure Appl. Geophys., vol. 165, pp. 1167–1193, 2008. doi:10.1007/s00024-008-0352-6

[27] S. Baffelli, O. Frey, and I. Hajnsek, "Geostatistical analysis and mitigation of the atmospheric phase screens in Ku-Band Terrestrial Radar Interferometric observations of an Alpine Glacier," IEEE Trans. Geosci. Remote Sens., vol. 58, no. 11, pp. 7533–7556, Nov. 2020. doi:10.1109/tgrs.2020.2976656

[28] E. Smith and S. Weintraub, "The constants in the equation for atmospheric refractive index at radio frequencies," Proc. IRE, vol. 41, no. 8, pp. 1035–1037, Aug. 1953. doi:10.1109/JRPROC.1953.274297

[29] R. Iglesias, X. Fabregas, A. Aguasca, et al., "Atmospheric phase screen compensation in ground-based SAR with a multiple-regression model over mountainous regions," IEEE Trans. Geosci. Remote Sens., vol. 52, no. 4, pp. 2436–2449, April 2014. doi:10.1109/TGRS.2013.2261077

[30] L. Pipia, "Polarimetric differential SAR interferometry with ground-based sensors," Doctoral dissertation, Universitat Politècnica de Catalunya, 2009.

[31] C. Michael and M. Byron, Imaging through turbulence, Florida, USA: CRC Press, 2018.

[32] A. Kolmogorov, "The local structure of turbulence in incompressible viscous fluid for very large reynolds numbers," Proc. R. Soc. vol. 434, 1991.

[33] H. Wackernagel, Multivariate geostatistics, Berlin, Germany: Springer-Verlag, 1998; ISBN 978-3-642-07911-5.

[34] J. Chilès and P. Delfiner, Geostatistics: Modeling spatial uncertainty: Second edition, Hoboken, NJ, USA: Wiley, 2012; ISBN 9780470183151.

[35] L. Noferini, M. Pieraccini, D. Mecatti, et al., "Permanent scatterers analysis for atmospheric correction in ground-based SAR interferometry," IEEE Trans. Geosci. Remote Sens., vol. 43, no. 7, pp. 1459–1470, July 2005. doi:10.1109/TGRS.2005.848707

[36] L. Pipia, X. Fàbregas, A. Aguasca, et al., "Atmospheric artifact compensation in ground-based DInSAR applications," IEEE Geosci. Remote Sens. Lett., vol. 5, no. 1, pp. 88–92, Jan. 2008. doi:10.1109/LGRS.2007.908364

[37] ITU, "ITU-R P.453-11: The Radio Refractive Index: Its Formula and Refractivity Data," ITU-R, Geneva, 2015.

[38] J. Chiles, Geostatistics : Modeling spatial uncertainty, 2nd ed., Hoboken, NJ: John Wiley & Sons, Inc., 2011; ISBN 9780470183151.

[39] Y. Izumi, L. Zou, K. Kikuta, et al., "Iterative atmospheric phase screen compensation for near-real-time ground-based InSAR measurements over a mountainous slope," IEEE Trans. Geosci. Remote Sens., vol. 58, no. 8, pp. 5955–5968, Aug. 2020.doi:10.1109/TGRS.2020.2973533

[40] M. Costantini, "A novel phase unwrapping method based on network programming," IEEE Trans. Geosci. Remote Sens., vol. 36, vol. 3, pp. 813–821, March 1998.

5 Applications

5.1 IMAGING OF VEGETATION USING POLARIMETRIC GB-SAR

5.1.1 System Specification

Trees are composed of leaves, branches, and trunks, but it is not easy to visualize each component by radar imaging due to the limitations of radar resolution. In this section, we demonstrate examples of three-dimensional tree imaging using ground based synthetic aperture radar (GB-SAR) with polarimetric information analysis. This approach is not only imaging of the objects but aims to characterize the objects, whose size is smaller than radar resolution.

Figure 5.1 shows the polarimetric GB-SAR system used in this study [1]. The technical specification of the system is summarized in Table 5.1. This system can be operated at frequencies between 400 MHz and 6 GHz and is based on a vector network analyzer (VNA). The antennas are mounted on an antenna positioner with a scanning aperture of 20 m horizontally and 1.5 m vertically. Broadband polarimetric radar images can be reconstructed from scattering data using wave migration algorithms, as discussed in [2]. Because a three-dimensional reflectivity image can be formed by synthesizing the two-dimensional aperture, a diffraction stacking algorithm was applied to reconstruct the radar image from the scattering data.

5.1.2 Algorithm for 3D Imaging

A three-dimensional SAR image can be reconstructed by synthesizing the data set acquired using a two-dimensional aperture. Migrations are often used for image reconstruction by wave field extrapolation, which has been discussed by many authors [3–13]. The goal of migration is to make the stacked section appear similar to the real location. A migration method of diffraction stacking has been employed to reconstruct the image [3, 14]. This is a very convenient and traditional method that is commonly used for seismic data processing. This approach is an integral solution to the wave equation. It can extrapolate the wave fields of an observation aperture to a larger 3D space. Diffraction stacking migration produces one migrated sample (a pixel) at a time by computing the diffraction shape for a scatter point at that location, summing and weighing the input energy along a diffraction path, and placing the summed energy at the scatter point location on the migrated section. The process is repeated for all migrated samples. During summation, the amplitudes of the input data are weighted. This migration algorithm focuses the time-domain scattering signal on the target in space by computing (5.1).

$$P(x,y,z) = \iint f(\tau, x_m, y_m, z_m)\Big|_{y_m=0} A(\theta)dz_m dx_m \tag{5.1}$$

where $P(x,y,z)$ is the imaging area.

DOI: 10.1201/9781003316312-5

FIGURE 5.1 Polarimetric GB-SAR system.

TABLE 5.1
Specifications of the Developed Polarimetric GB-SAR System

Vector network analyzer	HP 8720ES/HP 8753E	0.05~20.05 GHz
Antenna	Dual polarized diagonal horn: ETS-EMCO 3164-03 # 1044	0.5~6.3 GHz
Antenna positioner	DEVICE Inc. DX3151BV1/O	Accuracy 0.1 mm
Aperture length	Horizontal	20 m
	Vertical	1.5 m

$f(\tau, x_m, y_m, z_m)$ defines the acquired data in the time domain.

$$\tau = \frac{\sqrt{(x_t - x)^2 + (y_t - y)^2 + (y_t - y)^2} + \sqrt{(x_r - x)^2 + (y_r - y)^2 + (y_r - y)^2}}{v} \quad (5.2)$$

v is the electromagnetic wave propagating speed in media.

$T_x(x_t, y_t, z_t)$ is the transmitting antenna position.

$R_x(x_r, y_r, z_r)$ is the receiving antenna position.

$A(\theta)$ is the antenna directivity compensation coefficient.

θ is the compensation angle, where the maximum value is taken as half of the antenna beam width.

These variables are described in Figure 5.2. In this algorithm, antenna radiation pattern compensation $A(\theta)$ is included. It is used to weight the time domain data. It makes the data synthesis consistent with the real conditions of the radar system measurement.

5.1.3 IMAGING OF THREE DIFFERENT TREES

We acquired three data sets at the Kawauchi campus of Tohoku University, Sendai, Japan, during different seasons. The experimental site has three different types of trees. The configuration of the target area and the coordinate system are shown in Figure 5.3. The main targets are denoted by T1, T2, and T3 and represent three different types of trees. Tree T1 is a Japanese zelkova (scientific name: *Zelkova serrata* (Thunb). Makino), a deciduous tree that has no leaves in spring and exuberant broad leaves during early summer up to mid-autumn, after which the leaves fall off. Tree T2 is a Japanese cedar (scientific name: *Cryptomeria japonica* (L. fil.) D. Don), an evergreen with needles. It almost does not change from spring to winter. Tree T3 is a shrub of the Azalea genus (scientific name: *Rhododendron quinquefolium* Bisset

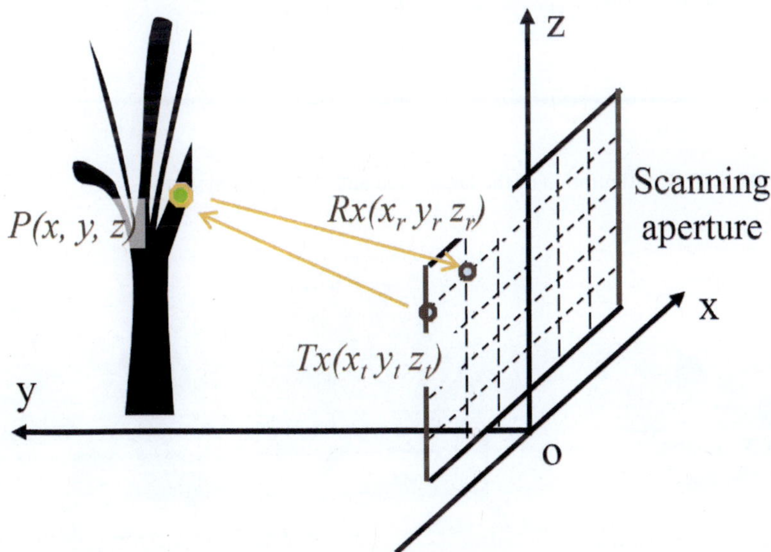

FIGURE 5.2 Geometry of image reconstruction using diffraction stacking with antenna directivity compensation.

et Moore), which is a short bush surrounded by some plants: Japanese honeysuckle (scientific name: *Lonicera japonica* Thunb.). From spring to summer, they have very dense foliage, whereas there are some stems and branches after autumn. The tree leaves are shown in Figure 5.4.

GB-SAR data include the HH, VH, and VV polarization signals. Each polarization data set is processed separately. First, acquired frequency domain data are filtered by a broadband-pass filter, shown in Figure 5.5, where the low frequency to

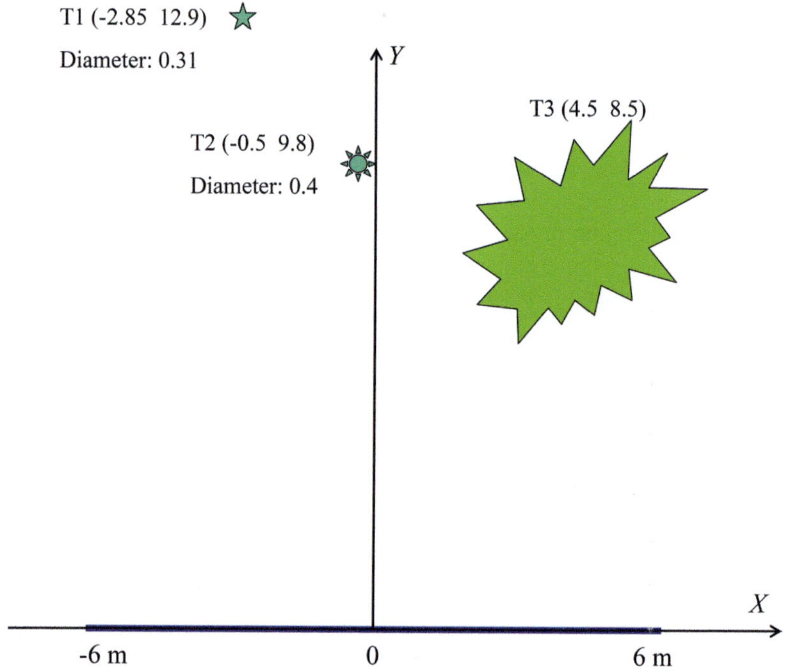

FIGURE 5.3 Configuration of the target area and coordinate system.

FIGURE 5.4 Leaves: (a) broad leaves of tree T1: Japanese zelkova, (b) needles of tree T2: Japanese cedar, (c) leaves of the short tree: azalea, and a plant: Japanese honeysuckle.

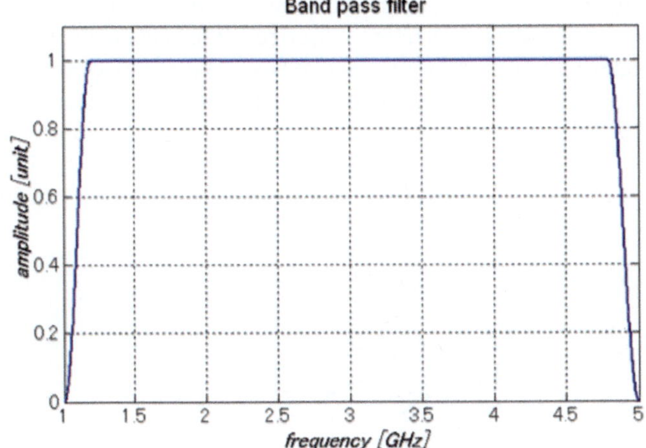

FIGURE 5.5   Band-pass filter used to obtain time domain signal.

pass is 1.2 GHz and the high frequency to pass is 4.8 GHz, to retain the broadband information and smooth the data to reduce the resonances of the time domain signal caused by inverse Fourier transformation. After processing using the inverse Fourier transform, time gating is applied to the time domain data. Because the main targets are located in the range of 6–14 m, a minimum time of 30 ns and a maximum time of 120 ns are selected for time gating. It can remove the antenna direct coupling, near-range ground reflections, and effects from far-range reflections.

The time domain data are then back-transformed to the frequency domain by FFT. The frequency domain data were calibrated using radar system calibration coefficients, which were calculated during system performance evolution. The calibrated data are then compressed by a matched filter, in which the reference signal is the reflection from an aluminum plate measured by the same radar system. After using the inverse FFT again, the time domain data are obtained. The migration of diffraction stacking is applied to migrate the time-domain data, while the antenna radiation directivity compensation is performed simultaneously, for which the compensation angle is 30°, and the special domain data are obtained. Here the spatial domain resolutions are set as 0.05 × 0.05 × 0.05 m for diffraction stacking calculations. The spatial region of the reconstructed images is −6 to 6 m in the azimuth (X) direction, 5 to 15 m in the range (Y) direction, and 0 to 6 m in the vertical (Z) direction according to the coordinate system in Figure 5.3.

Using the 3D spatial domain data, the 3D visualization images are obtained. Figures 5.6, 5.7, and 5.8 show the 3D reconstructed images of the HH, VH, and VV polarimetric returns for the three experiments, respectively. Figure 5.9 shows the trees in three different seasons when polarimetric GB-SAR data were acquired.

By analyzing the three polarimetric images of each measurement, we can identify the differences among the different polarimetric returns. The HH image shows reflections from trunks, some horizontal branches, and ground clutter; the VH image indicates strong reflections from leaves and some slanting branches; and the VV image shows reflections from vertical trunks and branches.

The rectangular box and circles marked in the 3D images show good polarimetric correspondences with the components of the trees. Blue rectangular boxes and circles show the image of tree T1, yellow rectangular boxes and circles show the image of tree T2, and red circles show the image of tree T3 in Figures 5.6, 5.7, and 5.8, respectively.

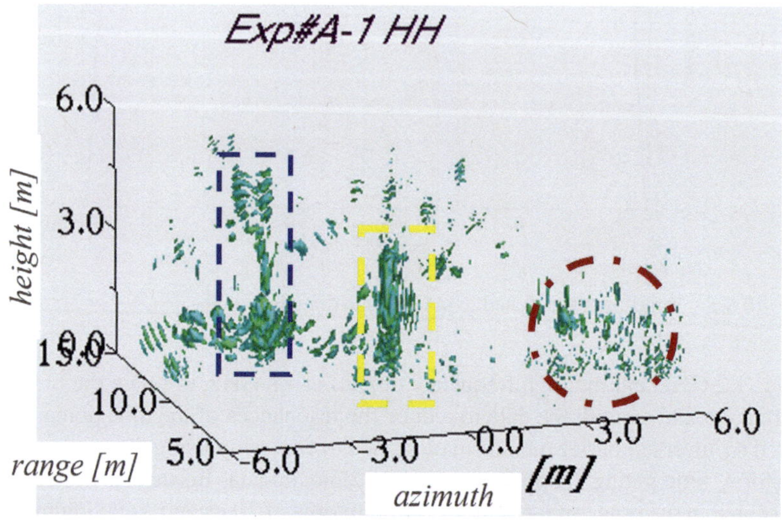

(a) amplitude: 5.5x10^{-3}

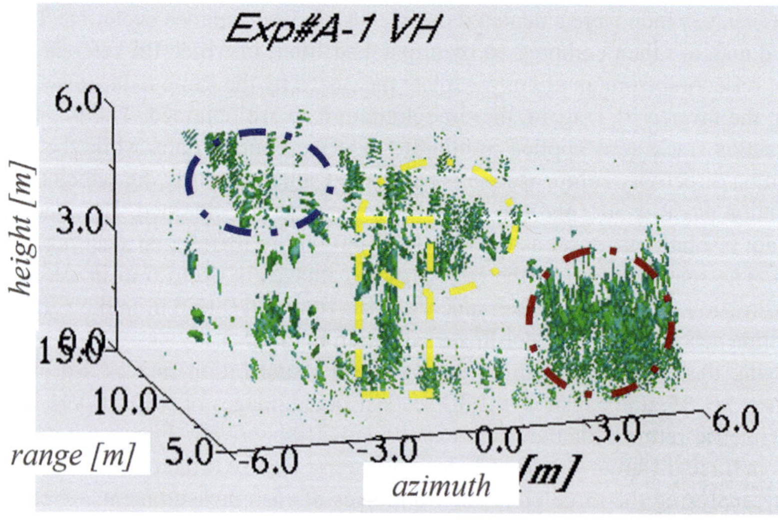

(b) amplitude: 1.2x10^{-3}

FIGURE 5.6   3D polarimetric images of trees in spring: (a) HH, (b) VH, and (c) VV.

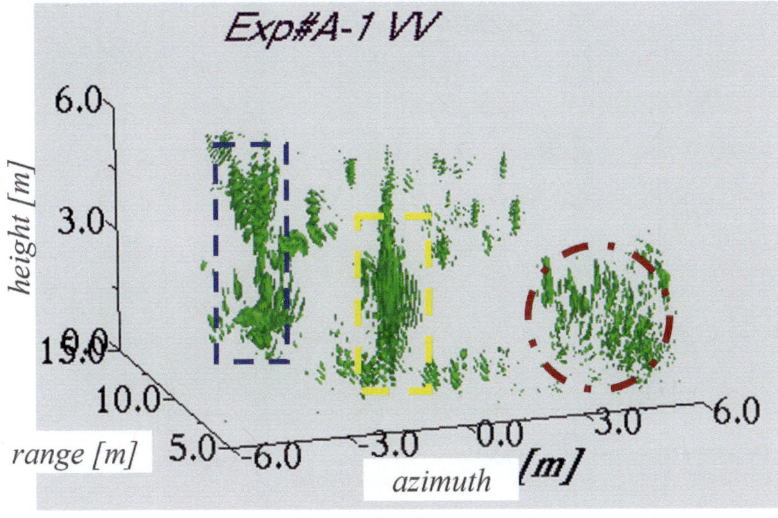

(c) amplitude: 5.5x10⁻³

FIGURE 5.6 (Continued)

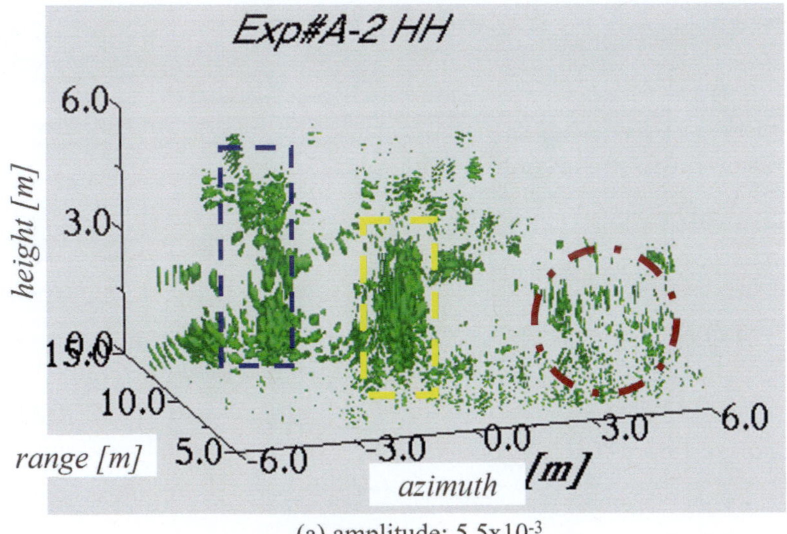

(a) amplitude: 5.5x10⁻³

FIGURE 5.7 3D polarimetric images of trees in summer: (a) HH, (b) VH, and (c) VV.

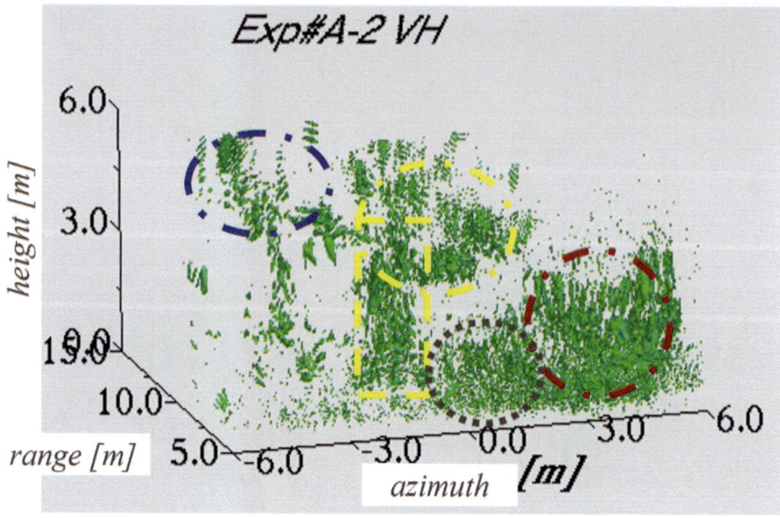

(b) amplitude: 1.2×10^{-3}

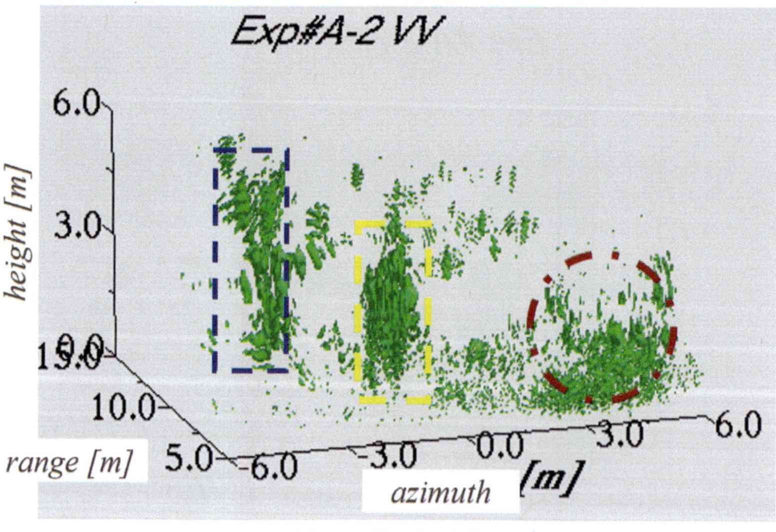

(c) amplitude: 5.5×10^{-3}

FIGURE 5.7   (Continued)

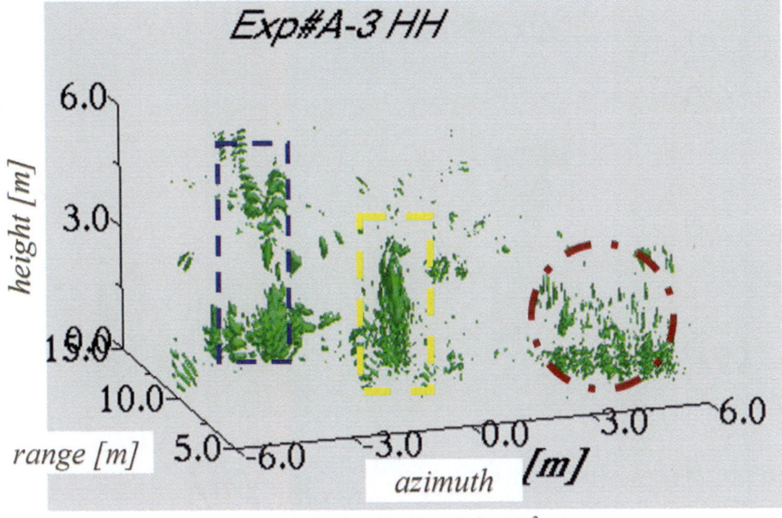

FIGURE 5.8 3D polarimetric images of trees in autumn: (a) HH, (b) VH, and (c) VV.

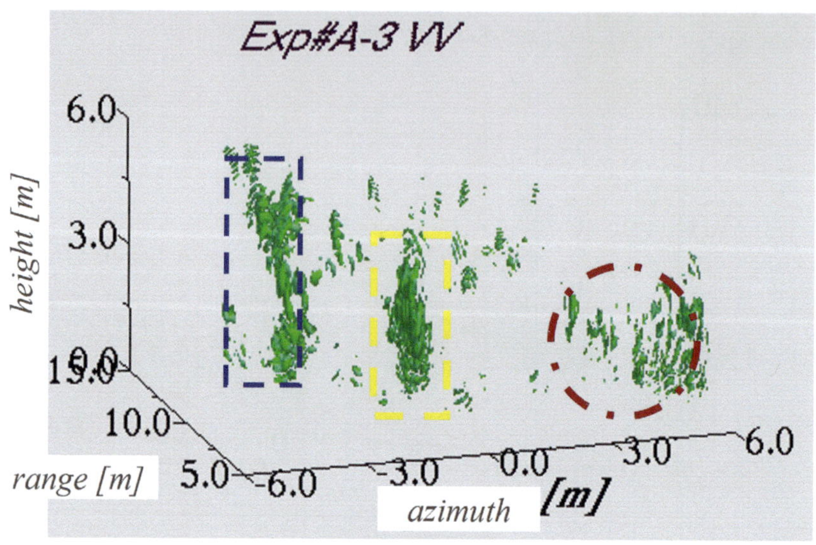

(c) amplitude: 5.5x10^{-3}

FIGURE 5.8 (Continued)

(a)

FIGURE 5.9 Trees in three different seasons when polarimetric GB-SAR data were acquired. (a) Spring, (b) summer, (c) autumn.

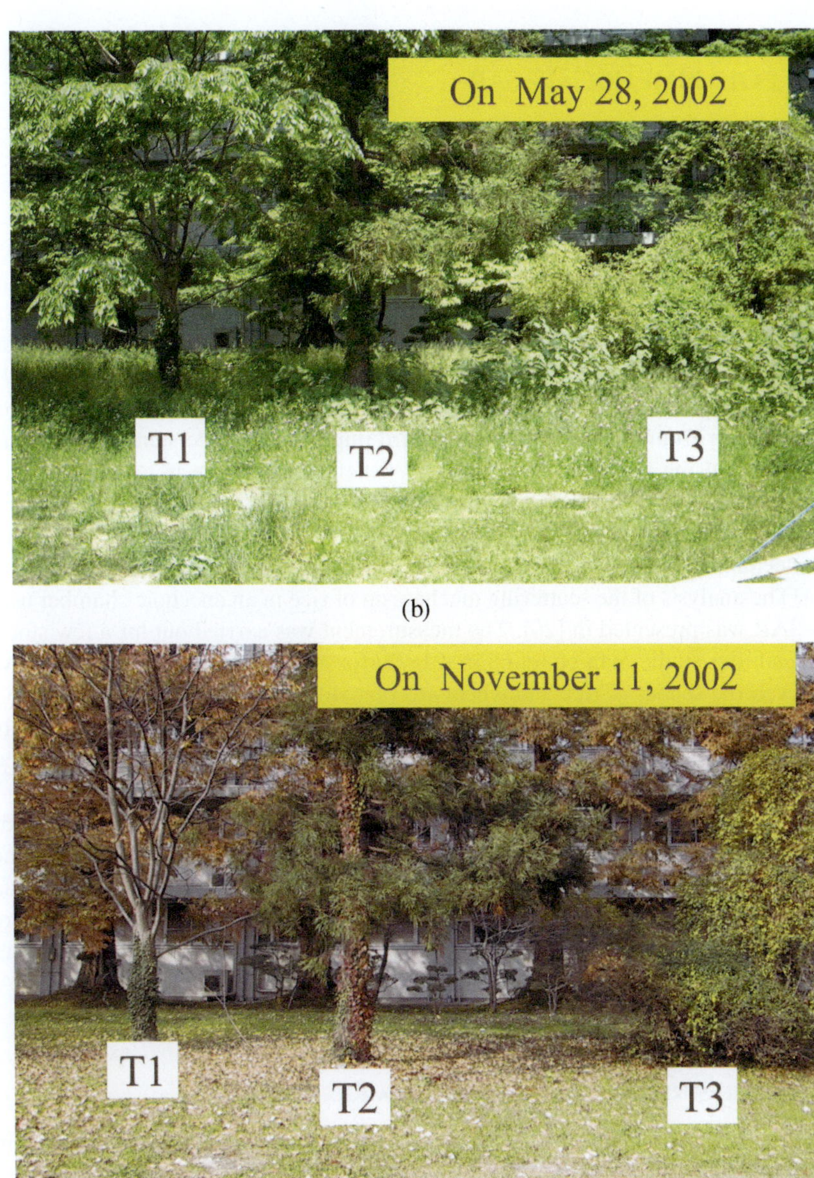

FIGURE 5.9 (Continued)

By using the same polarimetric GB-SAR system, we demonstrated that polarimetric information identified the distribution of cherry blossoms, trunks, and branches in a similar manner to the example shown in [1].

5.2 RICE PADDY FIELD MONITORING USING POLARIMETRIC GB-SAR

5.2.1 OBSERVATION OF VEGETATION BY SAR

It is important to monitor the growth of plants for agricultural purposes. Rice is an important food in many countries. Since those countries, especially tropical areas, have heavy rainfall, microwave remote sensing is more effective than optical sensors in monitoring rice growth because radar can be used under all weather conditions.

So far, there have been several studies reported on the monitoring of rice paddy fields by microwave remote sensing. One method is to use polarimetric satellite SAR [15–17] employing the C- or X-band. Although these studies were carried out together with ground measurements, it is still important to understand the scattering mechanism using a ground-based system. One such example is to use a ground-based polarimetric scatterometer [18, 19]. The system provides the backscattering coefficient of the rice paddy field. However, it is difficult to analyze the scattering mechanism because the system only provides the averaged value over certain area. The analysis of the scattering mechanism of rice in an anechoic chamber using GB-SAR was presented in [20]. The measurement was carried out for a few stocks. Some similar studies can be found in [21–23] for monitoring wheat using GB-SAR. However, the shapes of rice and wheat are very different, especially after ear emergence. While ears of wheat grow vertically, ears of rice grow not only vertically but also bow down if rice grows enough. This difference causes differences in the scattering mechanism between them.

This section presents the measurements of a rice paddy field using polarimetric GB-SAR and an analysis of the results. We focused on polarization and frequency range to understand the status of rice growth. The measurements were carried out in different seasons to detect seasonal changes in the rice field. This research has two aspects: application of polarimetric GB-SAR to real field and fundamental analysis of electromagnetic scattering for L-band satellite polarimetric SAR sensors.

5.2.2 MEASUREMENT CONFIGURATION IN A RICE FIELD

The test field is located in the Miyagi Prefecture Furukawa Agricultural Station in Japan. Six measurements for rice paddy field monitoring were carried out at sites #1 and #2 in 2008 and 2009, respectively. We used a VNA-based GB-SAR equipped with dual ridged horn antennas mounted on a rail to move the radar unit, which is described in Section 5.1. The measurement parameters are summarized in Table 5.2. The measured frequency range is 1 to 10 GHz, with 1601 sampling points. The scan aperture was 8 m with scanning intervals of 1 cm. These parameters were used for all measurements. Figure 5.10 shows the measurement configuration. Incident angles were 50° and 60° in sites #1 and #2, respectively.

Figure 5.11 shows the conditions of the rice crops. The measurement dates of (a), (b), and (d) were June 4, July 6, and August 20, 2009, respectively, and were located

TABLE 5.2

Measurement Parameters for Rice Field Monitoring

Frequency	1–10 GHz
Frequency step	5.625 MHz
Incident angle	50° (site 1, 2008), 60° (site 2, 2009)
Scan length	8 m
Scan step	1 cm

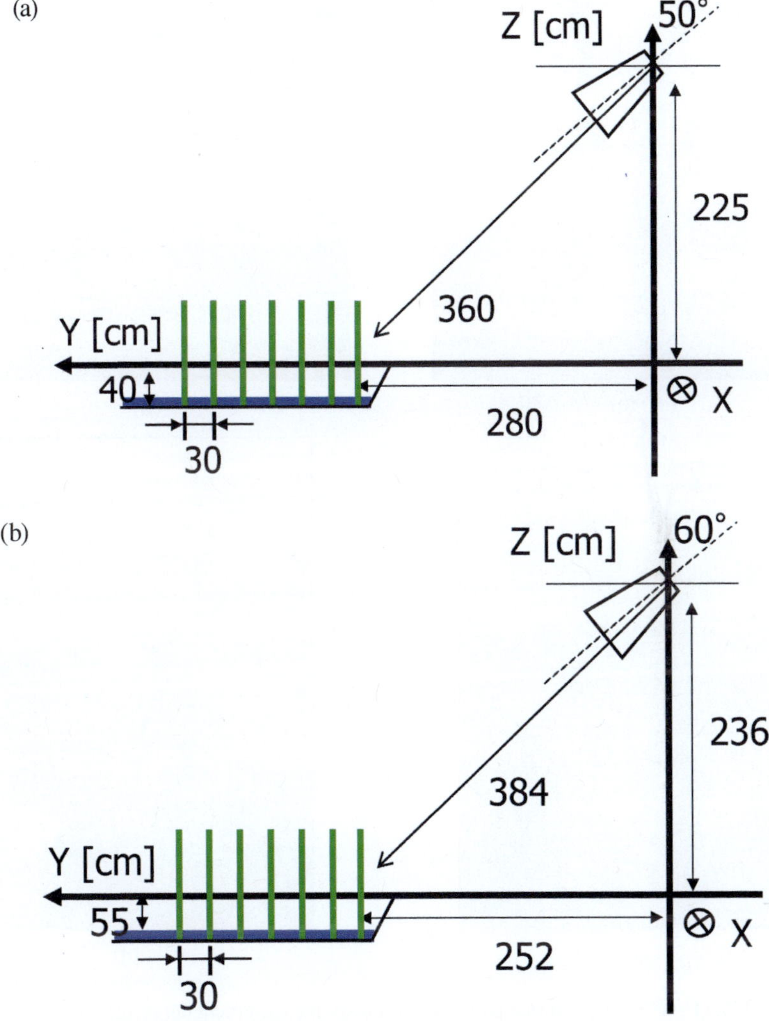

FIGURE 5.10 Measurement configuration in the rice paddy field. The measurements were carried out at (a) site #1 in 2008 and (b) site #2 in 2009 at different field and incident angles. The separation of each plant was approximately 30 cm in the ground range and 15 cm in the azimuth direction.

at site #2. The measurement dates of (c), (e), and (f) were August 3, September 4, and September 19, 2008, at site #1, respectively. Because the two sets of measurements were carried out in different years, at different locations, and in different configurations, the data were analyzed separately.

(a)   (b)

(c)   (d)

FIGURE 5.11 Condition of rice paddy fields on six measurement dates. (a), (b), and (d) were at site #2 on June 4, July 6, and August 20, 2009, respectively. (c), (e), and (f) were at site #1 on August 3, September 4, and September 19, 2008, respectively. (a) Two weeks after the rice was planted (June 4, 2009). (b) 1.5 months after the rice was planted (July 6, 2009). (c) Before ear emergence (August 3, 2008). (d) Just after ear emergence (August 20, 2009). (e) After ear emergence (September 4, 2008). (f) Just before harvest (September 19, 2008).

(e)

(f)

FIGURE 5.11   (Continued)

5.2.3   EAR EMERGENCE DETECTION

First, we analyzed the data acquired at site #1. The rice conditions correspond to those in Figures 5.11(c), (e), and (f). The ear of the rice can be seen in (e) and (f), but not in (c). The ear becomes horizontal as it grows. The main difference between (c), (e), and (f) is the existence of the ear.

Figure 5.12 shows the SAR image of the HH component in the rice field acquired on August 3. The frequency range was selected from 1 to 4 GHz by applying the Hanning window. To suppress antenna coupling, we subtracted the averaged signal from the raw signals. The averaged signal was acquired by averaging in the azimuth direction for all raw data traces. In Figure 5.12, the rice plants appear from around 3.5 m in the range direction. The row of planted rice, which is almost parallel to the scan axis, can be clearly seen. The SAR images for the other polarizations and other measurement data were reconstructed in the same manner. Figure 5.13 shows the color composite image of the three measurement data sets, where the |HH|, |HV+VH|/2, and |VV| components correspond to red, green, and blue, respectively. Since the radar is a monostatic radar, theoretically, HV and VH must be the same. However, in a realistic case, due to antenna differences, there is a small difference. Therefore, in the following analysis, (HV+VH)/2 is used as the HV component.

We observe that strong VV component appears more in the near range. This can be caused by the trunk of the rice plant. Comparing (a) with (b) and (c) of Figure 5.13, the HH components appear to increase after ear emergence. Figure 5.14 shows the amplitude of the averaged signal of the HH component. First, we took the average of the SAR image in the azimuth direction for three measurement data. Subsequently, we normalized the averaged signal by the averaged signal of August 3. Therefore, the result reflects the increase in the HH component before and after ear emergence. Red and blue lines correspond to normalized values for September 4 and 19, respectively. It is obvious that

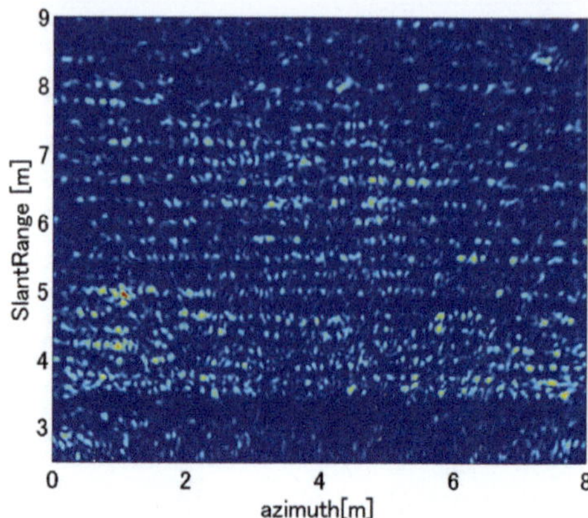

FIGURE 5.12   SAR image of the HH component in the rice field on August 3 at site 1. The frequency range is 1–4 GHz.

the HH components are clearly increased after ear emergence. This can be caused by the horizontal structure of the ear. The value of September 19 is slightly smaller than that of September 4. This is because the rice is drier, and reflectivity is decreased.

Since we measured a wide frequency bandwidth, we carried out frequency analysis to determine the frequency which gives a clearer difference. First, averaged subtraction is applied to the raw data set. Then, we use the inverse Fourier transform to convert sth. to a time domain signal without any band pass filtering. Then, a time window is applied to obtain only reflection from the rice field. Finally, the signal is transformed back to the frequency domain by Fourier transform. Figure 5.15 shows the frequency spectrum obtained by the processing at the center of the scan aperture. Blue, green, and red lines correspond to the data acquired on August 3, September 4, and September 19, respectively. It is obvious that the difference between before and after emergence becomes large around the C-band, especially at 6 GHz, which is the wavelength of 5 cm.

Figure 5.16 shows the color composite image after SAR processing. The frequency was limited with a central frequency of 6 GHz and a bandwidth of 150 MHz. Although the row of rice plants cannot be seen clearly due to the low spatial resolution caused by the limited bandwidth, the color change is obvious. Before ear emergence, the VV component was dominant, but after ear emergence, the HH component increased and become dominant. To confirm this, the same processing as Figure 5.14 was applied. Figure 5.17 shows the amplitude of the averaged signal of the HH component normalized by the data acquired on August 3. It is obvious that the HH component after ear emergence drastically increases.

Even if the resolution of the image is not sufficient to visualize each target, we can detect the clear changes on the target by using polarimetric analysis with an effective frequency range.

5.2.4   RICE GROWING UNTIL EAR EMERGENCE

This section presents the measurement results and analysis at site #2, acquired in 2009. The conditions of the rice crops are shown in Figure 5.11. The measurement

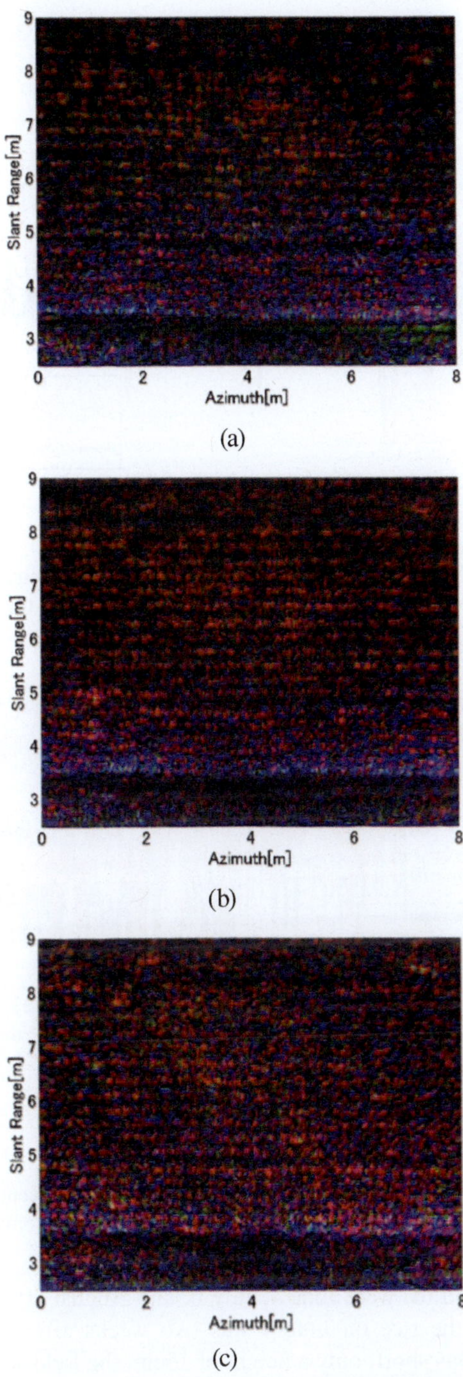

FIGURE 5.13 Color composite image acquired on three different dates at site 1. Red, green, and blue colors correspond to |HH|, |HV+VH|/2, and |VV|, respectively. Frequency was selected from 1 to 4 GHz. (a) August 3, (b) September 4, (c) September 19.

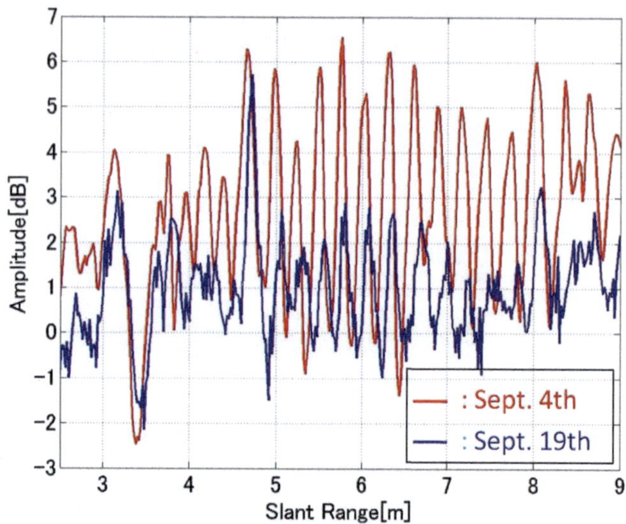

FIGURE 5.14 Amplitude of the averaged signal of the HH component normalized by the averaged data acquired on August 3. The positive sign represents the increase of HH components from August 3.

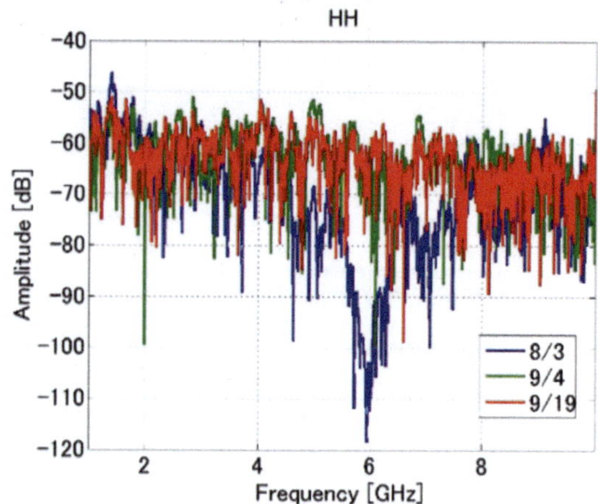

FIGURE 5.15 Frequency spectrum of the HH component at the center of the scan axis after average subtraction. Time gating was applied to obtain scattering only from rice.

dates of (a), (b), and (d) were June 4, July 6, and August 20 in 2009, respectively. The condition of the rice on June 4 was two weeks after the rice was planted. Because the rice was short, only a height of 15cm, the field was not covered by the rice. The condition on July 6 was one and half months after the rice was planted. The size of the rice increased, but the field was not completely covered with rice. In the condition on August 20, the ears of the rice can be seen, but it did not lie down. At this point, most of the field was covered with rice. The incident angle of

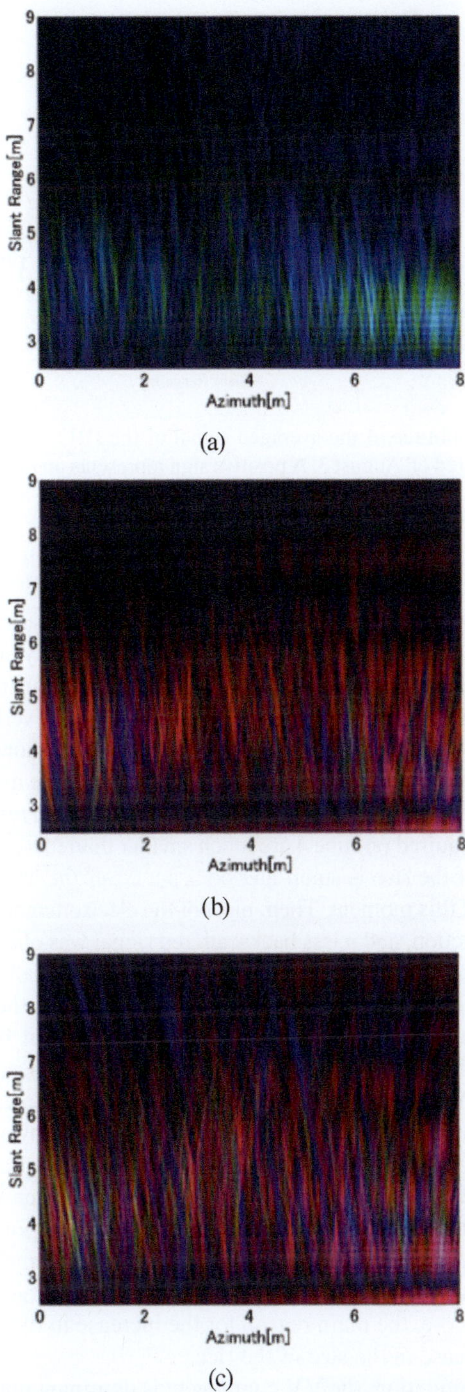

FIGURE 5.16 Color composite image acquired on three different dates at site 1. Red, green, and blue colors correspond to |HH|, |HV+VH|/2, and |VV|, respectively. The frequency was selected at 6 GHz with 150 MHz bandwidth. (a) August 3, (b) September 4, (c) September 19.

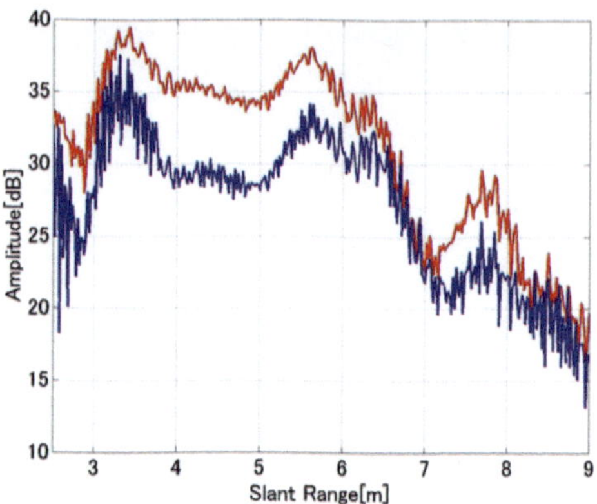

FIGURE 5.17 Amplitude of the averaged signal of the HH component normalized by the averaged data acquired on August 3. A positive sign represents an increase in HH components.

the electromagnetic wave is 60°, and the other measurement parameters are exactly the same as those at site #1.

Figure 5.18 shows a color composite image of three measurements. The same processing procedure was applied as used in site #1. The frequency range was selected from 1 to 4 GHz using the Hanning window. As for Figures 5.18 (a) and (b), the VV component is dominant. Since the structure of the rice spreads in the vertical direction, the VV component becomes the dominant polarization. Figure 5.19 shows the amplitude of the averaged SAR image along the azimuth direction for the three measurements. The HH, HV, and VV components correspond to red, green, and blue, respectively. All components in the data acquired on June 4 are much smaller than those in other measurements because the size of the rice is small and does not cover the entire area. The field was filled with water at this moment. Then, most of the electromagnetic wave was scattered in the forward direction, and a less backscattered signal was obtained.

As for the VV component, 2 weeks later (Figure 5.19(a)) and 1.5 months later (Figure 5.19(b)), the VV component was dominant. On the other hand, at just after ear emergence (Figure 5.19(c)), the VV component became smaller than at 1.5 months later in the far range, although it increased in the near range. The increase of the VV component in the near range was due to an increase in the size of the rice, which produces a larger vertical structure. However, in the far range, since the fields were covered with rice, the electromagnetic wave was scattered at the canopy and did not reach the trunk of the rice, which can contribute to the VV component. Therefore, in the far range, the VV component decreased.

As for HH components, the component at just after ear emergence became larger than that 1.5 months later. Because the ear of the rice emerged, it spread vertically at this time. Therefore, the main reason for the increase in the HH component could simply be the increase in the size of the rice.

From this interpretation, the VV component is dominant until the field is covered by rice. Therefore, we observed that by observing the ratio of the VV component to the other component, the degree of area covered with rice can be estimated and reflect the growth of the rice.

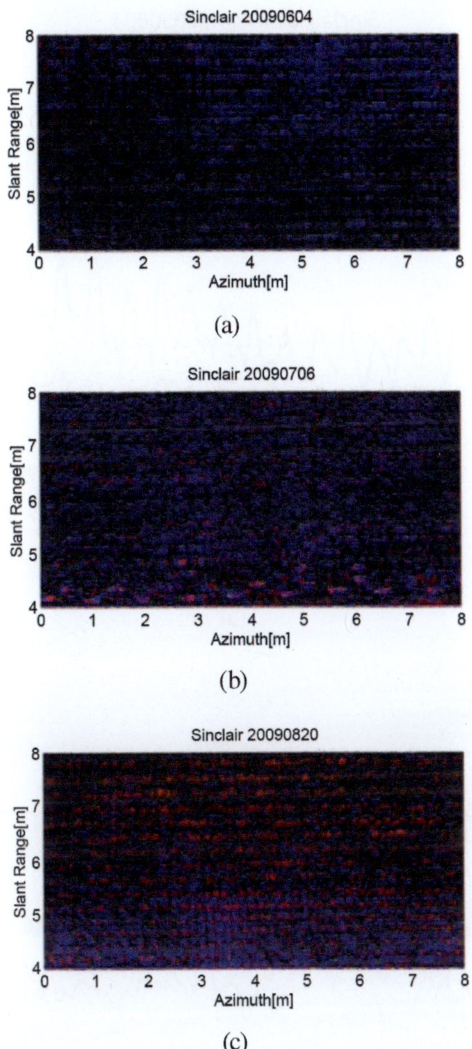

FIGURE 5.18 Color composite image acquired on three different dates at site 2. Red, green, and blue correspond to |HH|, |HV+VH|/2, and |VV|, respectively. Frequency was selected from 1 to 4 GHz. (a) June 4, (b) July 6, (c) August 20.

5.2.5 SUMMARY

Rice field monitoring by polarimetric GB-SAR is presented in this section. In all, we carried out six measurements. From the measurements at site #1, rice growth before and after the emergence of the rice ear was monitored. Although the frequency range of 1 to 10 GHz was used for the measurement, the analysis was carried out in the range of 1 to 4 GHz for L-band satellite SAR. Ear emergence was successfully detected by L-band. Focusing on the detection of ear emergence, the C-band is the most effective frequency range for detection according to frequency analysis. From measurements at site #2, rice growth from the

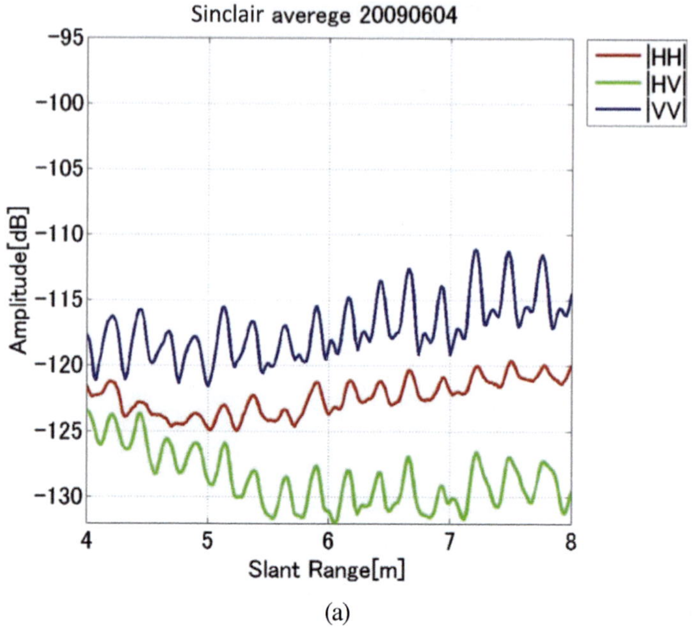

(a)

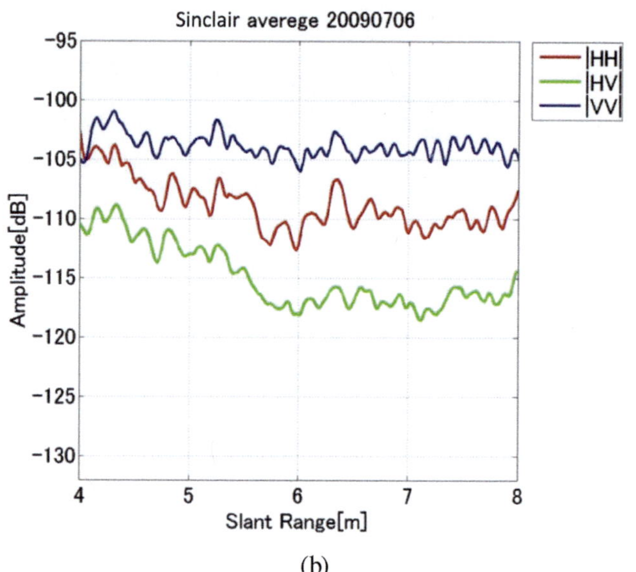

(b)

FIGURE 5.19 Amplitude of the averaged signal along the azimuth direction at site #2. Red, green, and blue correspond to |HH|, |HV+VH|/2, and |VV| components, respectively. (a) June 4, (b) July 6, (c) August 20.

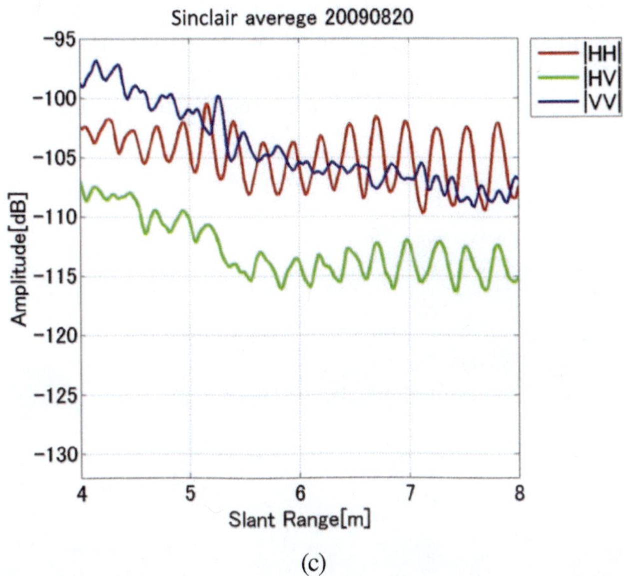

(c)

FIGURE 5.19   (Continued)

earlier stage of the rice to just after the emergence at which the plants were still vertically standing was monitored. From this interpretation, the degree of rice growth may be estimated by calculating the ratio of VV components to the other components.

5.3   SOIL EROSION ESTIMATION

We think GB-SAR can be used for quantitative estimation of soil erosion. This section demonstrates estimation of soil erosion where a large-scale landslide was caused by an earthquake.

5.3.1   LANDSLIDE MONITORING IN ARATO-ZAWA

The Iwate-Miyagi Nairiku Earthquake occurred on June 14, 2008 [24]. This earthquake occurred in the northern part of Japan, in the mountainous area of Miyagi and Iwate prefectures. A very large-scale landslide occurred at Arato-zawa, Kurikoma city. The research group of Tohoku University installed a 17-GHz GB-SAR system (IBIS-KU by IDS), shown in Figure 5.20(a) in November 2011, and long-term monitoring of Arato-zawa started in June 2012 [25]. The monitoring area, which can be seen in Figure 5.20(b), is about 200 m high and 1 km wide, and the distance from GB-SAR to the observation area is about 500 m. Figure 5.20(c) shows the displacement map of Arato-zawa obtained by GB-SAR interferometry. The most important purpose of this observation is prediction of landslides; however, we observed that the ground surface is stable several years after the earthquake, but small ground surface displacement is continuously measured.

Ground displacement is a natural phenomenon that is influenced by natural factors such as rainfall, topography, and wind. In the particular case of Arato-zawa, with high precipitation in some periods, we expected that the precipitation would affect the rate of the ground surface displacement, which is also an indication of

(a)

(b)

FIGURE 5.20   GB-SAR system installed in Arato-zawa. (a) GB-SAR system. (b) landslide, and (c) displacement map.

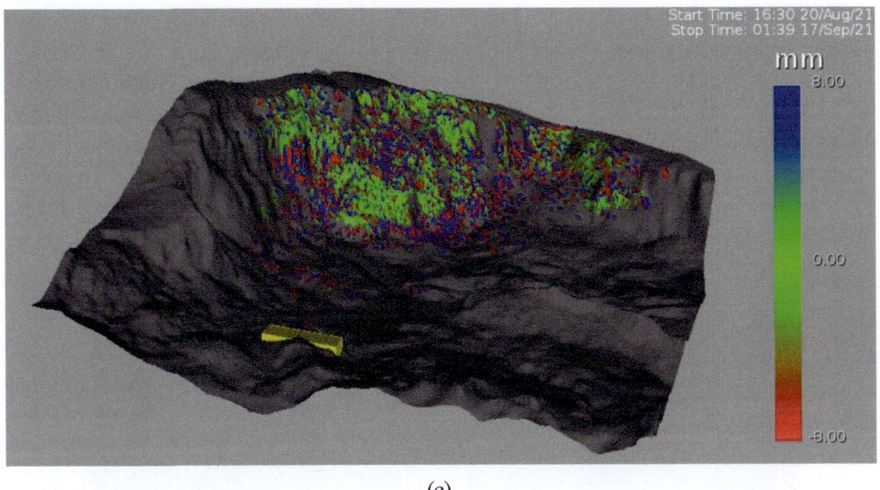

(c)

FIGURE 5.20   (Continued)

erosion occurrence. The surface fluctuated, with varied displacement values with the sampling rate data every 30 min.

This section investigates the natural factors causing displacements related to soil erosion, such as precipitation and slope steepness. Water flowing and absorbing into soil with extreme volume due to heavy rain and typhoons is expected to move soil particles to lower elevations and generate displacements. Meanwhile, SAR illumination is expected to be able to detect this movement.

5.3.2 EROSION FACTOR

Precipitation data used for this work were obtained from the Japan Meteorological Agency (JMA) website. Using the information available on the website, we could choose the rainfall stations near the observation area in Arato-zawa. The distance between the rainfall stations and the observation area is indicated by the GB-SAR location. According to the coordinate location provided by JMA, it was found that there are three nearest rainfall stations to the GB-SAR position: Komano-yu rainfall station, Arato-zawa rainfall station, and Fukayama-dake rainfall station.

This information is exploited to determine the distance between each rainfall station and the GB-SAR location. The distances of the Komano-yu, Arato-zawa, and Fukayama-dake rainfall stations were 4.1, 2.1, and 1.9 km, respectively. Figure 5.21 shows the location of each station in the topography.

Precipitation is defined as water that penetrates to the ground at a certain depth, generally defined in millimeters of depth. Precipitation data provided by JMA are served in hourly sampling rates [26]. The precipitation data of the Komano-yu, Arato-zawa, and Fukayama-dake rainfall stations are shown in Figure 5.22 with the time series selected from October 6–31. In Figure 5.22, the hourly data are shown in the accumulation of precipitation per day, which is well known as the daily rainfall intensity.

It is obvious that on October 12, 13, and 19, the precipitation reached the highest intensity compared to other days. These days corresponded to the days when a typhoon attacked Japan, which is known as the Hagibis Typhoon (Typhoon No.19 or Reiwa 1 East Japan Typhoon) on October 12–13.

Also, it should be recognized that the precipitation obtained within the three rainfall stations only differs slightly. Therefore, we will use only precipitation data from the Komano-yu rainfall station from here on. This selection considered data availability compared to the other two rainfall stations.

Precipitation is relatively high at a particular time in Arato-zawa. To examine this in actual circumstances, the precipitation distribution is explored from January

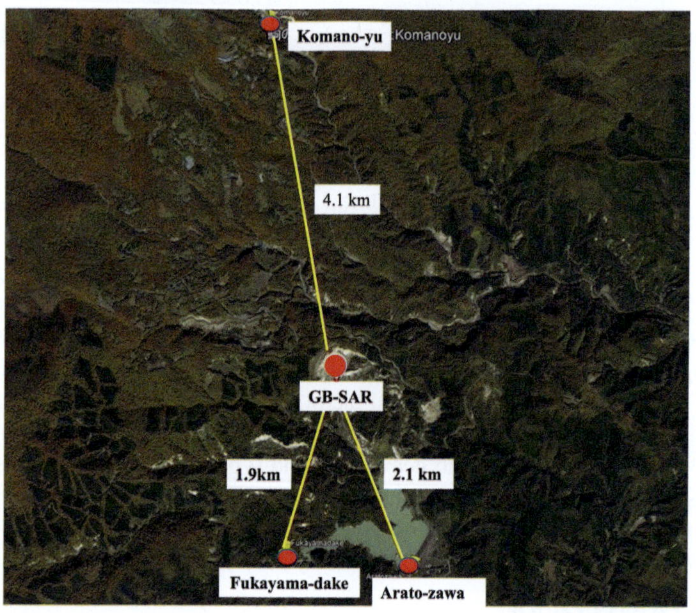

FIGURE 5.21 Location of rainfall stations near Arato-zawa site.

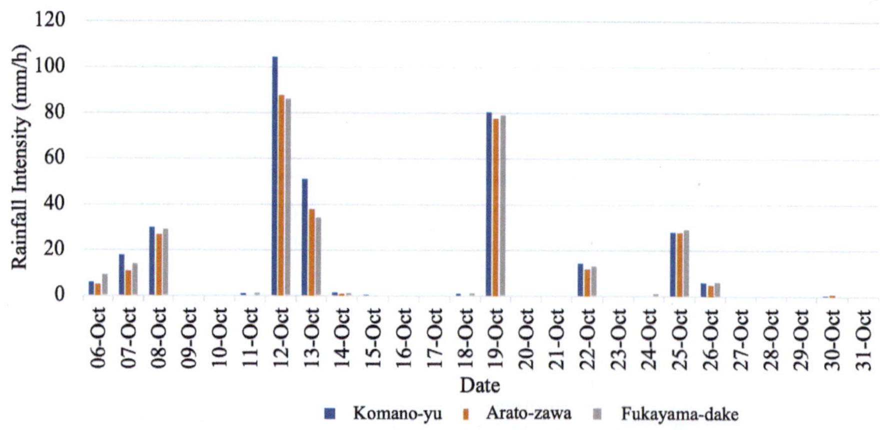

FIGURE 5.22 Monthly rainfall intensity of the three rainfall stations.

to October 2019. The precipitation is shown every hour in daily data, with the water volume counted in millimeter depth units. To show the rainfall distribution in Arato-zawa throughout the year, precipitation data provided by the Komano-yu rainfall station are shown in the monthly intensity defined as follows:

$$I = \sum_{k=1}^{N} \frac{(\Delta V)_k}{\Delta t} \tag{5.3}$$

where I is the accumulated rainfall intensity in a month, k defines the index of precipitation events, $(\Delta V)_k$ is the precipitation depth in the k-th precipitation event, N is the total amount of precipitation events, and Δt is the rainfall duration in which precipitation is considerably constant in an hour (h). The monthly rainfall intensity obtained in October from the Komano-yu rainfall station is shown in Figure 5.22.

From Figure 5.23, the intensity in Arato-zawa within 10 months fluctuated, reaching 70.5 mm/month in September to 450 mm/month in October. From January to May, the monthly intensity was relatively stable when the water penetration volume was less than 200 mm. However, due to the changing season, which is usually recognized with the rainy days in June, the rainfall intensity approached 406 mm/month. This finding also indicates that the intensity in October is the highest rainfall intensity among the months. The October intensity will be used for erosion analysis in the next section.

The erosivity factor [27, 28] is defined as the ability of water to erode soil into particles and gradually move particles from one place to another. The general erosivity factor is processed using years of data sets. However, this work approached the erosivity factor R from monthly precipitation data with daily processing. The erosivity factor R is defined as

$$R = \frac{1}{N} \left[E I_{30} \right]_k \tag{5.4}$$

where N is the total number of days during rainfall; E is the storm kinetic energy (MJ ha^{-1}); k is the index of each rainfall storm, which means every hour; and I_{30} is the

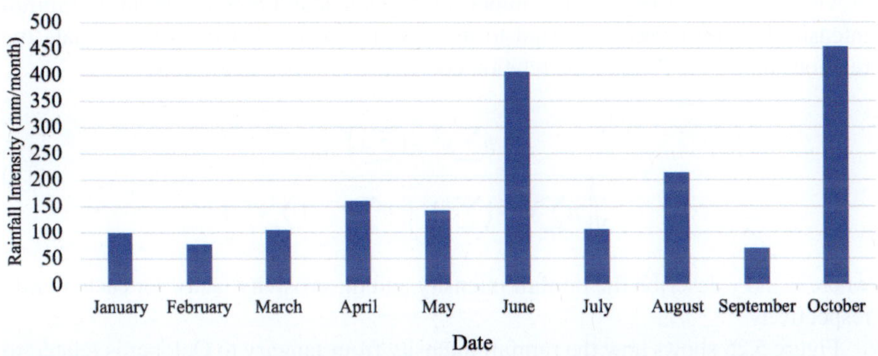

FIGURE 5.23 Monthly rainfall intensity of Komano-yu rainfall station in 2019.

30-min maximum rainfall intensity (mm h^{-1}). Meanwhile, the intensity in 30 min I_{30} used for the R calculation is obtained by:

$$I_{30} = \frac{[\Delta V_r]}{0.5h} \tag{5.5}$$

where $0.5h$ is normalization in 30 min, and ΔV_r is the precipitation depth defined in mm. Also, the kinetic energy E is defined by

$$E = \sum\nolimits_{j=1}^{N} e_r \Delta V_r \tag{5.6}$$

which is given by the rainwater with the unit of MJ ha^{-1}; j is the index of the precipitation events; N is the total number of precipitation events; and e_r is the rainfall energy per unit depth, which is defined by

$$e_r = 0.29\left[1 - (0.72\,\exp(-0.05I_r)) \right) \tag{5.7}$$

where I_r is the intensity of each precipitation data. In this work, the precipitation is obtained every hour; thus, the precipitation is equal to the rainfall depth, which is described as follows:

$$I_r = \Delta V_r \tag{5.8}$$

Figure 5.24 shows the erosivity factor R for each month in Arato-zawa, from January to October. It shows that the erosivity factor varies throughout the time period, from 0.6 megajoule mm/ha/hour/month in January to 54.8 megajoule mm/ha/hour/month in October. The trend shows that from January to April, rainwater had a slight effect on erosion, while in May and June, the erosivity factor increased sharply. Figure 5.24 also shows that rainwater has the strongest capability to erode the soil in October.

The ability of rainwater is expressed as the erosivity factor R, as explained in the previous section. However, it is important to understand how the monthly rainfall intensity in Arato-zawa is related to its erosivity factor R. We try to estimate the relationship by the Pearson correlation coefficient r, which is given by

$$r = \frac{\left[n\sum x^2 - \left(\sum x\right)^2 \right]}{\sqrt{\left[n\sum x^2 - \left(\sum x\right)^2 \right]\left[n\sum y^2 - \left(\sum y\right)^2 \right]}} \tag{5.9}$$

where x and y describe the rainfall intensity and the erosivity factor for each month, respectively.

Figure 5.25 shows how the rainfall intensity from January to October is related to the value of the erosivity factor. The Pearson correlation coefficient was 0.9494, and this value describes a strong positive correlation, indicating that the higher the rainfall intensity, the more excessive the erosivity factor. The more rain falls in Arato-zawa, the higher the possibility that erosion occurs.

Topography plays an important role in the erosion phenomenon. The steepness in mountainous areas such as the Arato-zawa region means a high possibility of soil erosion. Rainwater falling in higher places with steeper slopes highly triggers soil erosion [24]. In this work, the selection of the observation area to estimate the soil-loss volume corresponds to the slope steepness.

In Section 5.2.2, the precipitation and erosivity factors from January to October are discussed. The precipitation in October is the highest compared with the other 9 months' precipitation data. Therefore, to examine the distribution of precipitation in October, the rainfall intensity is shown as daily data. We examines the days with the

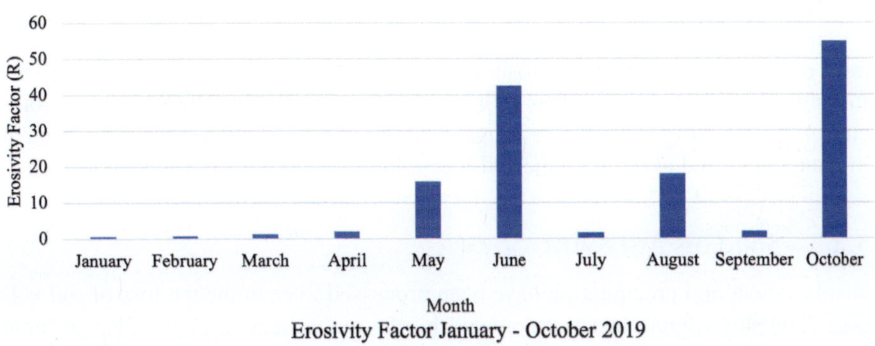

FIGURE 5.24 Erosivity factor R of the Komano-yu rainfall station.

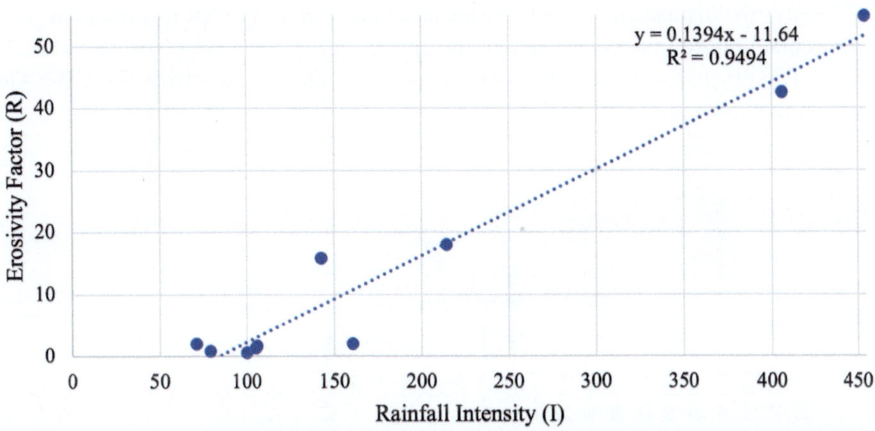

FIGURE 5.25 Correlation between rainfall intensity I and erosivity factor R in Arato-zawa.

most precipitation in October, which will be used for the soil-loss estimation. The daily rainfall intensity I_{day} is defined as

$$I_{day} = \sum_{j=1}^{J} \Delta V_j \text{ mm/day} \qquad (5.10)$$

where J is the number of storms, and ΔV is the rainfall depth.

This clearly confirms that the rainy days in October occurred in several days with fluctuating amounts. It also shows a more excessive amounts on several days, such as on October 4, 12, 13, and 19. The amount of rainfall intensity went from 0.5 mm/day on October 30 to 104.5 mm/day on October 12. This also supports the fact of the Hagibis typhoon occurrence on 12 October.

The erosivity factor of each day in October is shown in Figure 5.26. These two figures confirm that the erosivity factor corresponds to the rainfall intensity of each day, particularly the high rainfall intensity on October 4, 12, 13, and 19.

The erosivity factor R on October 4 is 35.1789 MJ mm/ha/hour/day. Meanwhile, most days in October show an erosivity factor less than 5 MJ mm/ha/hour/day. This finding will be discussed together with the displacement obtained by SAR interferometry in the following section to investigate the role of precipitation in the displacement in Arato-zawa and to estimate the soil-loss volume due to the erosion phenomenon.

5.3.3　Soil-Loss Assessment

Displacement and precipitation have been processed to examine the loss of soil volume. The soil volume lost is calculated by the multiplication of the displacement value by the size area of the pixel where the displacement occurs, represented by

$$Soil-loss_{total} = \sum_{n=1}^{N} d_n A_n \qquad (5.11)$$

where d is the displacement, and A is the size of the area. As SAR illumination generates pixel-based images, the size of the area used for this work is the size of each pixel. The pixel size of the SAR image depends on the location from the GB-SAR sensor, as shown in Figure 5.27(a).

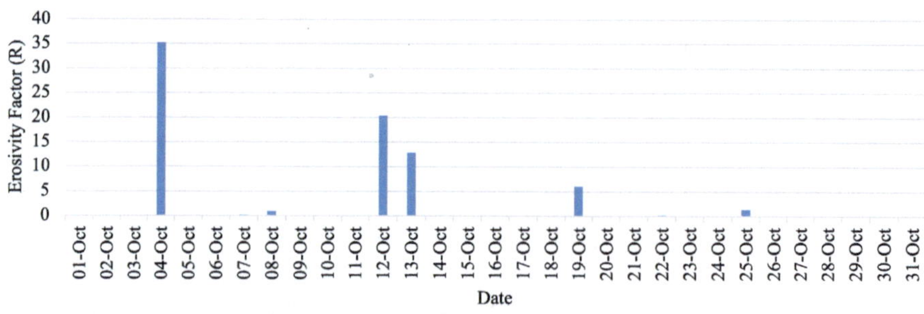

FIGURE 5.26　Erosivity factor R of the Komano-yu rainfall station in October 2019.

The point displacement shown in Figure 5.27(a) is located at different pixels. The azimuth angle of radiation pattern of GB-SAR θ_i is represented by

$$\theta_i = \frac{\lambda}{2L} \tag{5.13}$$

where λ is the wavelength (0.0174 m), L is length of the rail (2 m), and we obtain $\theta_i = 0.004$ rad. In addition, from Figure 5.27(b), the size of A_1 is obtained by

$$A_1 = \frac{\theta_i}{2}(R_1)^2 \tag{5.14}$$

By using (5.14), the size of each pixel in one SAR image can be given as follows:

$$A_n = \frac{\theta_i}{2}\left(R_n^{\;2} - R_{n-1}^{\;2}\right) \tag{5.15}$$

where n is the index of the ranges.

The displacement in Arato-zawa detected by SAR interferometry was used to examine soil erosion. To examine volume loss, it is essential to note that topography may cause different behavior in the movement of soil particles. Meanwhile, the observation area in Arato-zawa is mostly covered by a mountainous area with up and down hills and diverse steepness levels. In other words, this condition disputes the estimation of soil loss due to fluctuating surface steepness.

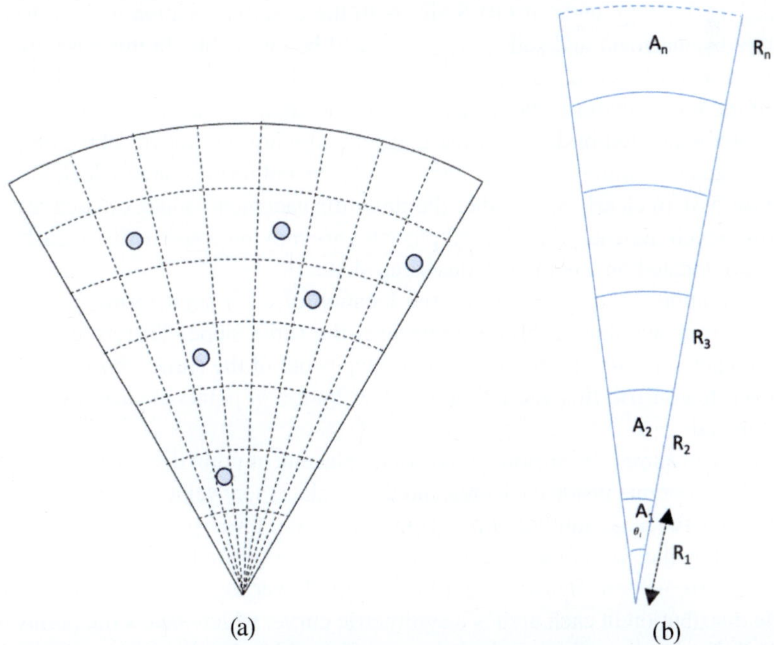

FIGURE 5.27 SAR image-based pixel size. (a) SAR image, (b) pixel in one-column array.

As the displacement represents in points, ten points were selected for real-time monitoring. This aims to show the continuous movement of particular points at various elevations, distances to the sensor, and historical locations from previous displacement results.

Regarding the estimation of soil-loss volume, these observation points were used to define the area dimension. Each point is connected to another point to create a specific area. The point connection was then approached by applying Delaunay triangulation, which maximizes the minimum angle to form the triangle. The result of this step is shown in Figure 5.27(a).

The displacement detected by GB-SAR monitoring every 30 min is accumulated to obtain the displacement as

$$Displacement_{total} = \sum_{i=1}^{N} d_i \tag{5.16}$$

where d is the displacement in temporal time, i is the time sequence, and N is the number of data acquisitions of the entire SAR image processing for soil-loss estimation. In this case, SAR processing is performed daily to correspond to precipitation data and to see the movement more accurately. The area selection located in the SAR illumination is shown in Figure 5.28(b).

The triangles resulted in a huge area and hence contained diverse slope steepness. It can be seen in Figure 5.29, which shows the real imagery of the area under this study, and Figure 5.30, which shows the displacement distributions in Areas #6, #7, and #11. Also, since the displacement is based on the LOS of the sensor, the triangle did not consider the looking angle of GB-SAR. With the unstable location of each triangle area, the displacement and soil-loss volume will be unreliable. In this case, the area selection for soil-loss estimation should be refined. The area was then selected on the basis of the homogenous slope distribution in one area, which was selected from the topography generated by DEM in the real-time monitoring system. Thus, 38 points were selected to continuously interconnect the area and form 24 area selections.

Figure 5.31(b) clearly shows that the daily displacement values of each area are uniform, which indicates that the soil particles are moving consistently to each other as they are located on a homogeneous steep slope.

As validation, Figure 5.32 shows the location of each point forming the selection area. It shows that each area represents the same slope. Therefore, it can be said that each area also represents various slopes out of the entire observation area. Moreover, to confirm this validation, the distribution of points in each area per day was analyzed.

Figure 5.33 shows the various trends of displacement point distribution related to the number of points inside each area and the displacement values of each point. The figures show the huge number of points in Area 5 and fewer points in Area 23. This was caused by the size of the area, which influenced the number of points inside the area. Also, from these figures, regardless of the lower number of points, it shows that the distribution in each area is a symmetric curve, which means the points have a normal distribution. The occurrence was affected by the slope steepness of each area, which caused the displacements to move uniformly.

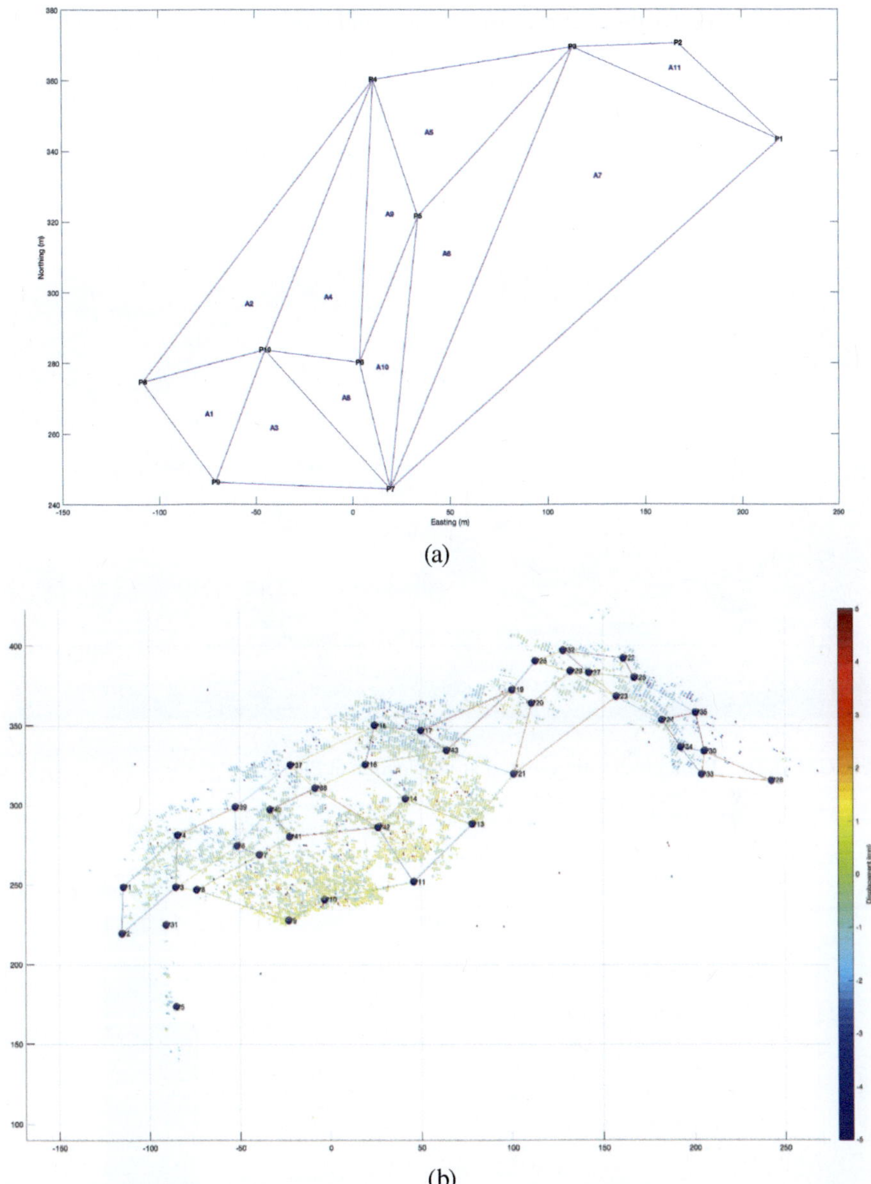

FIGURE 5.28 Area selection based on observation points. (a) Point-based area. (b) Slope-based area under SAR illumination.

Then, we will explain the displacement and soil-loss volume in relation to the precipitation and erosivity factors in each time series of each area. To achieve this aim, the displacement in 24 areas was examined on each day corresponding to the rainfall intensity and erosivity factor. The displacement corresponds to the rainfall intensity, while the soil-loss volume corresponds to the erosivity factor R.

Figure 5.34 shows the average displacement and rainfall intensity on the left side and the average soil-loss volume and the erosivity factor on the right side of each area daily based in October. The displacement is a result of the InSAR processing applied to GB-SAR images every 30 min. In addition, the rainfall intensity is the result of the accumulation of precipitation data every hour. The displacement and precipitation are shown at mm scale and are shown daily from October 1 to 31.

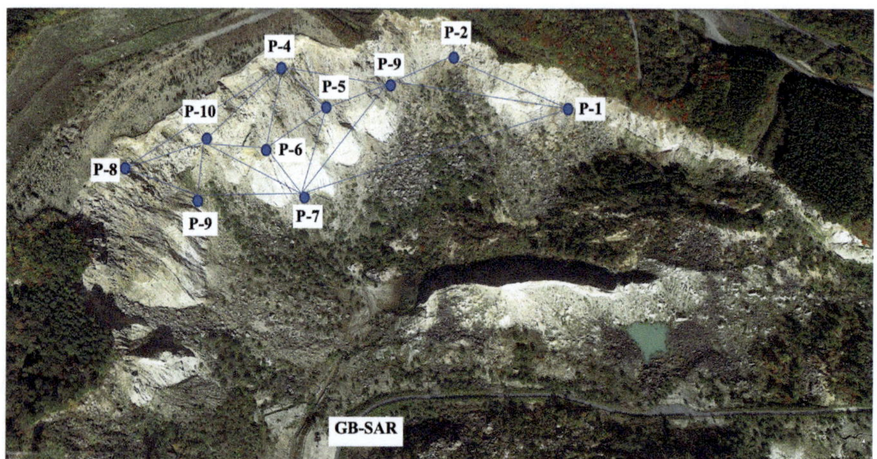

FIGURE 5.29 Point observation–based area selection at the site.

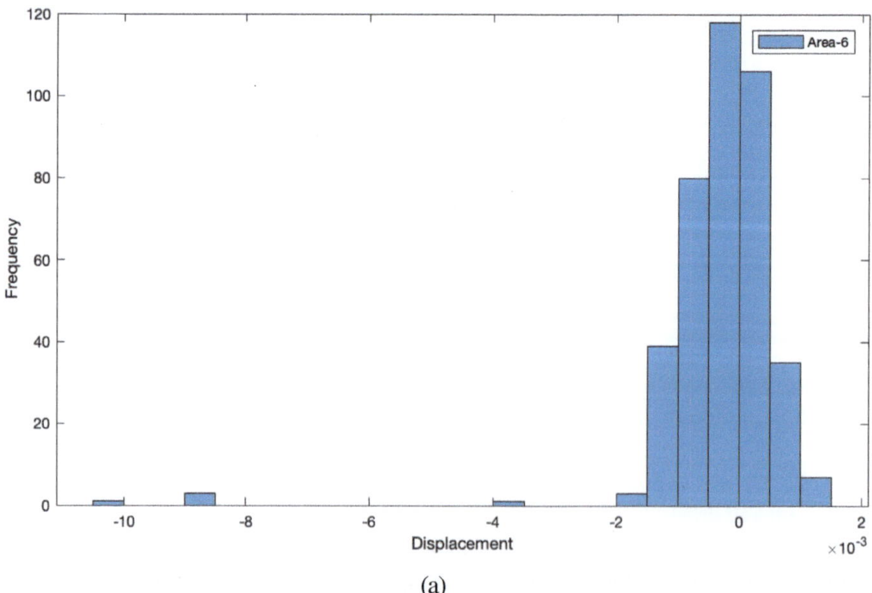

FIGURE 5.30 Displacement distribution on October 15: (a) Area #6, (b) Area #10, (c) Area #11.

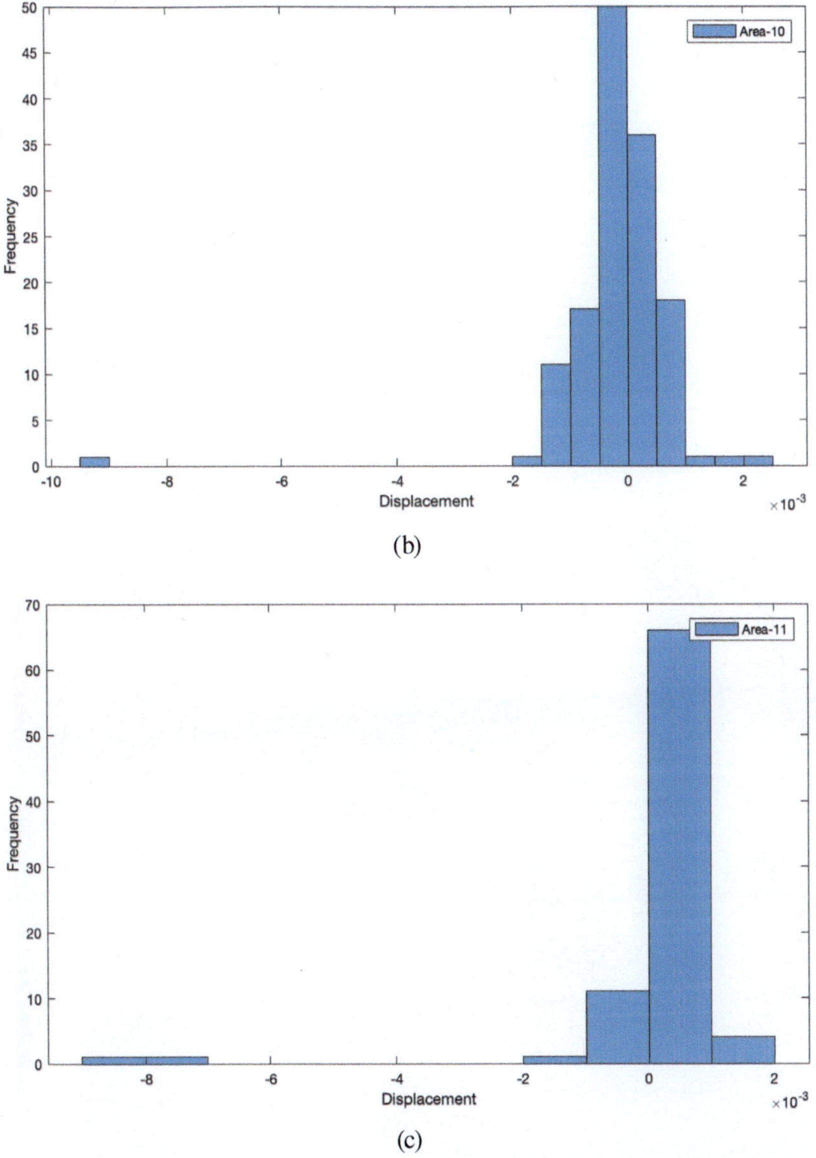

(b)

(c)

FIGURE 5.30 (Continued)

On the right side, the figures show the soil-loss volume and the erosivity factor, which were calculated in the previous section. Consequently, the soil-loss volume was also acquired every 30 min, while the erosivity factor was acquired every hour. The soil-loss volume is shown in m^3, and the erosivity factor is in MJ mm/ha h d. The left- and right-side figures show the figure in the same area, which straightforwardly compares the displacement and erosion phenomenon in daily based in each area.

From Figure 5.34, it is clear that the rainfall intensity occurred on particular days, which is shown by the sharp orange line. This phenomenon is also followed by the

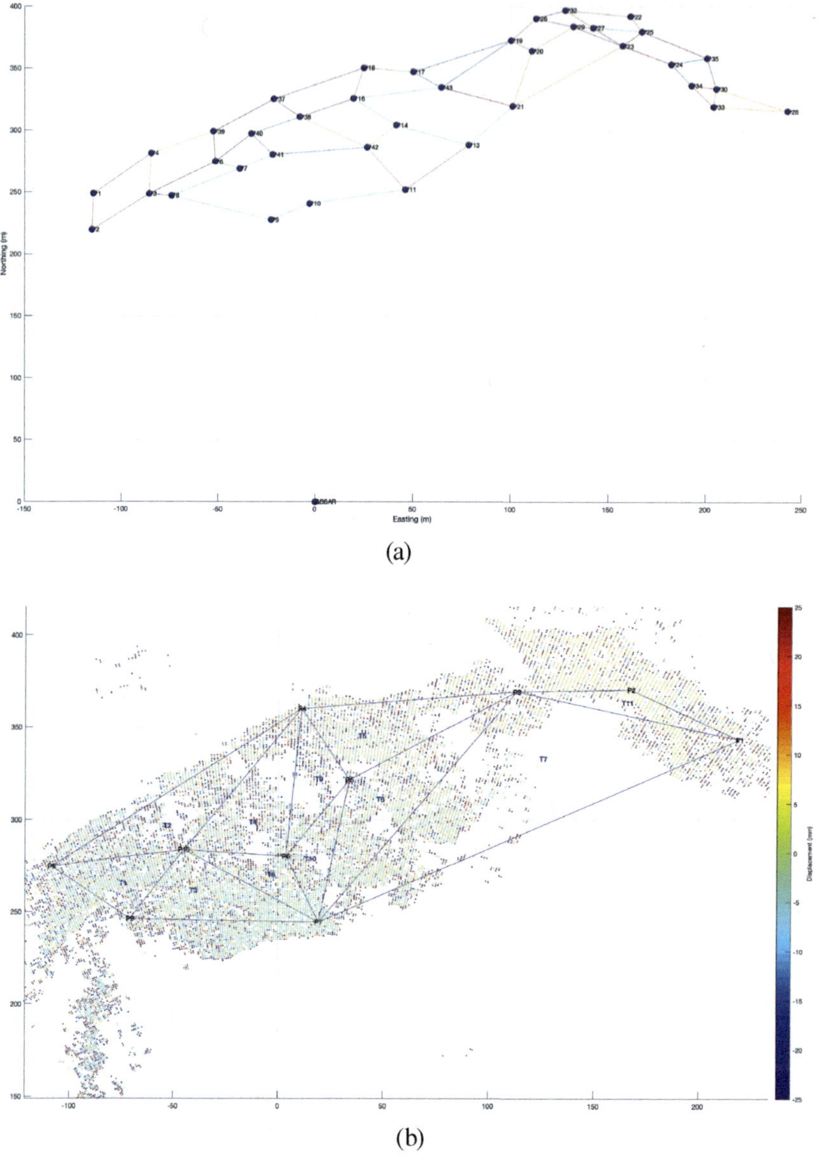

FIGURE 5.31 Area selection based on slope steepness. (a) Slope-based area. (b) Slope-based area under SAR illumination.

erosivity factor, with higher erosivity on specific days identical to high rainfall intensity days. Also, the displacement during the day shows fluctuating values, with specific days having higher amounts compared to other days. This phenomenon is also shown on the right side, where the soil-loss volumes fluctuating with particular days lost more soil mass than other days.

The displacement and soil loss show a corresponding trend in terms of the occurrence days observed in each area. However, in terms of displacement and rainfall intensity,

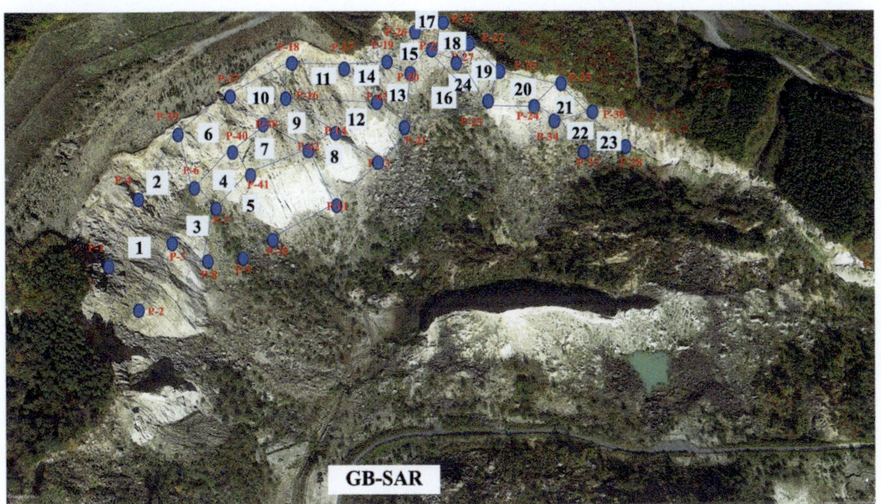

FIGURE 5.32 Slope observation–based area selection at the site.

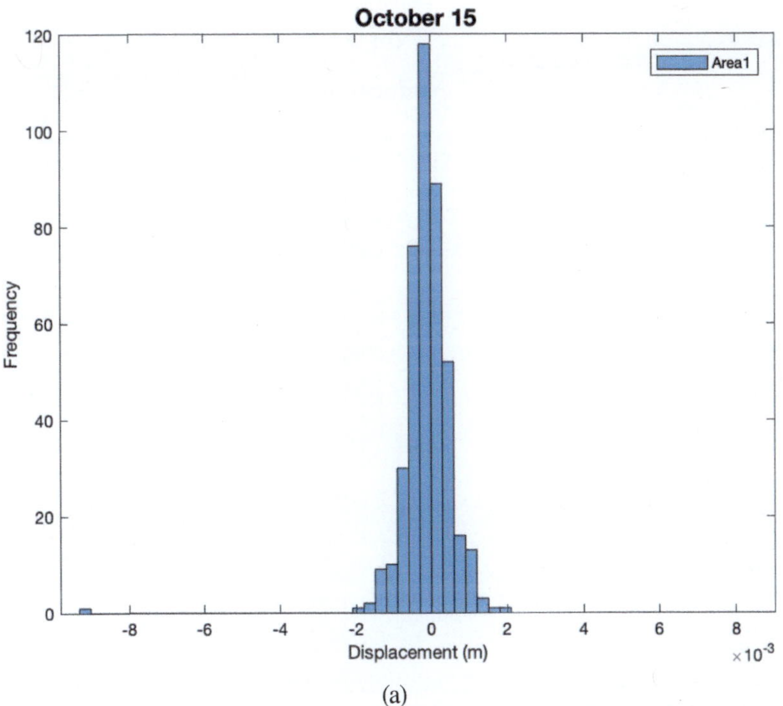

(a)

FIGURE 5.33 Displacement distribution on October 15. (a) Area #1, (b) Area #2, (c) Area #3, (d) Area #4, (e) Area #5, (f) Area #6, (g) Area #7, (h) Area #8, (i) Area #9, (j) Area #10, (k) Area #11, (l) Area #12, (m) Area #13, (n) Area #14, (o) Area #15, (p) Area #16, (q) Area #17, (r) Area #18, (s) Area #19, (t) Area #20, (u) Area #21, (v) Area #22, (w) Area #23, (x) Area #24.

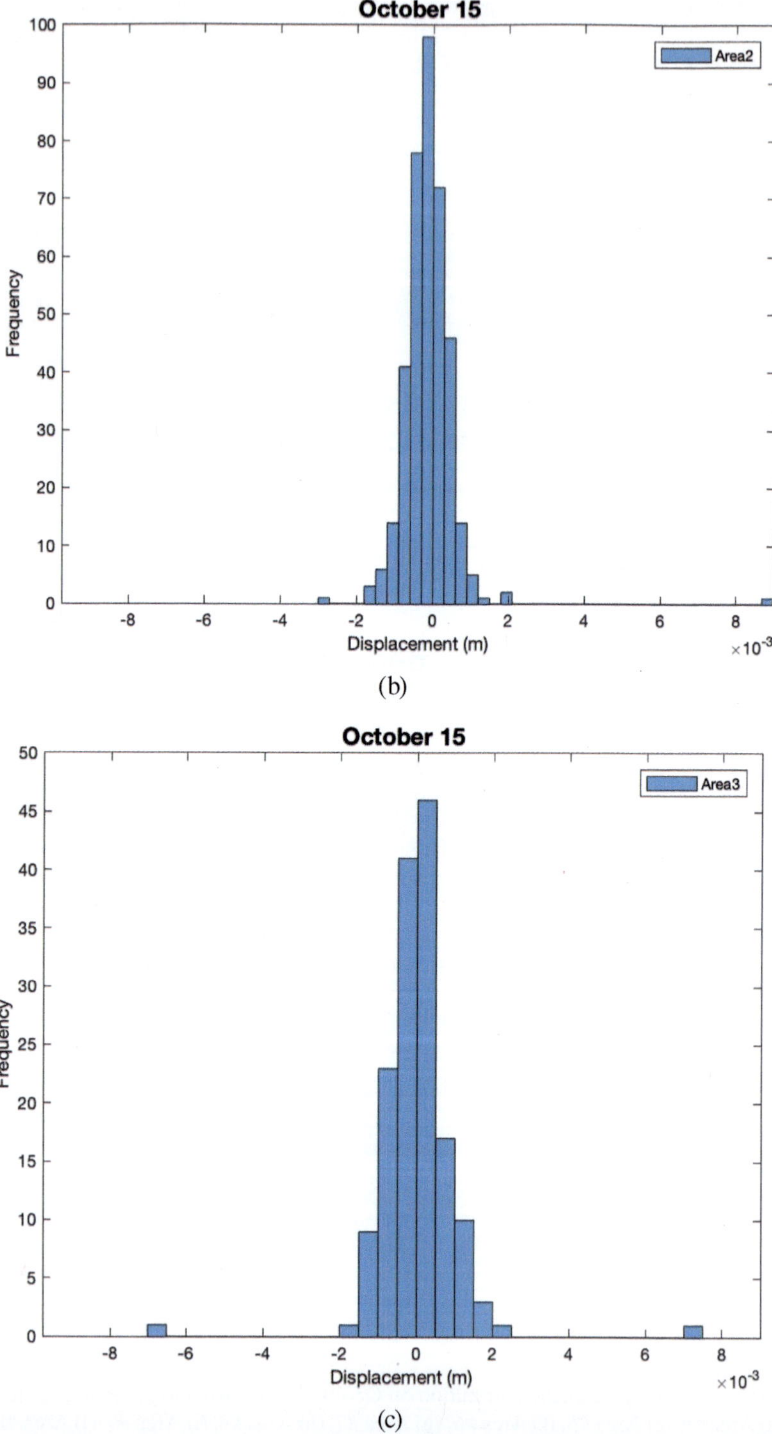

FIGURE 5.33 (Continued)

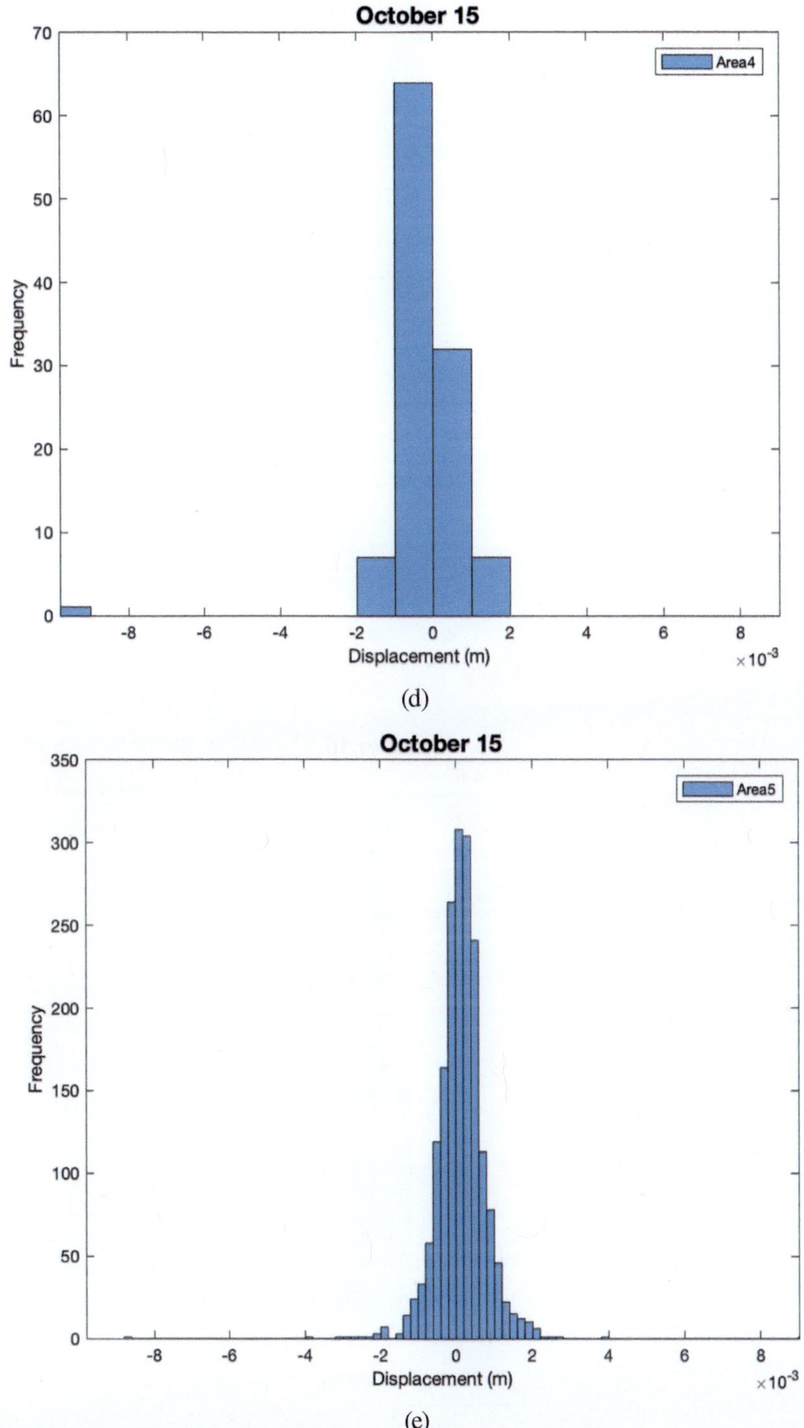

(d)

(e)

FIGURE 5.33   (Continued)

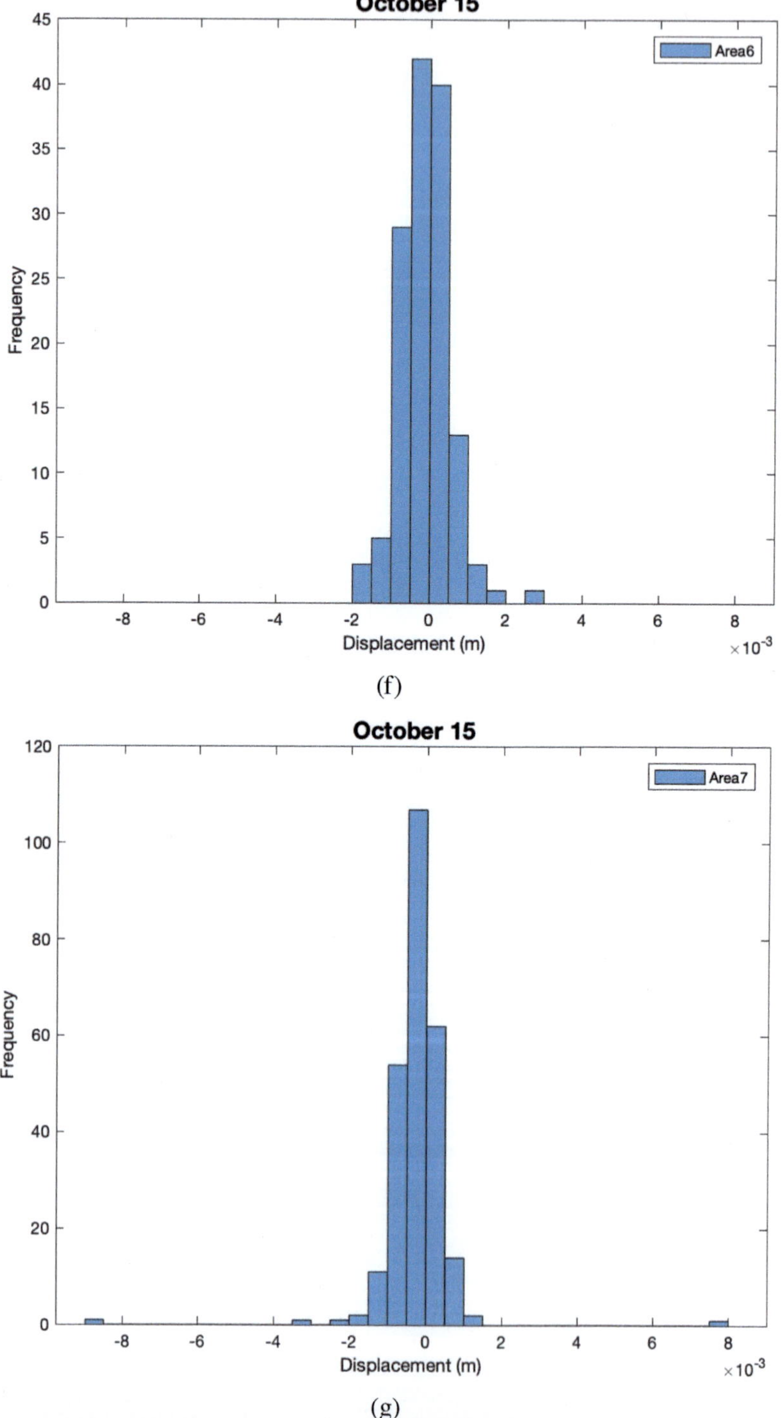

FIGURE 5.33 (Continued)

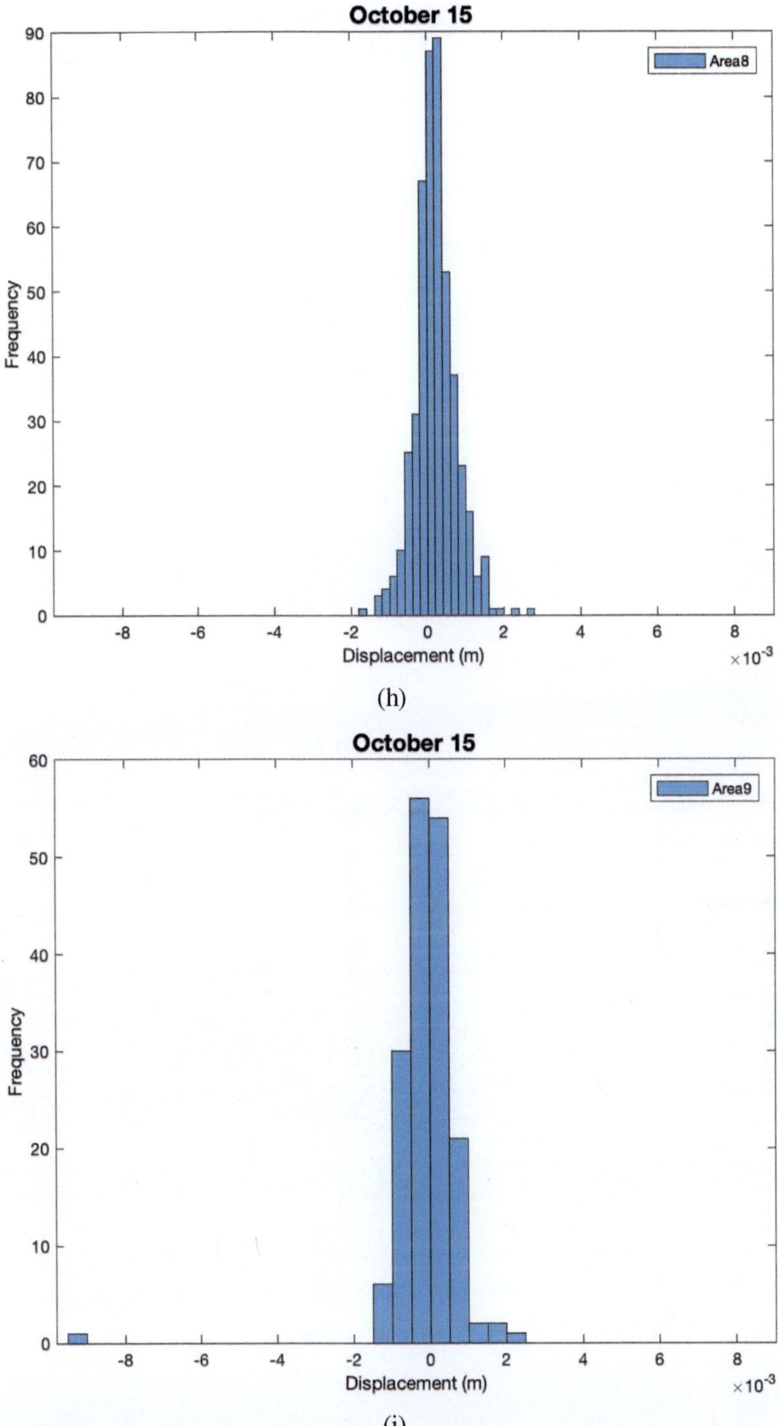

(h)

(i)

FIGURE 5.33 (Continued)

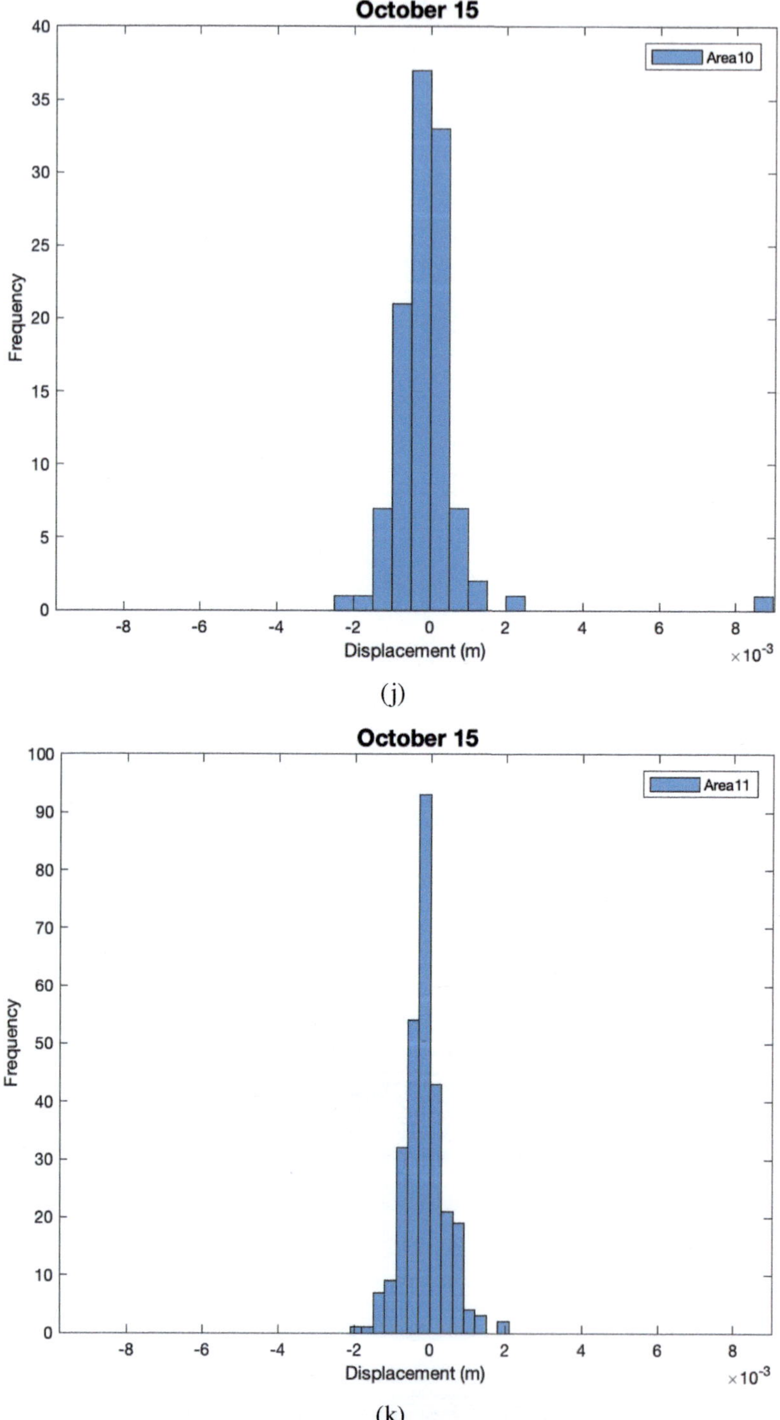

(j)

(k)

FIGURE 5.33 (Continued)

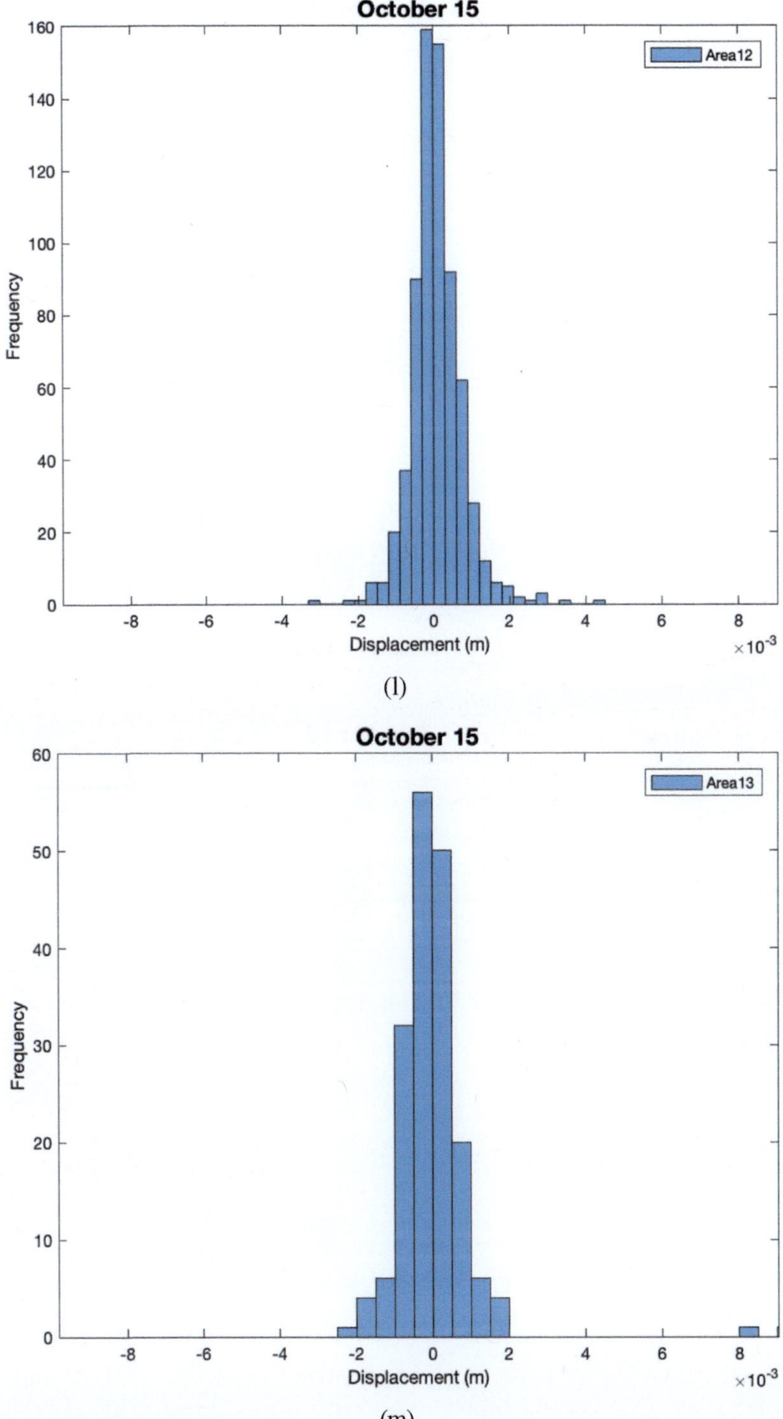

FIGURE 5.33 (Continued)

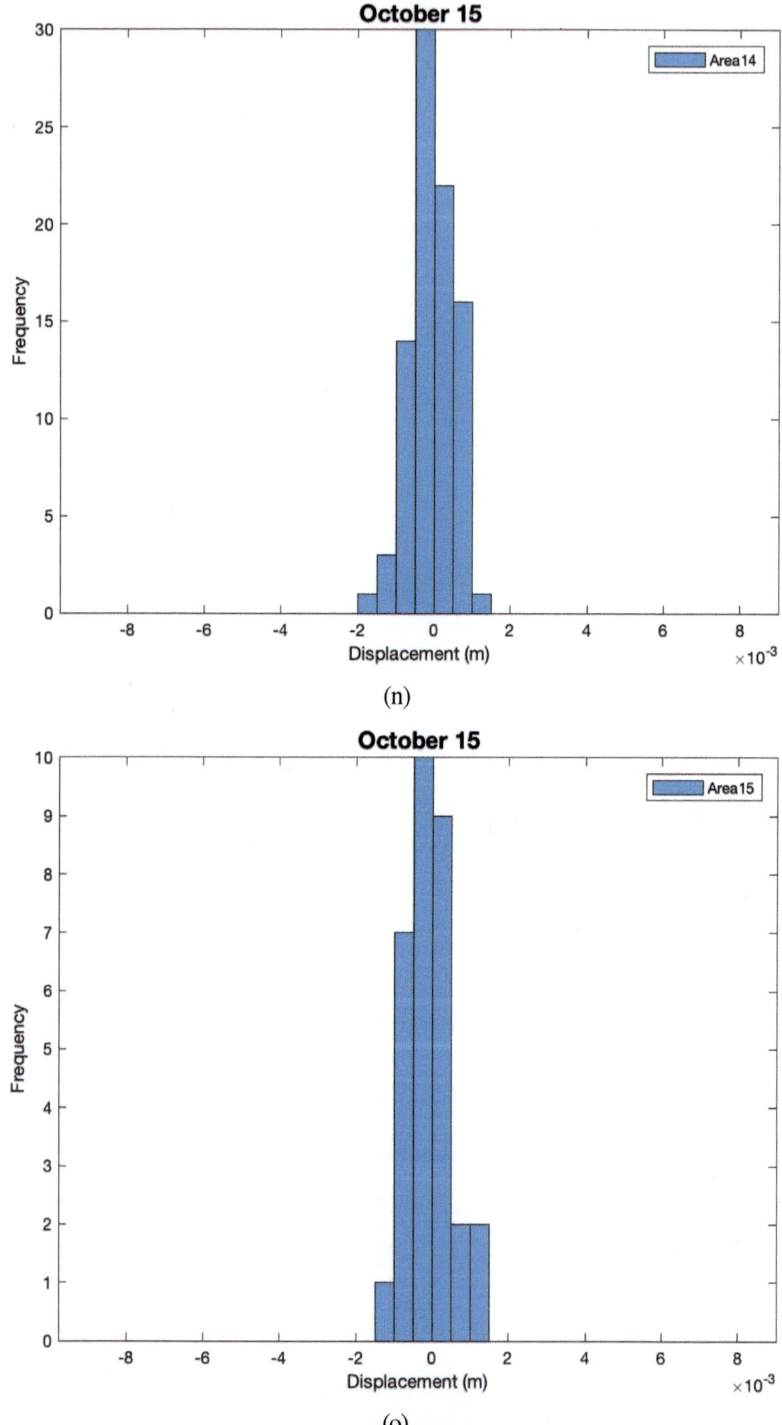

FIGURE 5.33   (Continued)

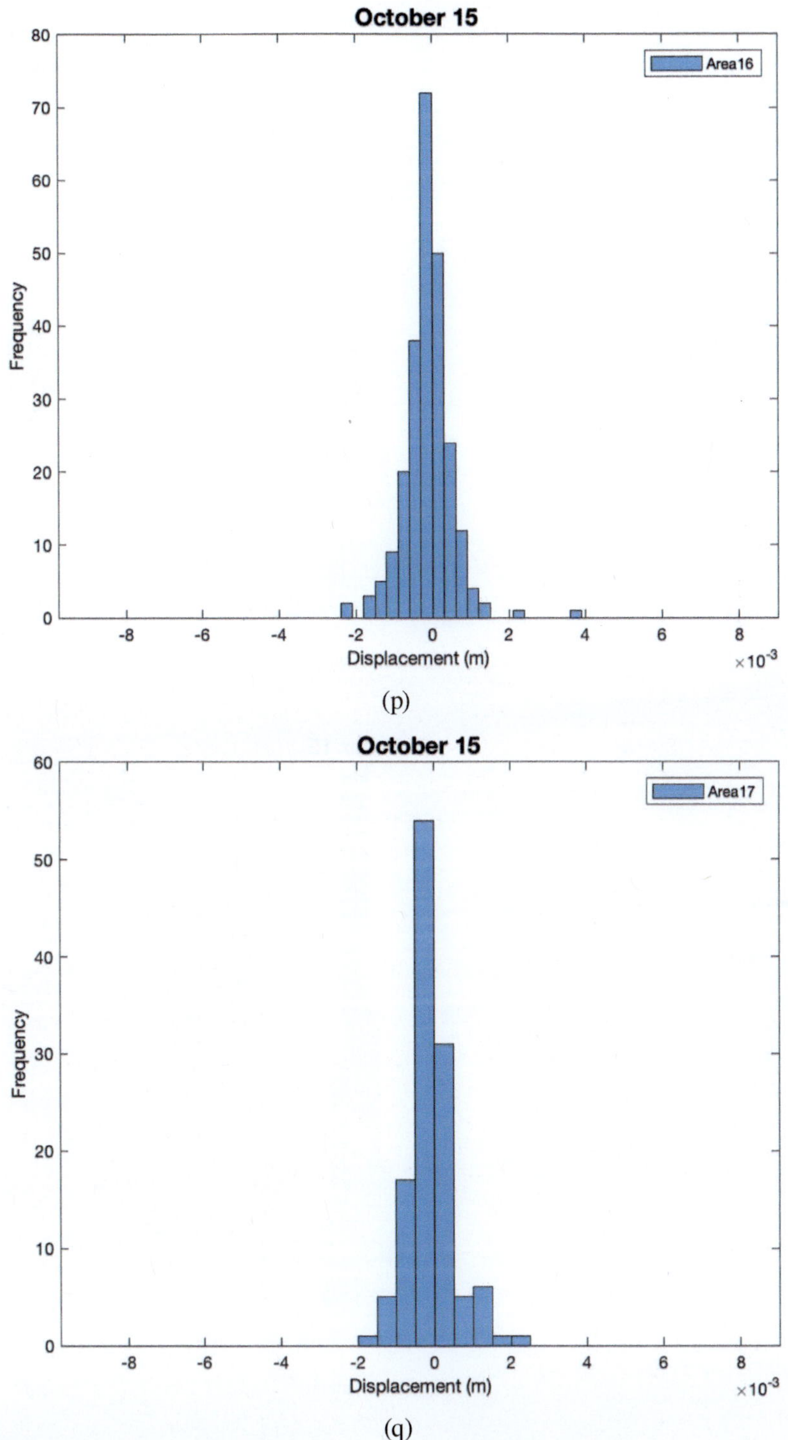

FIGURE 5.33 (Continued)

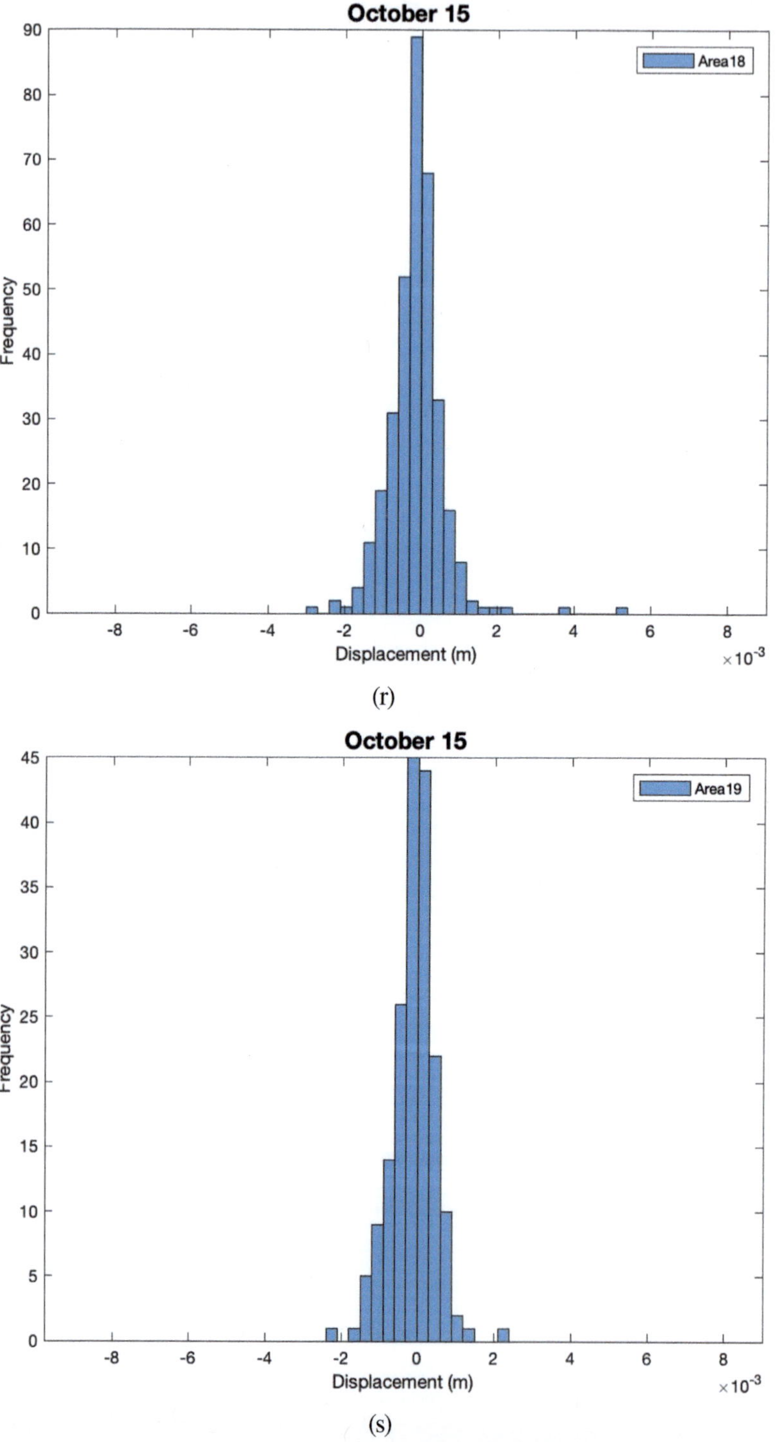

FIGURE 5.33 (Continued)

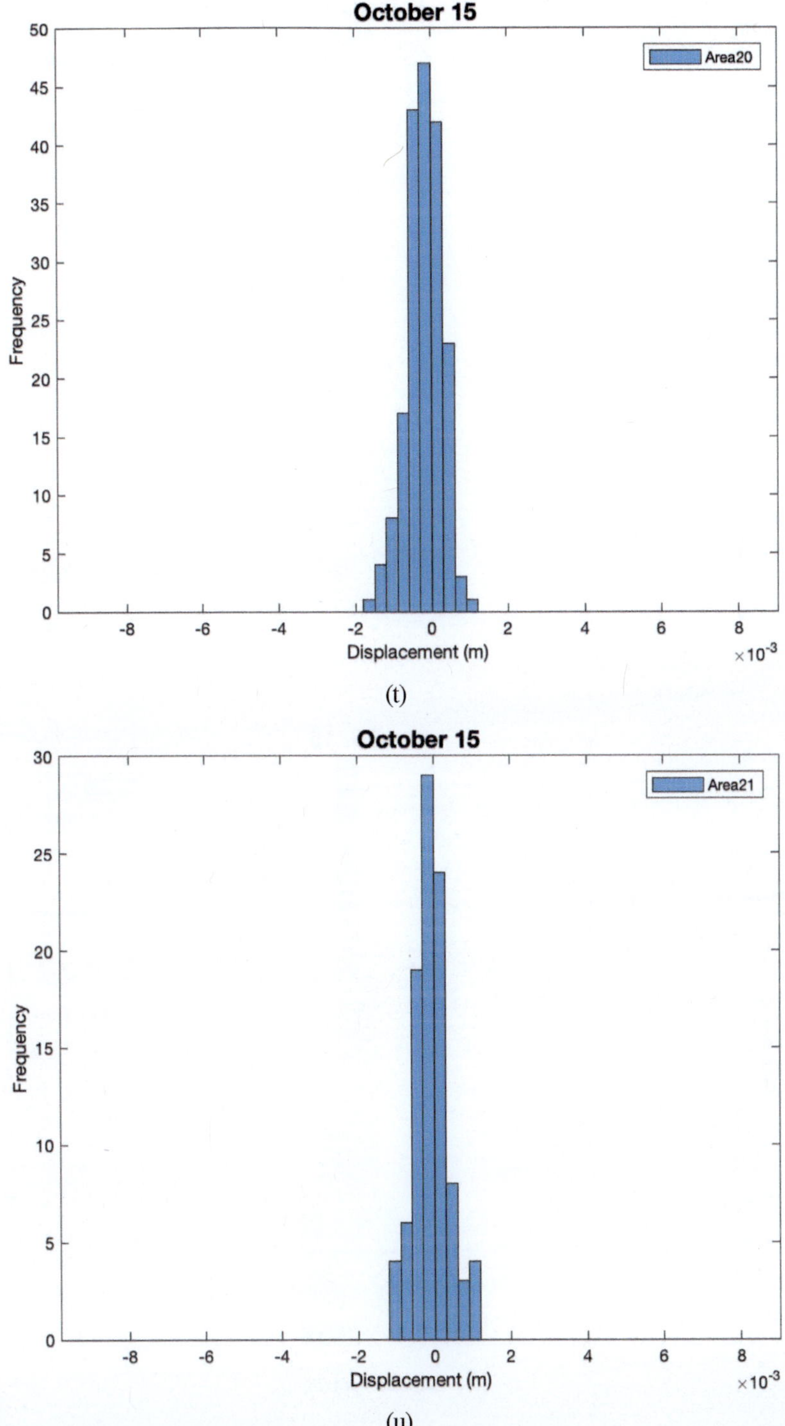

FIGURE 5.33 (Continued)

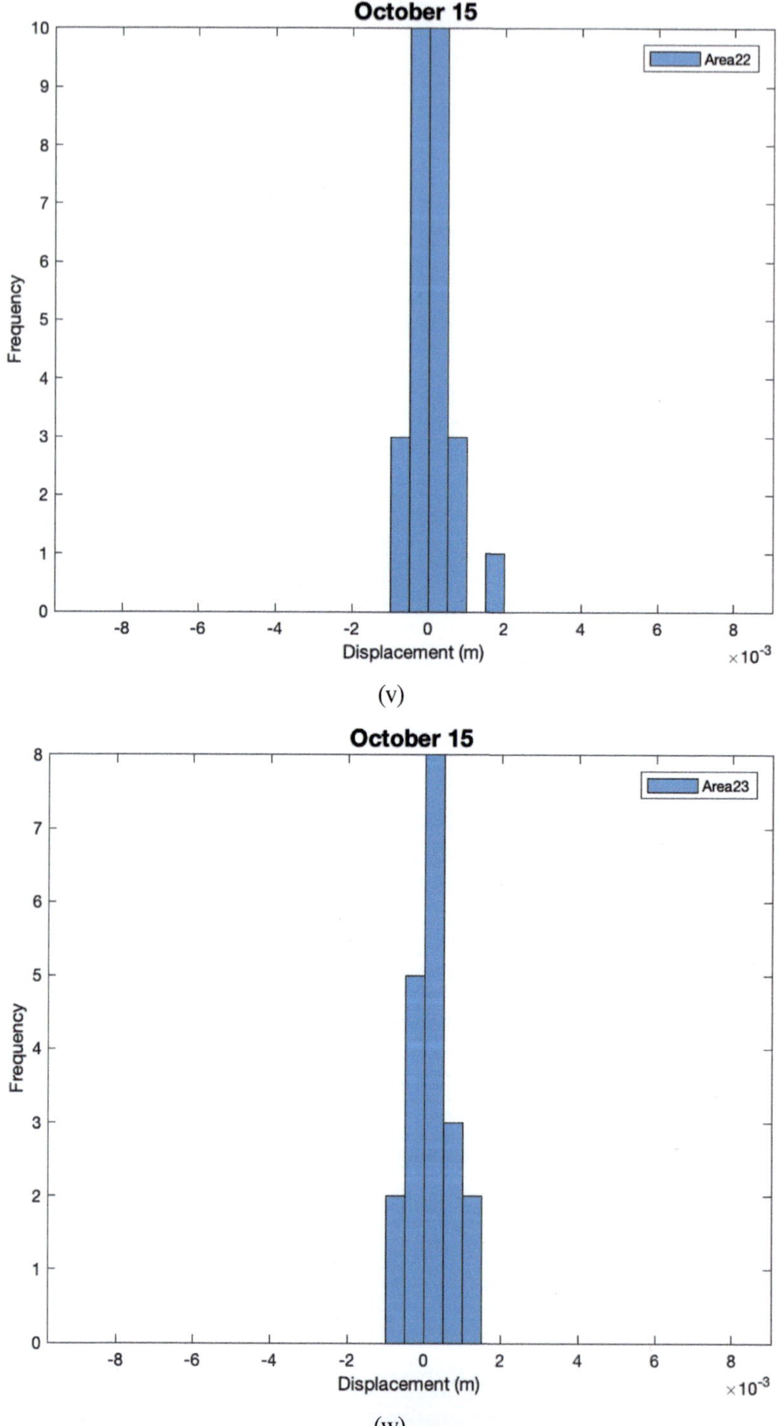

FIGURE 5.33 (Continued)

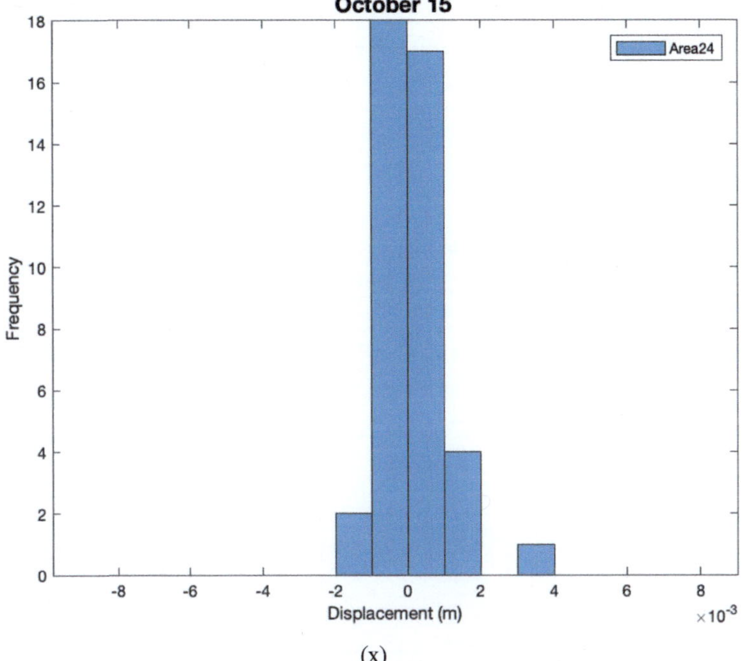

FIGURE 5.33 (Continued)

the relation does not show the corresponding days. Instead, during the higher amount, the displacement period lagged from the rainfall intensity period. Simultaneously, the soil-loss volume in the higher amount also lagged behind the erosivity factor.

Due to the different size of all 24 areas, the number of displacements varied, which caused the average values of each area to be different. Figure 5.34 shows that the displacement approached 7 mm in Area #15, Area #22, and Area #23. In addition, the soil-loss volume shows the highest amount in the same area. In accordance with the rainfall intensity and erosivity factor, an obvious lag occurred on October 5 and 12.

From the left-hand figure, it is shown that high rainfall intensity occurred on October 4 and 12. In each area, the displacement within these periods shows an increasingly high amount approaching approximately 6.5 mm in Areas #10 and #15. This trend is also shown in the soil-loss volume and erosivity factor in the right-hand figure. The maximum erosivity factor occurred on the same date as the rainfall intensity. Simultaneously, soil-loss volume occurred on the same date with the displacement, with a maximum amount of 0.05 m³ in the same area.

Also, in Area #12, the averaged displacement occurred mostly about 3 mm, with the movement relatively stable. From the data in Area #12, it is also obvious that the displacement and the rainfall intensity have a lag gap of several days, particularly on October 4, which lagged to several days on October 7.

Figure 5.35 shows the accumulated displacement and rainfall intensity on the left side and the accumulated soil-loss volume and the erosivity factor on the right side of each area daily based on October. The displacement shown in blue is the accumulated displacement points in each area. The displacement is described in mm with an

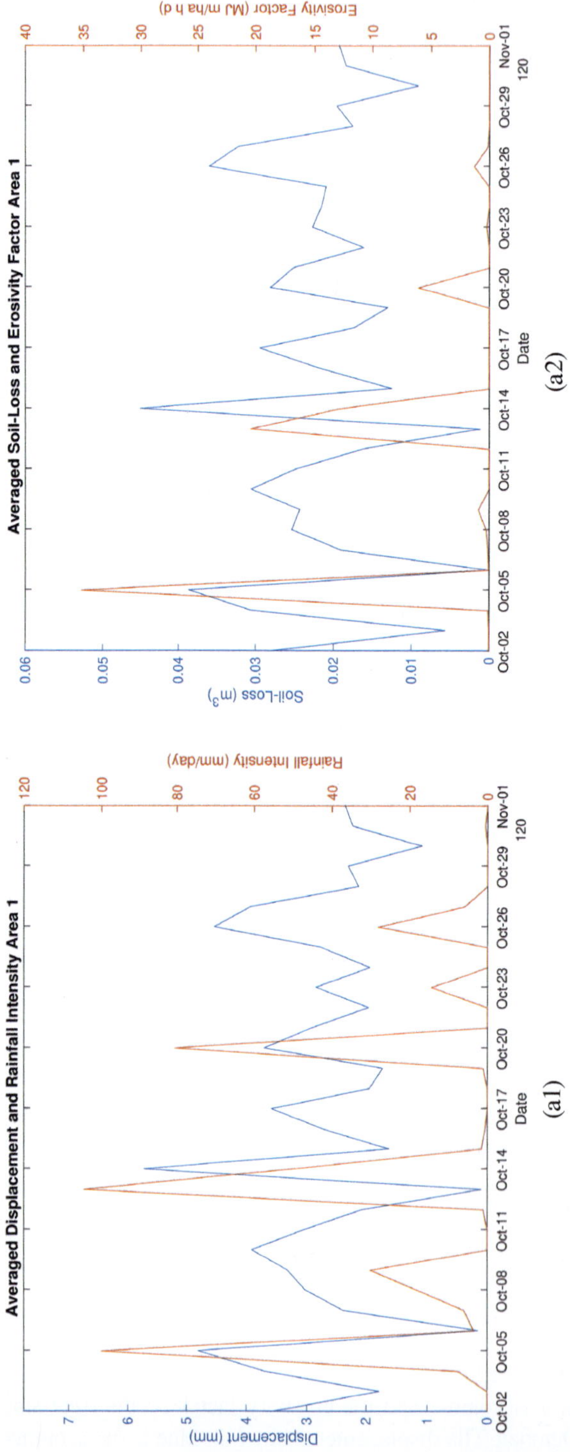

FIGURE 5.34 Averaged displacement and soil-loss volume in October. (a1) Displacement in Area #1. (b1) Displacement in Area #2. (c1) Displacement in Area #3. (d1) Displacement in Area #4. (e1) Displacement in Area #5. (f1) Displacement in Area #6. (g1) Displacement in Area #7. (h1) Displacement in Area #8. (i1) Displacement in Area #9. (j1) Displacement in Area #10. (k1) Displacement in Area #11. (l1) Displacement in Area #12. (m1) Displacement in Area #13. (n1) Displacement in Area #14. (o1) Displacement in Area #15. (p1) Displacement in Area #16. (q1) Displacement in Area #17. (r1) Displacement in Area #18. (s1) Displacement in Area #19. (t1) Displacement in Area #20. (u1) Displacement in Area #21. (v1) Displacement in Area #22. (w1) Displacement in Area #23. (x1) Displacement in Area #24. (a2) Soil loss in Area #1. (b2) Soil loss in Area #2. (c2) Soil loss in Area #3. (d2) Soil loss in Area #4. (e2) Soil loss in Area #5. (f2) Soil loss in Area #6. (g2) Soil loss in Area #7. (h2) Soil loss in Area #8. (i2) Soil loss in Area #9. (j2) Soil loss in Area #10. (k2) Soil loss in Area #11. (l2) Soil loss in Area #12. (m2) Soil loss in Area #13. (n2) Soil loss in Area #14. (o2) Soil loss in Area #15. (p2) Soil loss in Area #16. (q2) Soil loss in Area #17. (r2) Soil loss in Area #18. (s2) Soil loss in Area #19. (t2) Soil loss in Area #20. (u2) Soil loss in Area #21. (v2) Soil loss in Area #22. (w2) Soil loss in Area #23. (x2) Soil loss in Area #24.

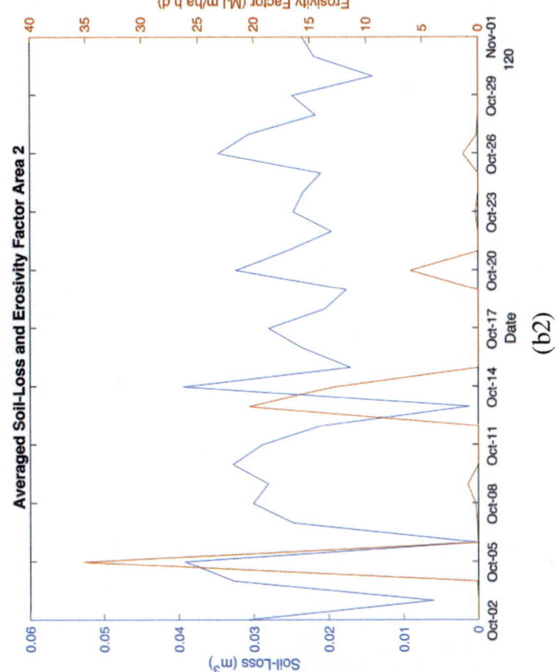

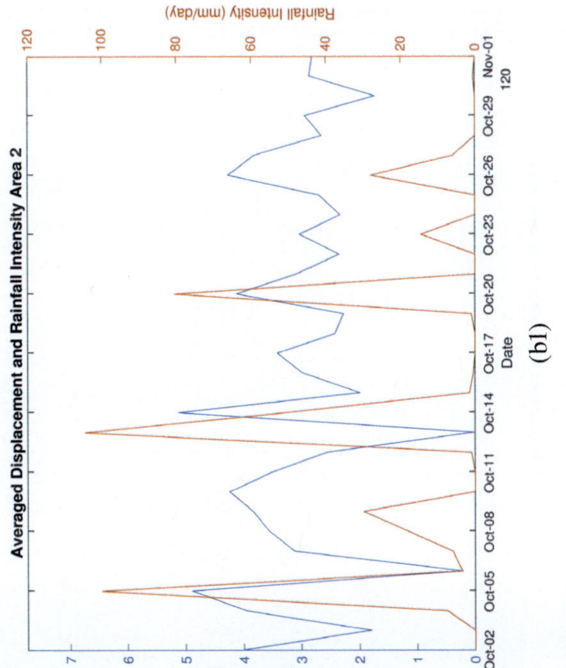

FIGURE 5.34 (Continued)

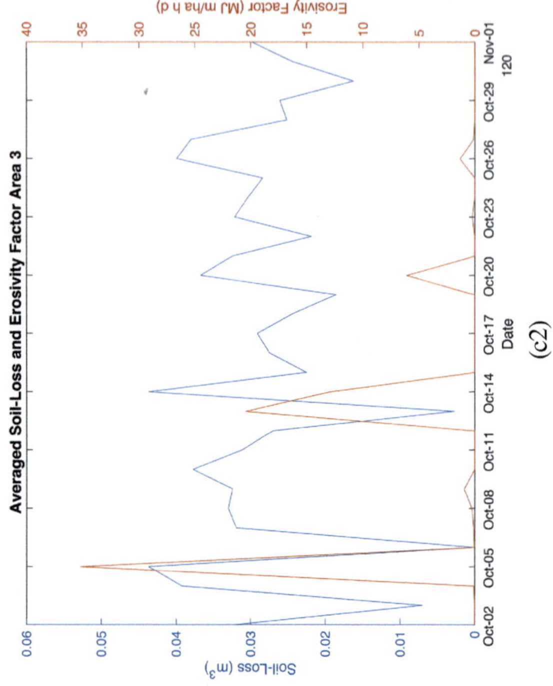

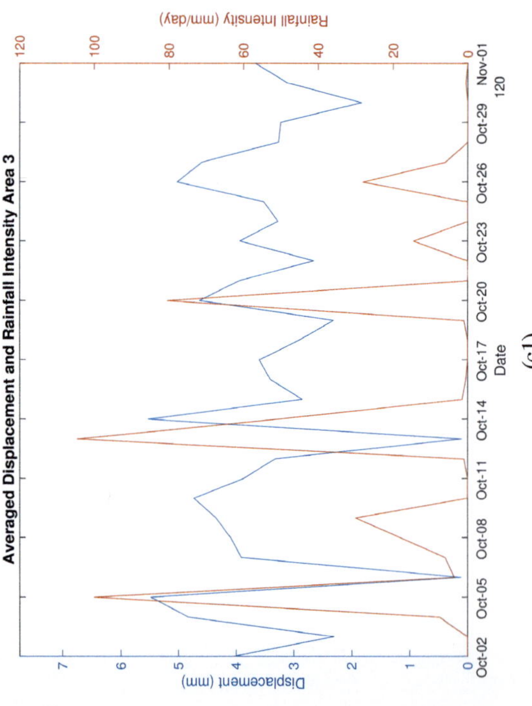

FIGURE 5.34 (Continued)

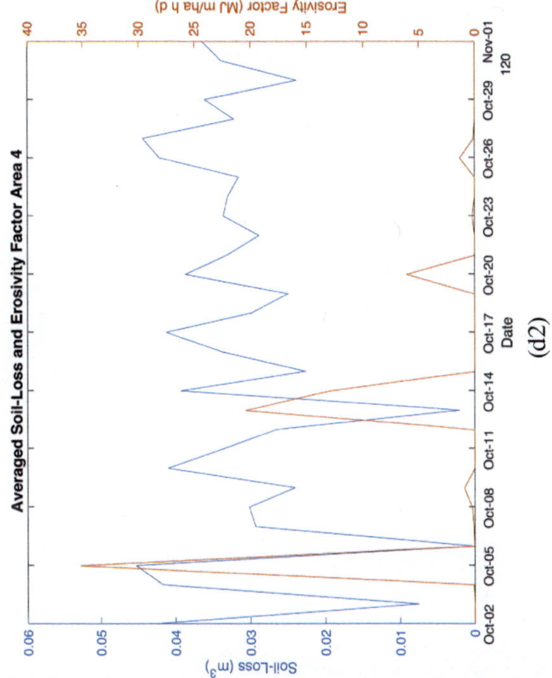

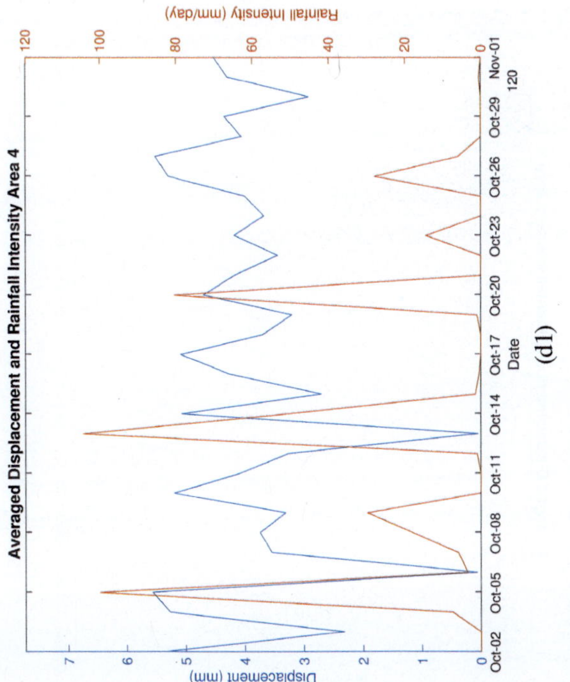

FIGURE 5.34 (Continued)

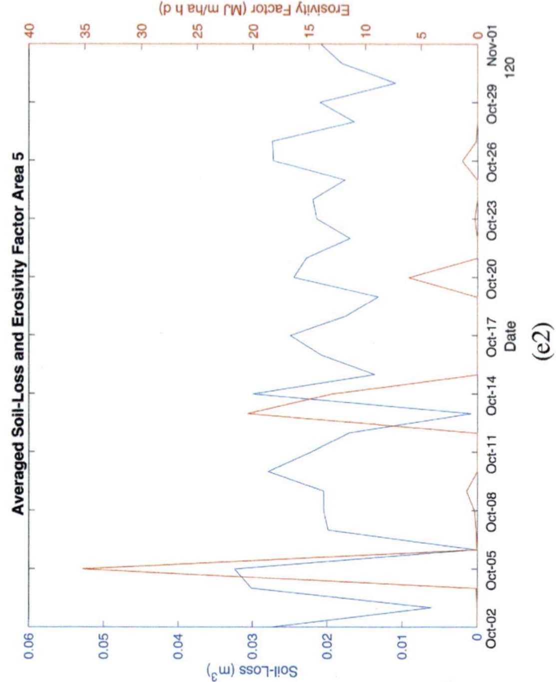

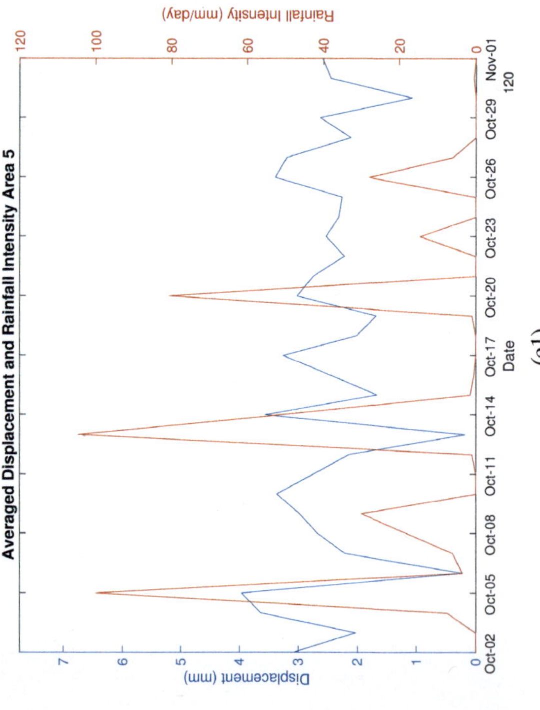

FIGURE 5.34 (Continued)

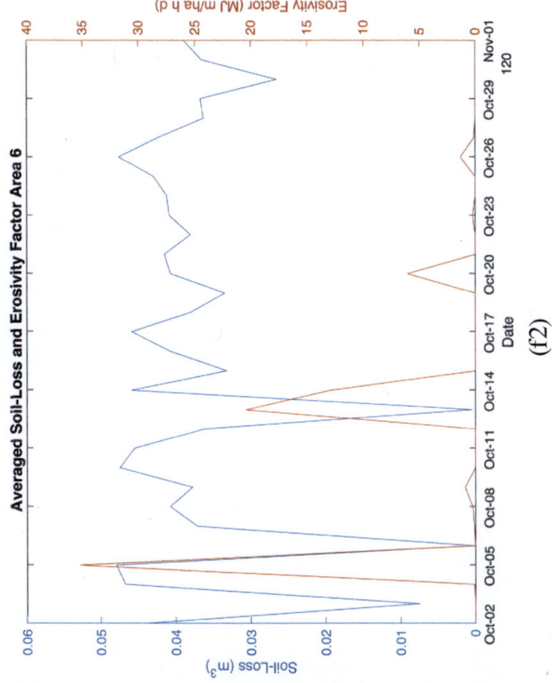

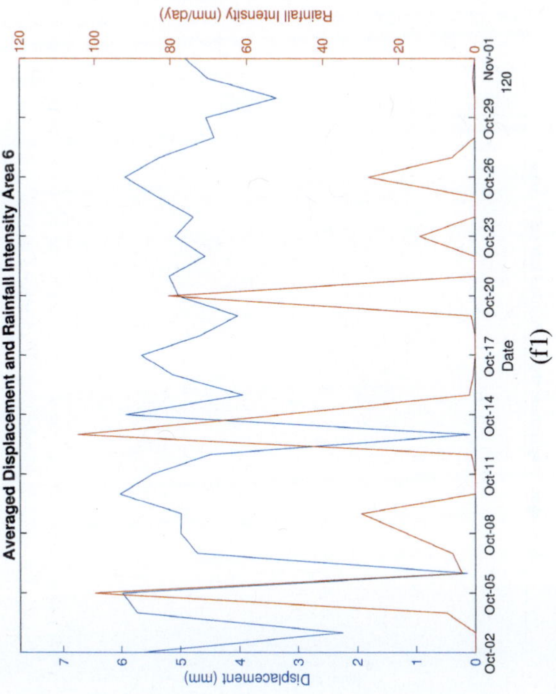

FIGURE 5.34 (Continued)

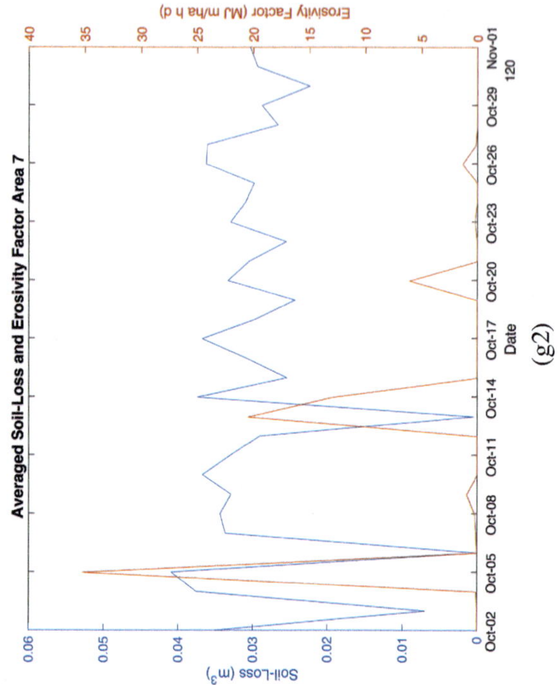

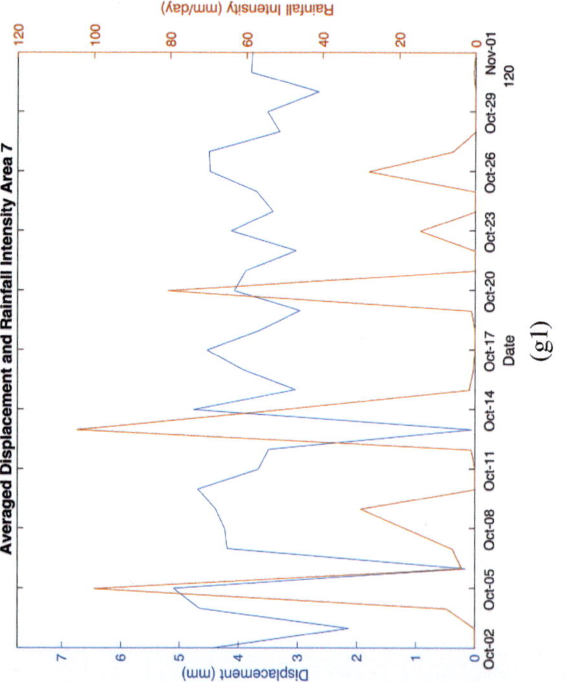

FIGURE 5.34 (Continued)

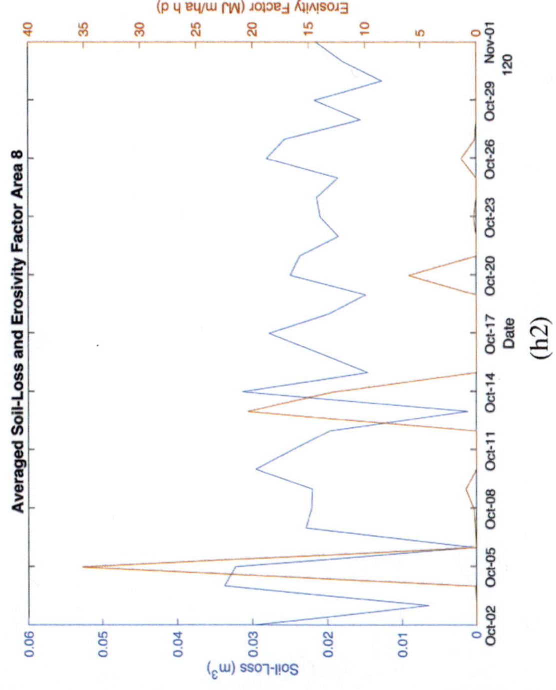

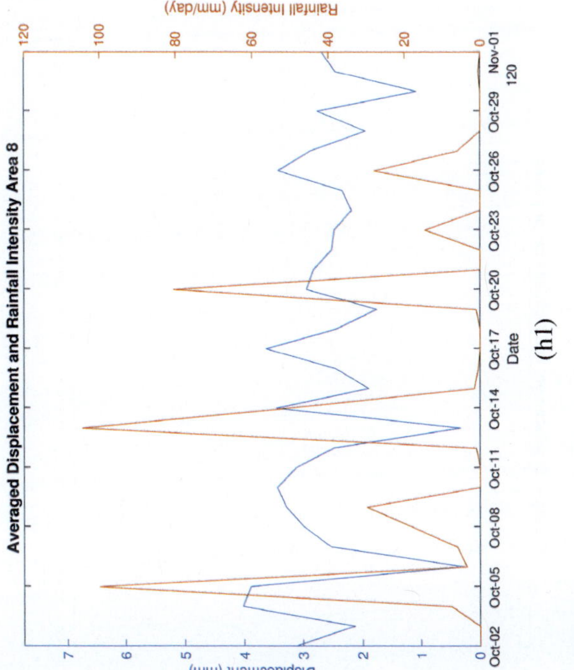

FIGURE 5.34   (Continued)

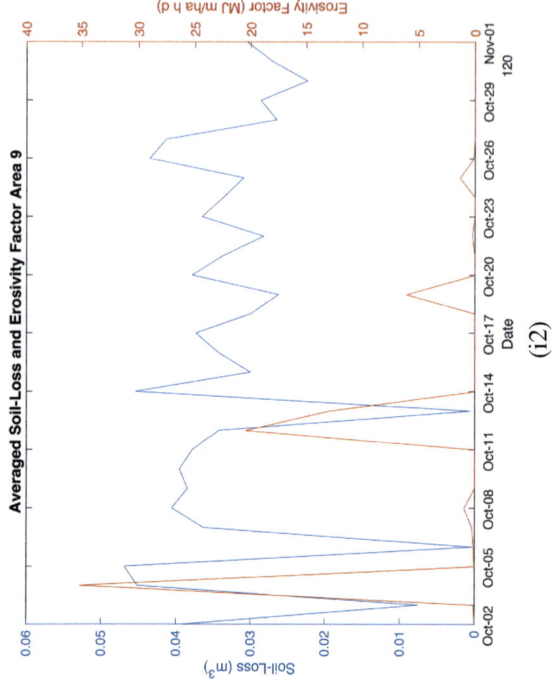

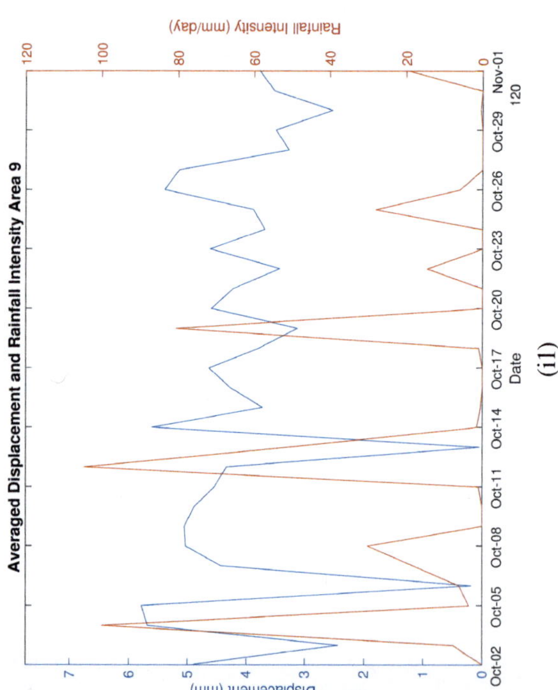

FIGURE 5.34 (Continued)

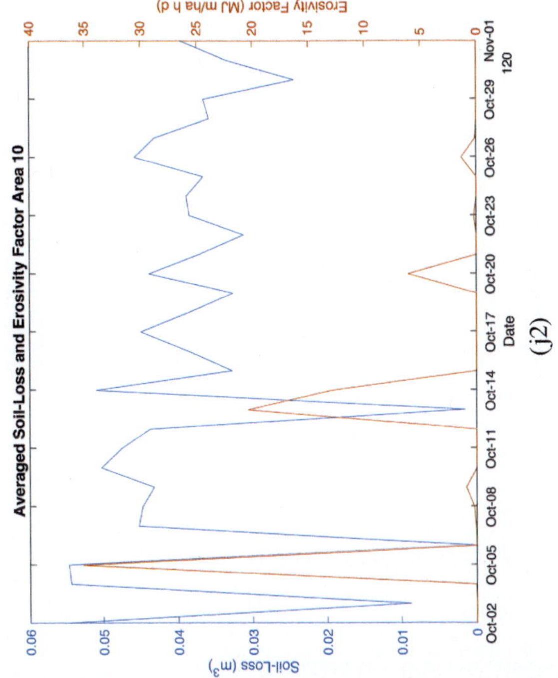

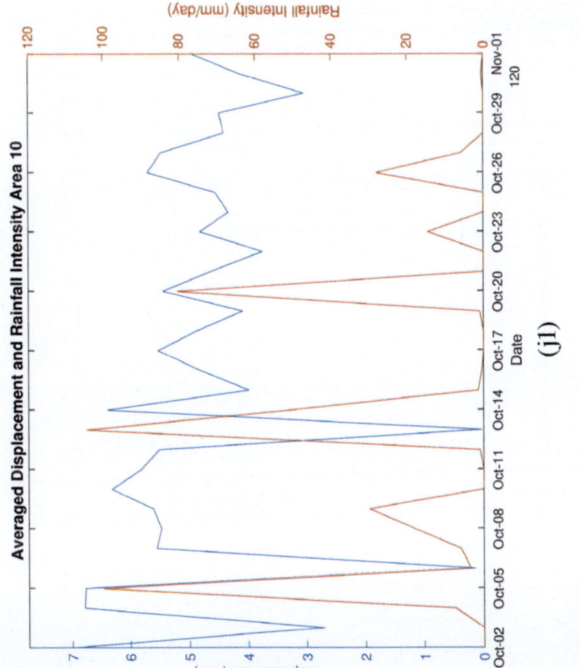

FIGURE 5.34 (Continued)

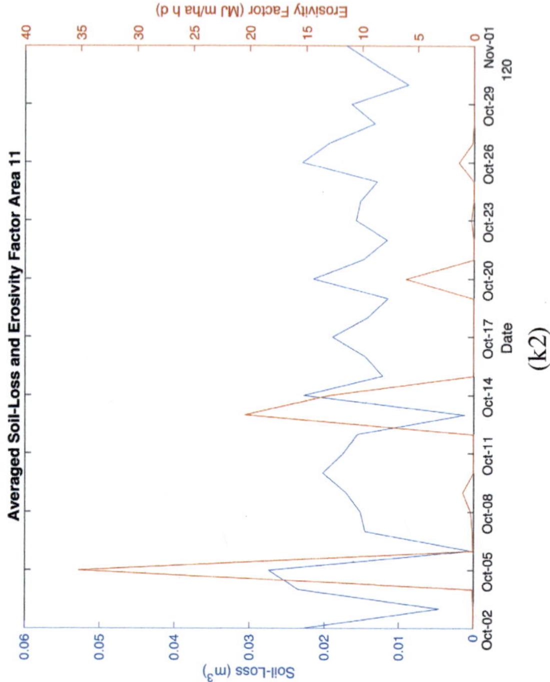

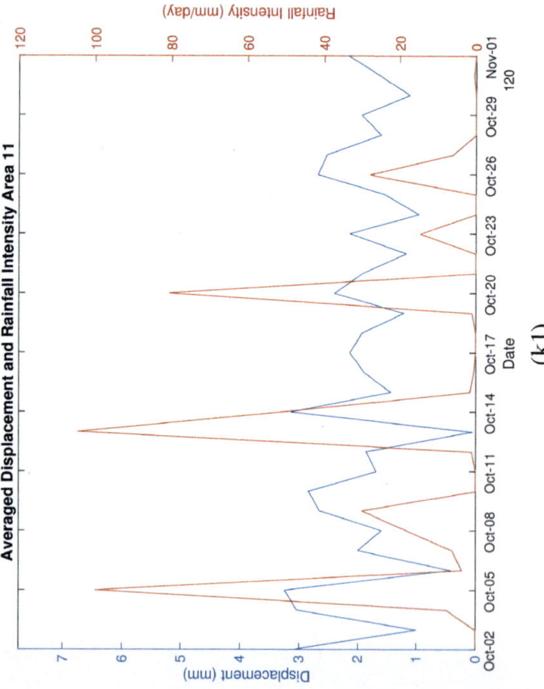

FIGURE 5.34 (Continued)

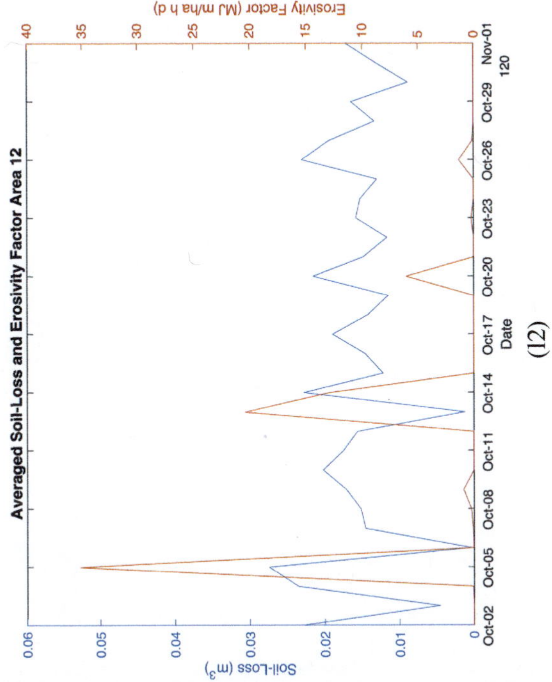

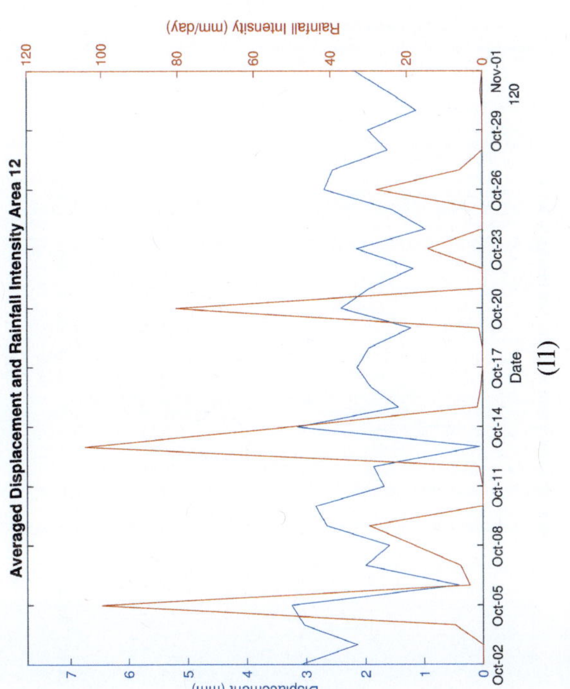

FIGURE 5.34 (Continued)

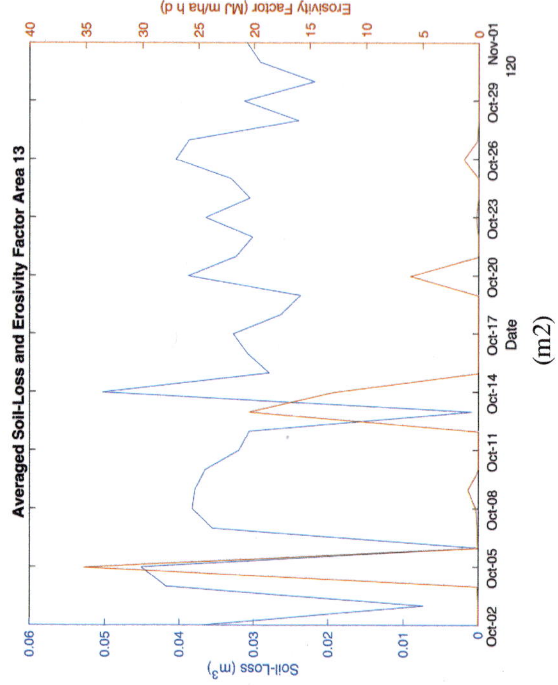

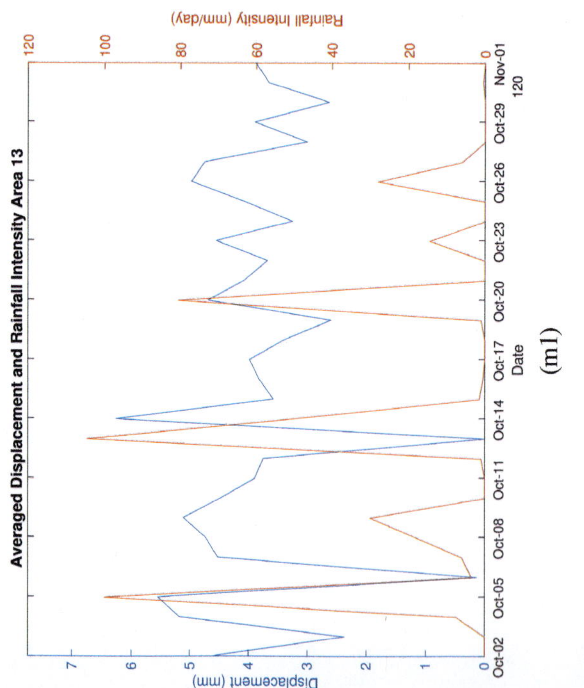

FIGURE 5.34 (Continued)

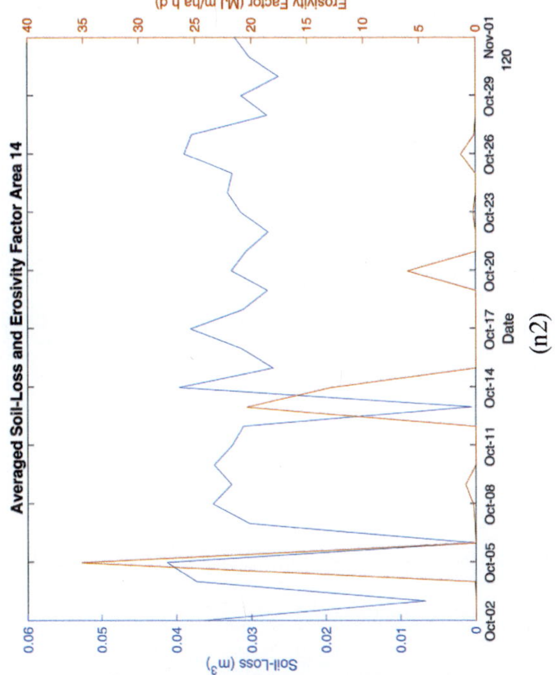

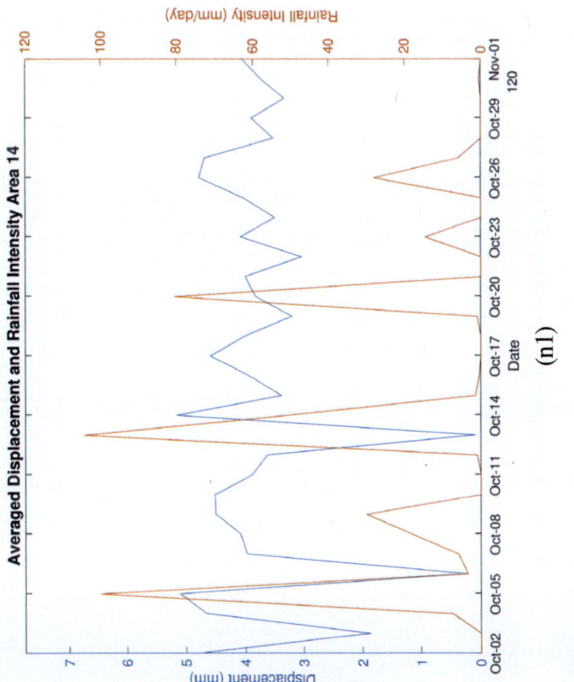

FIGURE 5.34 (Continued)

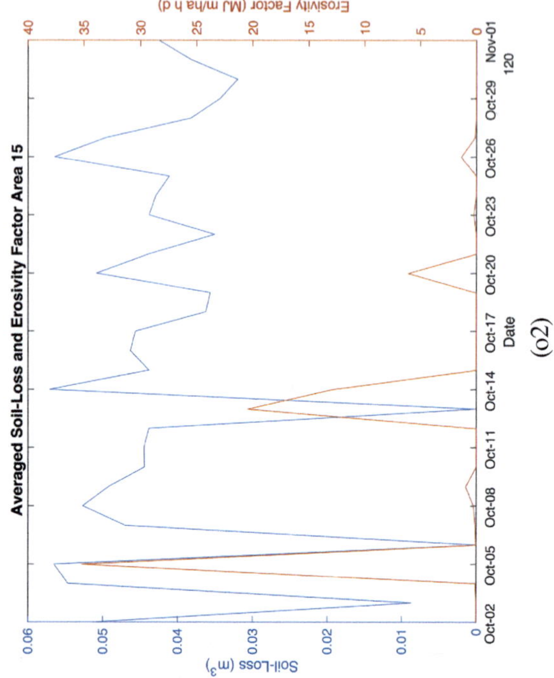

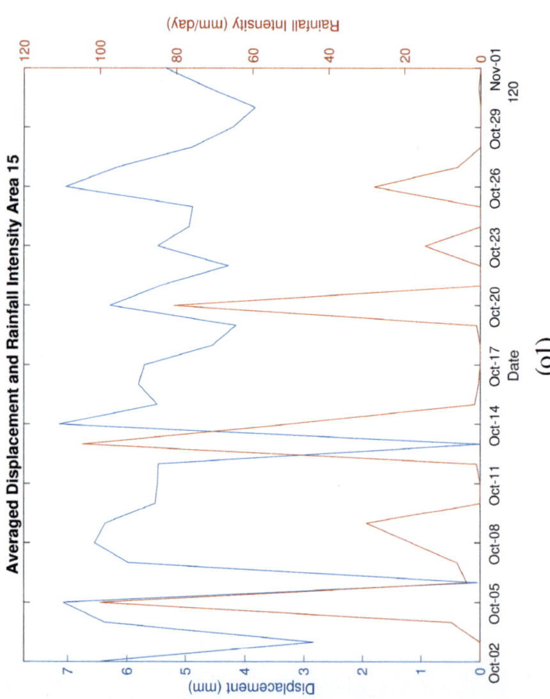

FIGURE 5.34 (Continued)

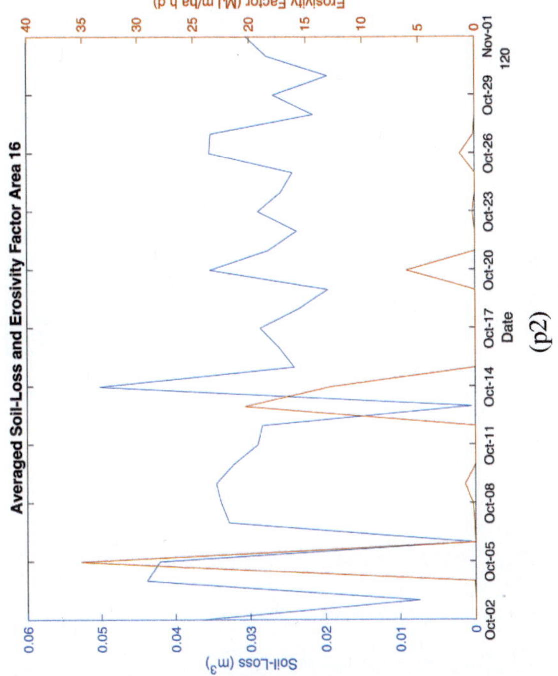

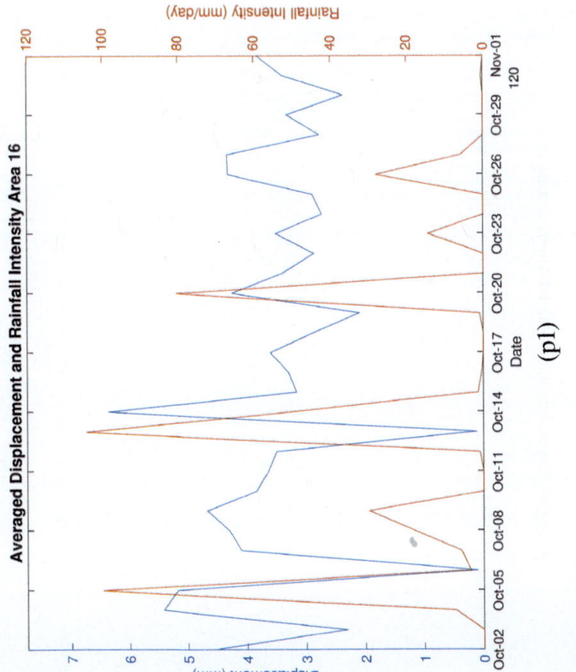

FIGURE 5.34 (Continued)

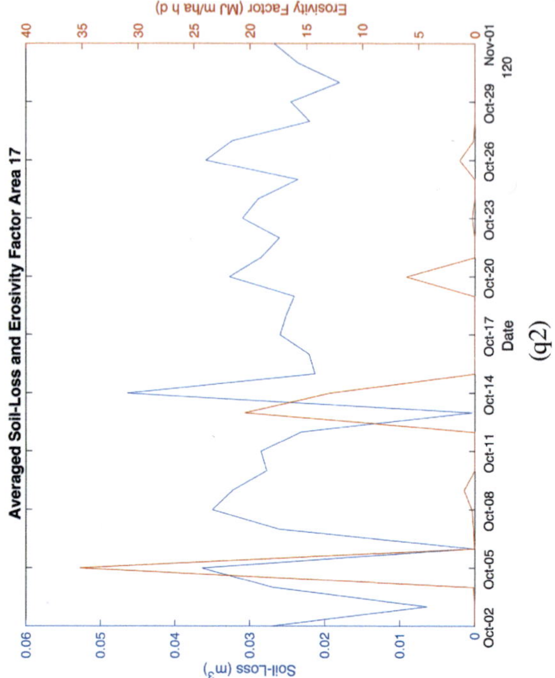

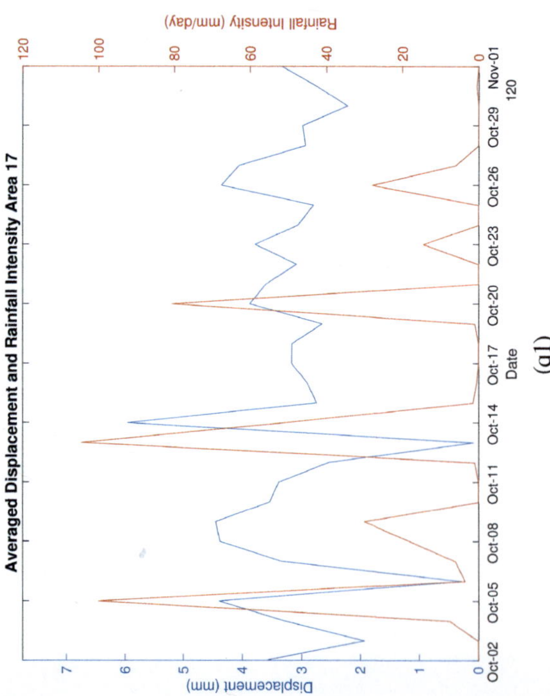

FIGURE 5.34 (Continued)

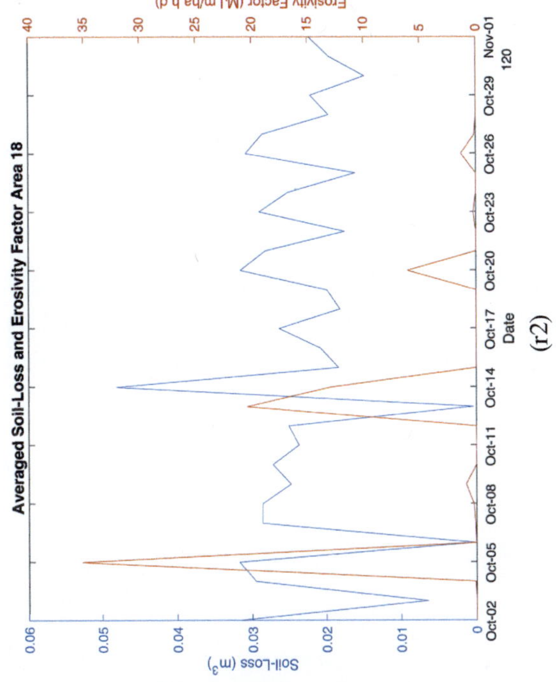

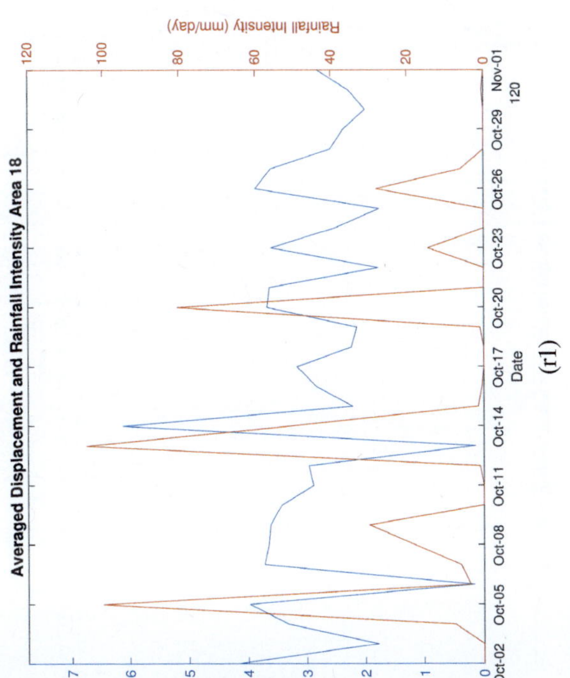

FIGURE 5.34 (Continued)

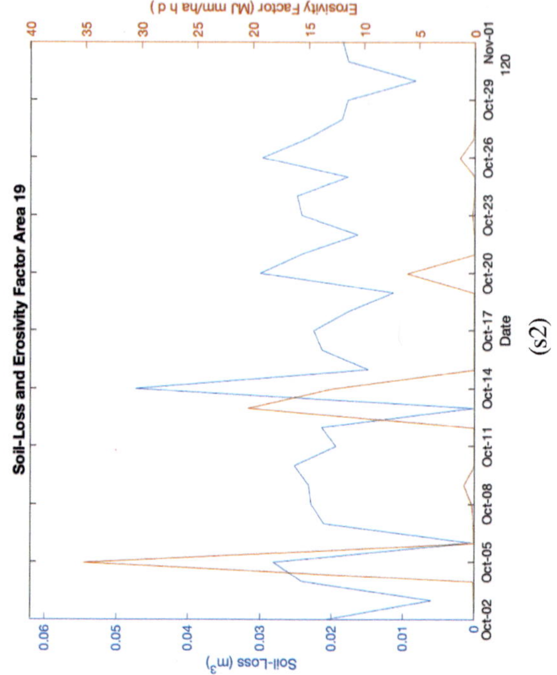

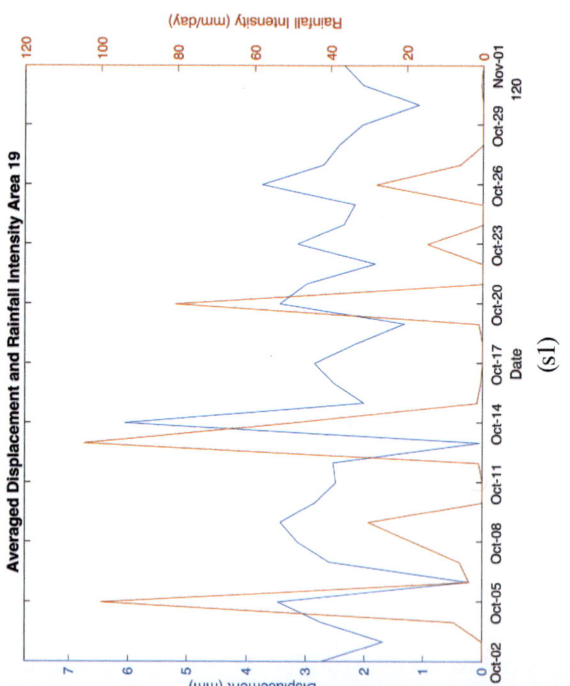

FIGURE 5.34 (Continued)

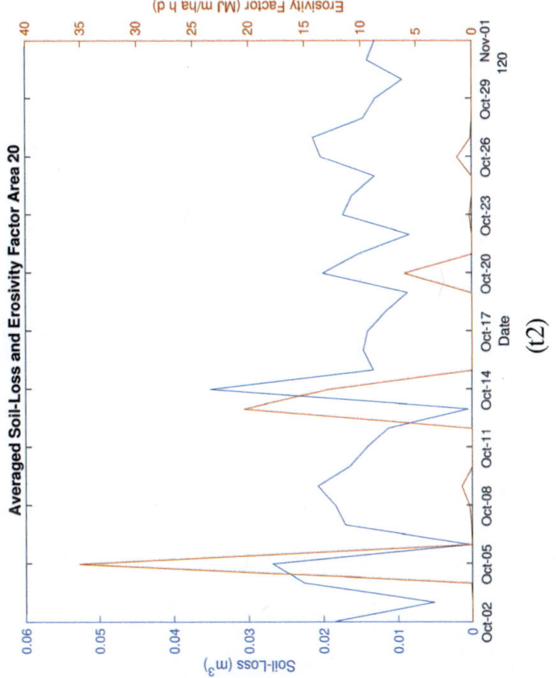

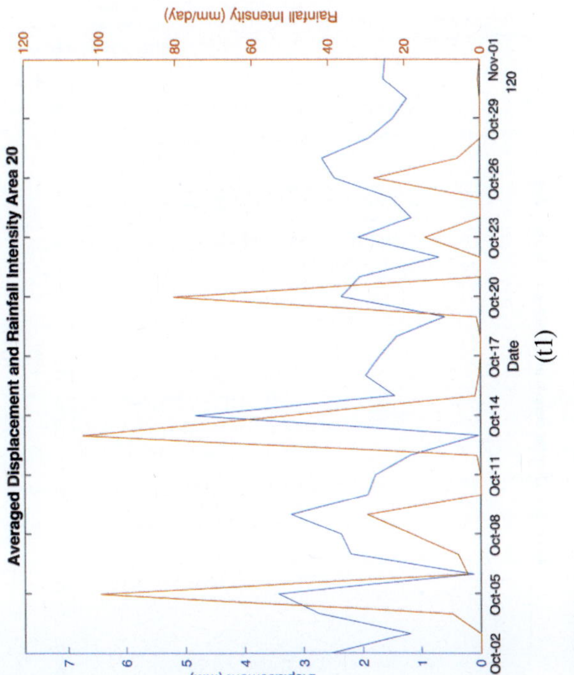

FIGURE 5.34 (Continued)

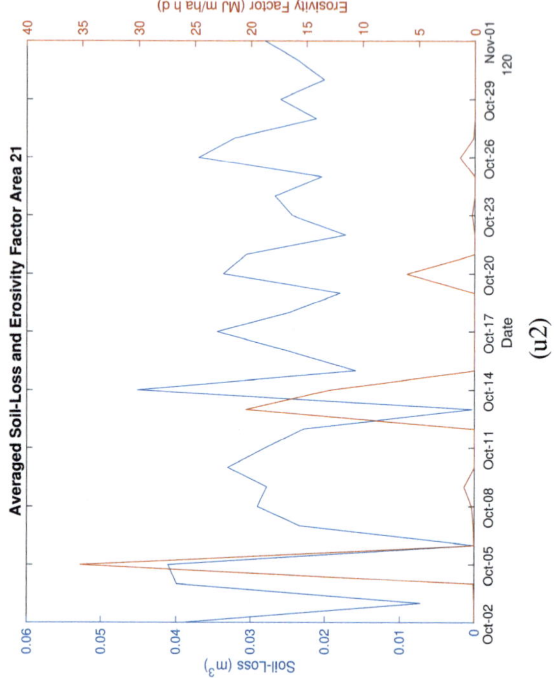

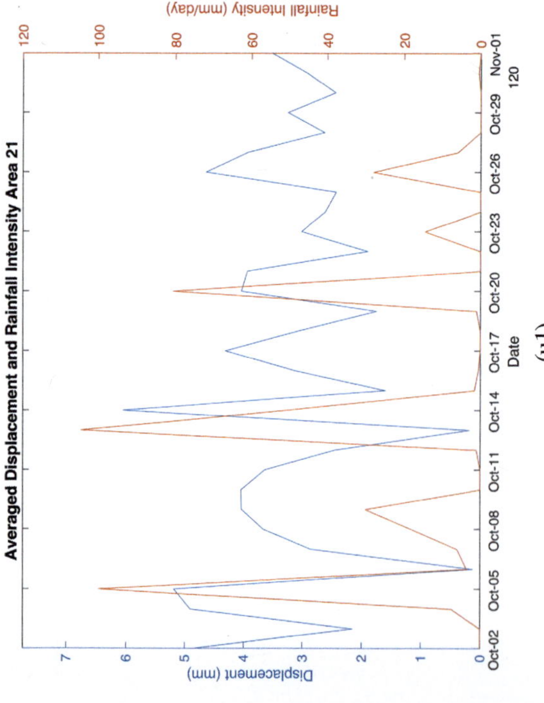

FIGURE 5.34 (Continued)

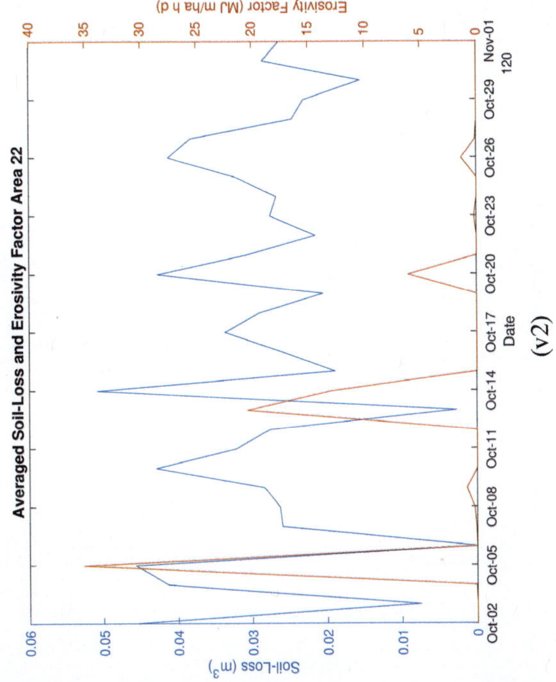

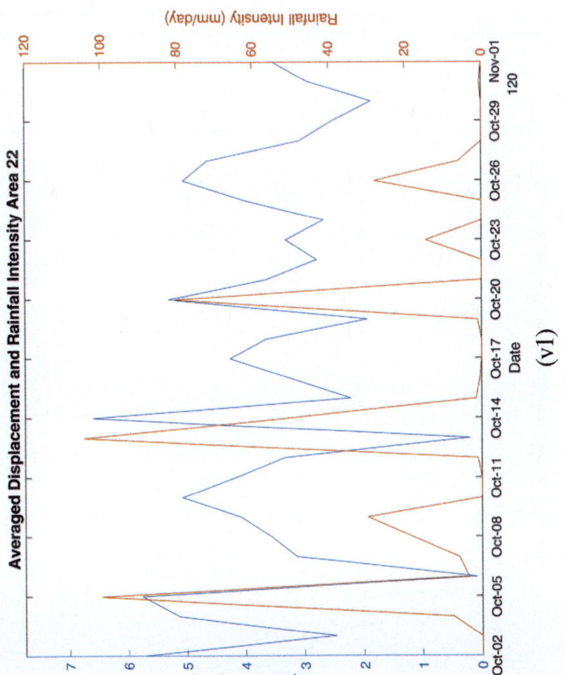

FIGURE 5.34 (Continued)

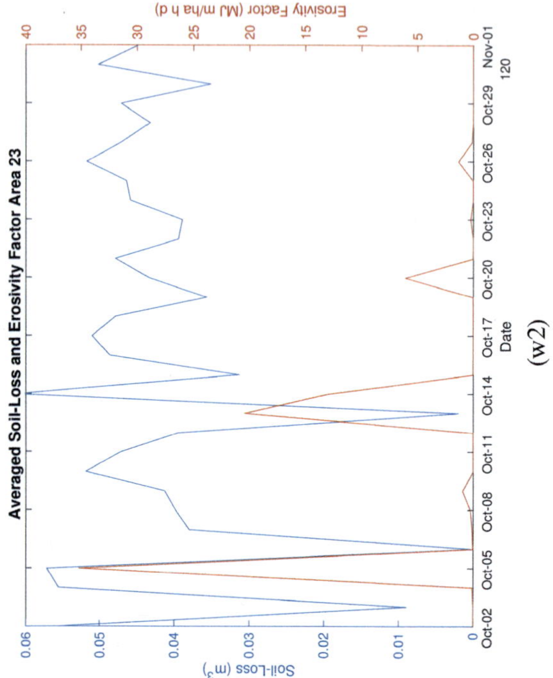

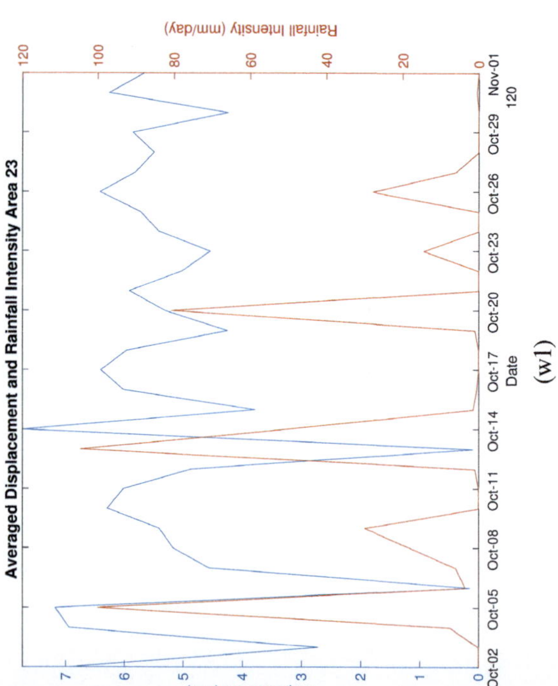

FIGURE 5.34 (Continued)

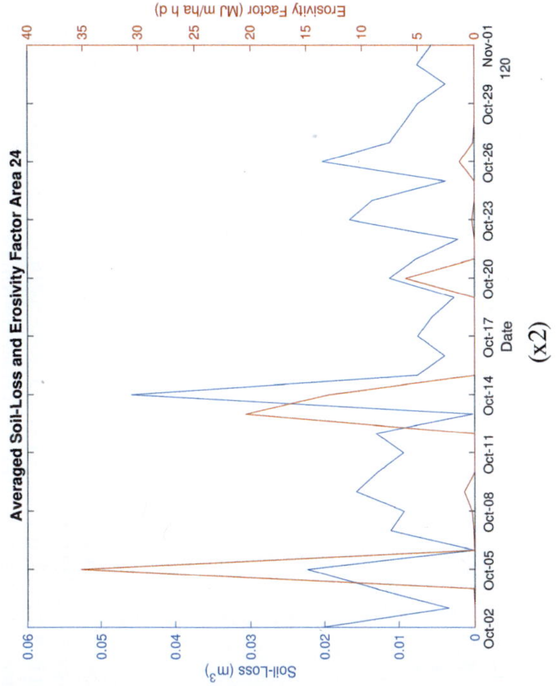

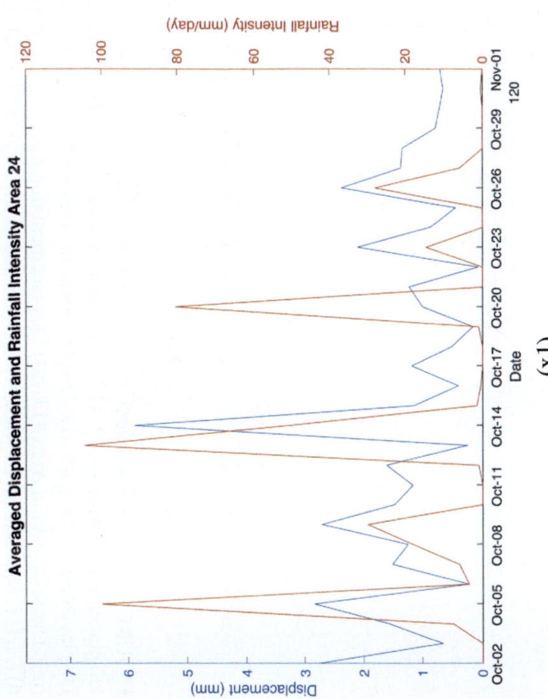

FIGURE 5.34 (Continued)

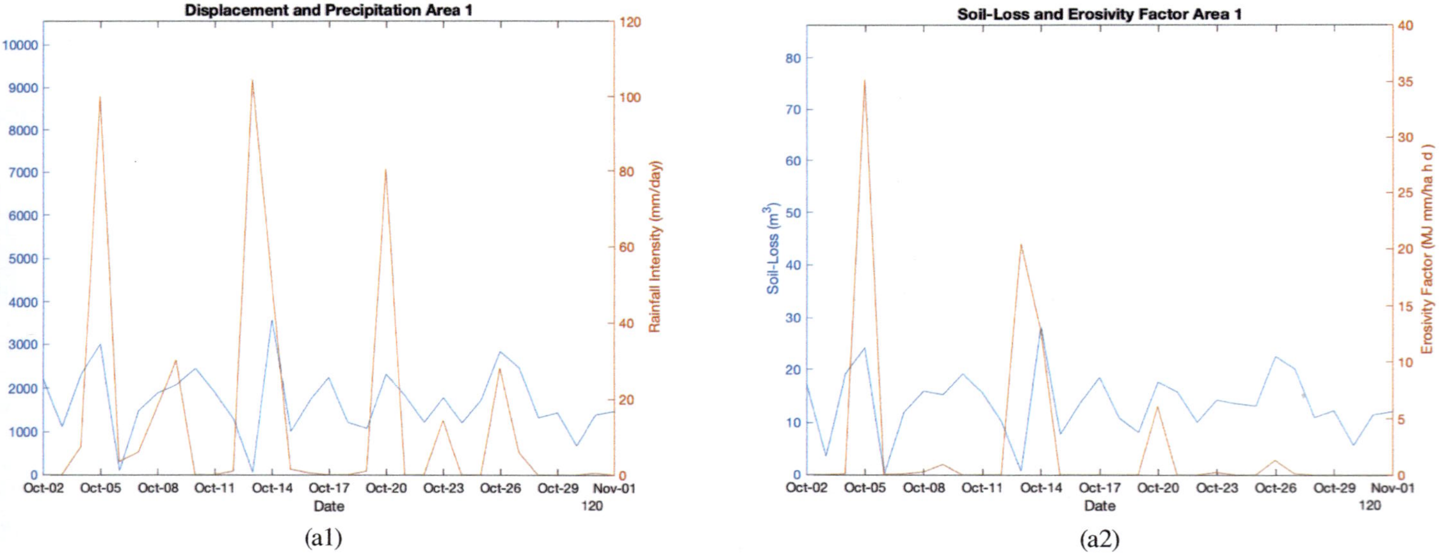

FIGURE 5.35 Accumulated displacement and soil-loss volume in October. (a1) Displacement in Area #1. (b1) Displacement in Area #2. (c1) Displacement in Area #3. (d1) Displacement in Area #4. (e1) Displacement in Area #5. (f1) Displacement in Area #6. (g1) Displacement in Area #7. (h1) Displacement in Area #8. (i1) Displacement in Area #9. (j1) Displacement in Area #10. (k1) Displacement in Area #11. (l1) Displacement in Area #12. (m1) Displacement in Area #13. (n1) Displacement in Area #14. (o1) Displacement in Area #15. (p1) Displacement in Area #16. (q1) Displacement in Area #17. (r1) Displacement in Area #18. (s1) Displacement in Area #19. (t1) Displacement in Area #20. (u1) Displacement in Area #21. (v1) Displacement in Area #22. (w1) Displacement in Area #23. (x1) Displacement in Area #24. (a1) Soil loss in Area #1. (b2) Soil loss in Area #2. (c2) Soil loss in Area #3. (d2) Soil loss in Area #4. (e2) Soil loss in Area #5. (f2) Soil loss in Area #6. (g2) Soil loss in Area #7. (h2) Soil loss in Area #8. (i2) Soil loss in Area #9. (j2) Soil loss in Area #10. (k2) Soil loss in Area #11. (l2) Soil loss in Area #12. (m2) Soil loss in Area #13. (n2) Soil loss in Area #14. (o2) Soil loss in Area #15. (p2) Soil loss in Area #16. (q2) Soil loss in Area #17. (r2) Soil loss in Area #18. (s2) Soil loss in Area #19. (t2) Soil loss in Area #20. (u2) Soil loss in Area #21. (v2) Soil loss in Area #22. (w2) Soil loss in Area #23. (x2) Soil loss in Area #24.

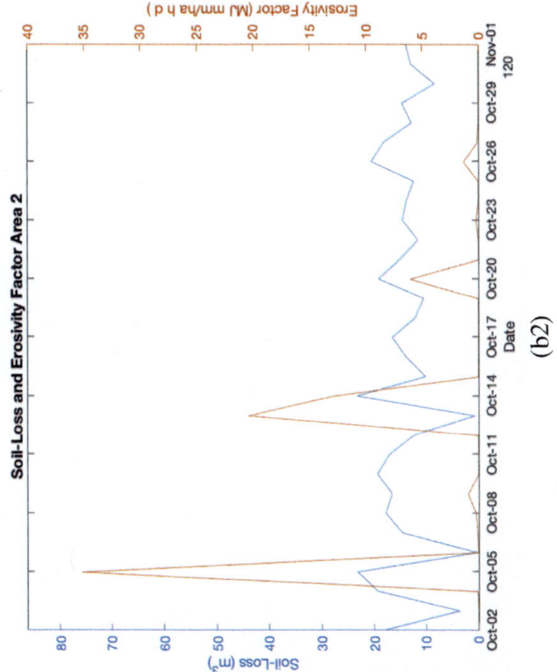

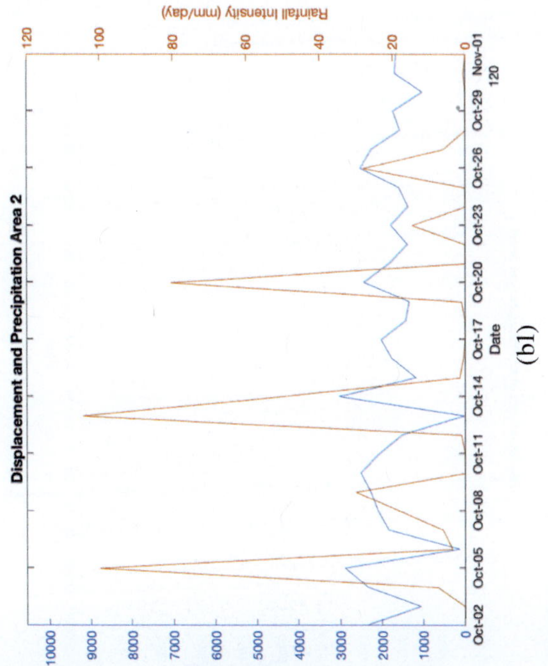

FIGURE 5.35 (Continued)

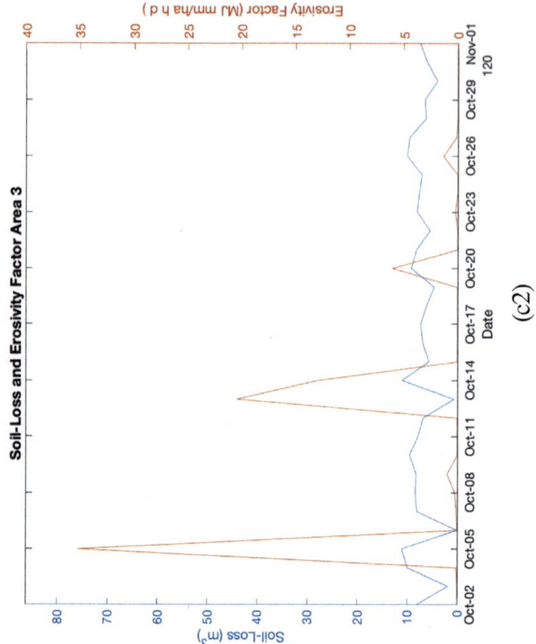

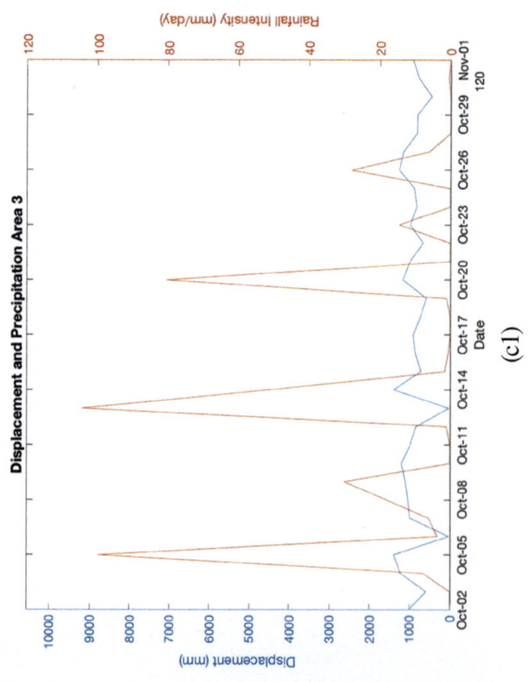

FIGURE 5.35 (Continued)

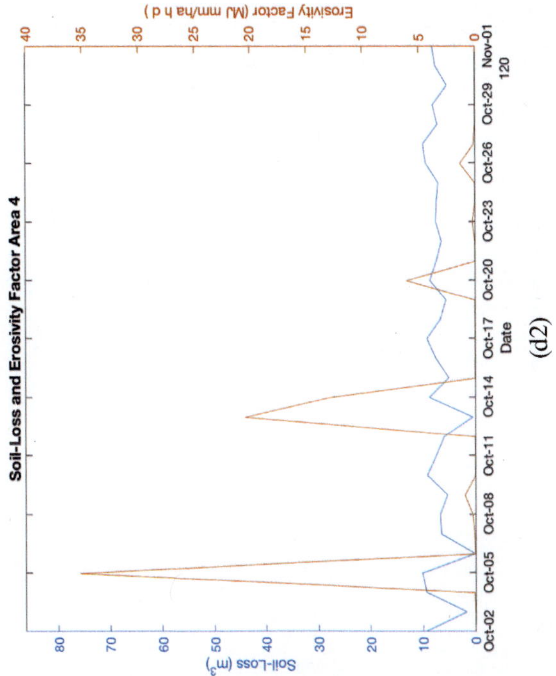

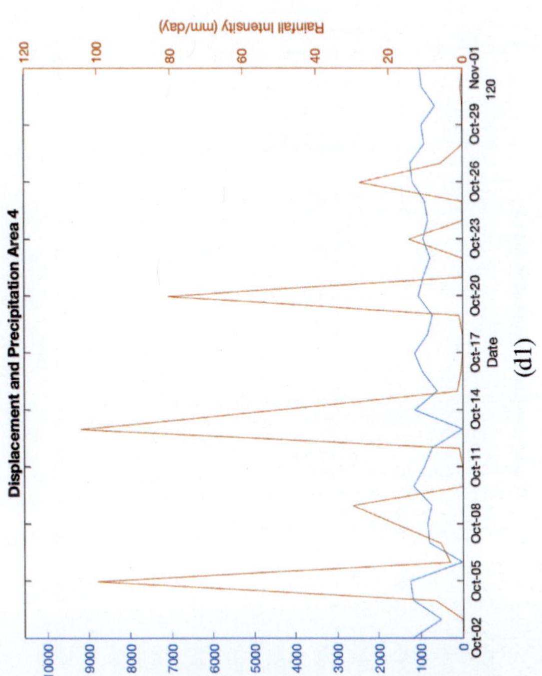

FIGURE 5.35 (Continued)

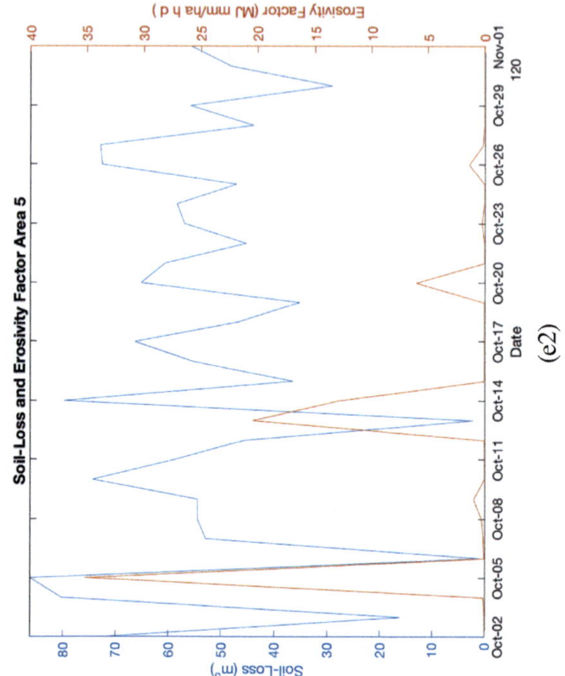

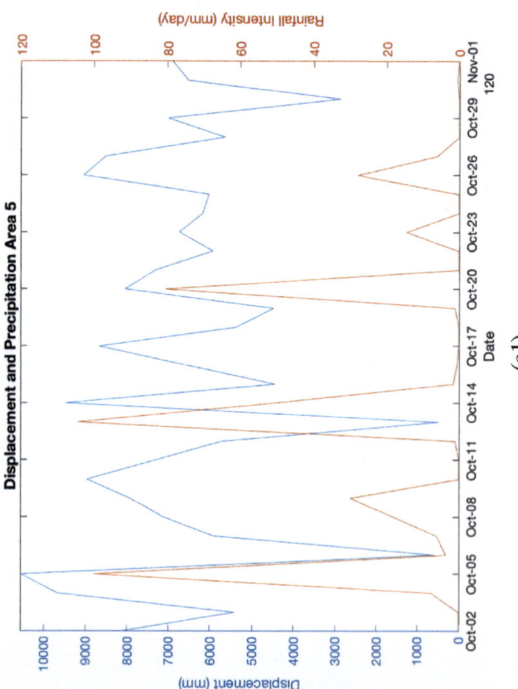

FIGURE 5.35 (Continued)

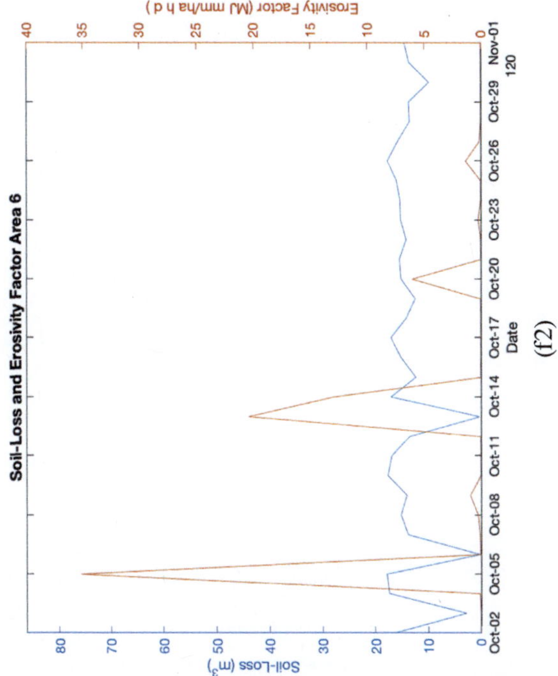

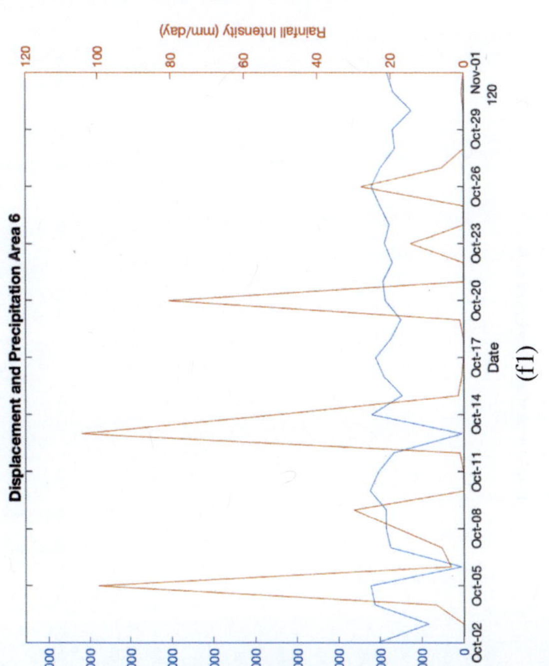

FIGURE 5.35 (Continued)

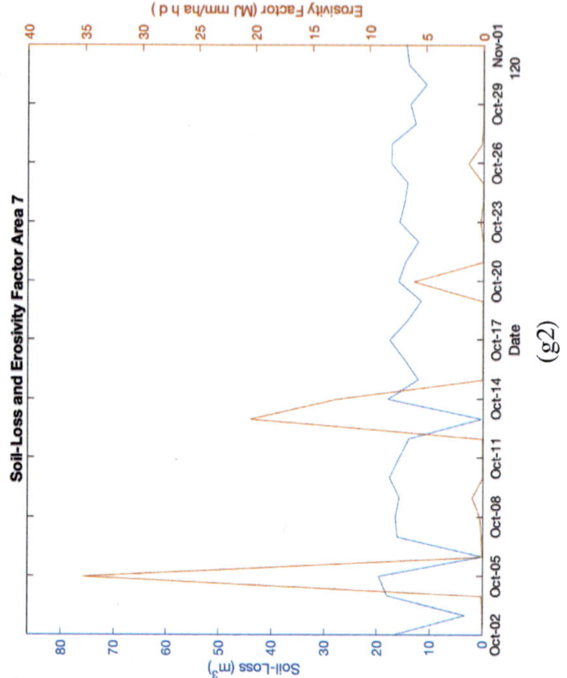

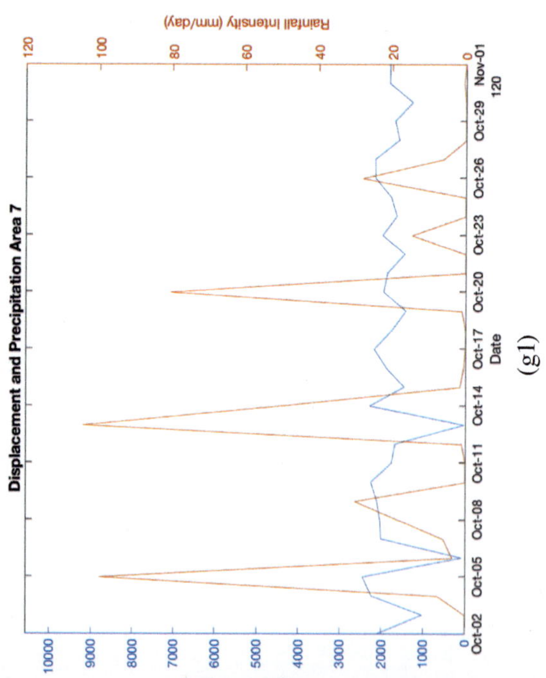

FIGURE 5.35 (Continued)

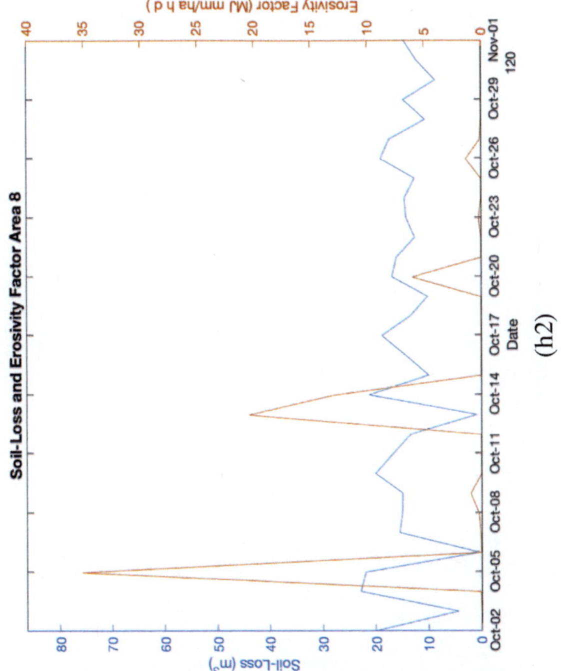

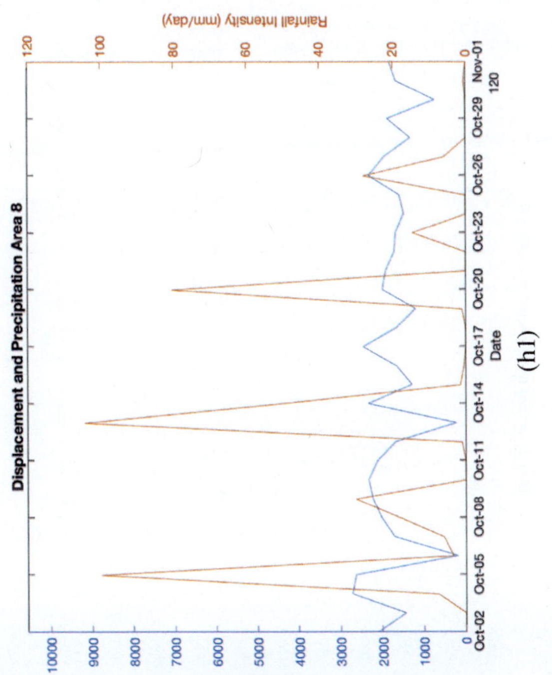

FIGURE 5.35 (Continued)

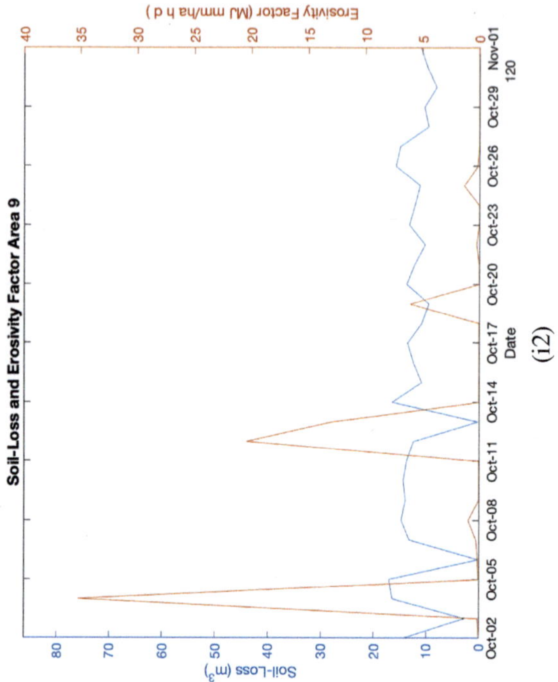

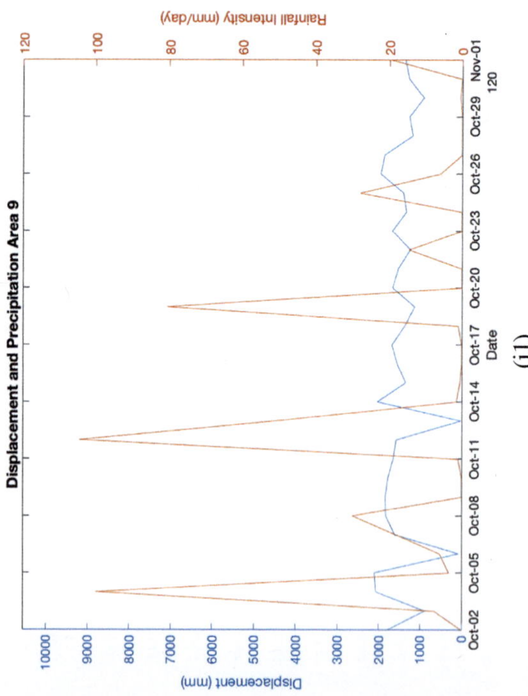

FIGURE 5.35　(Continued)

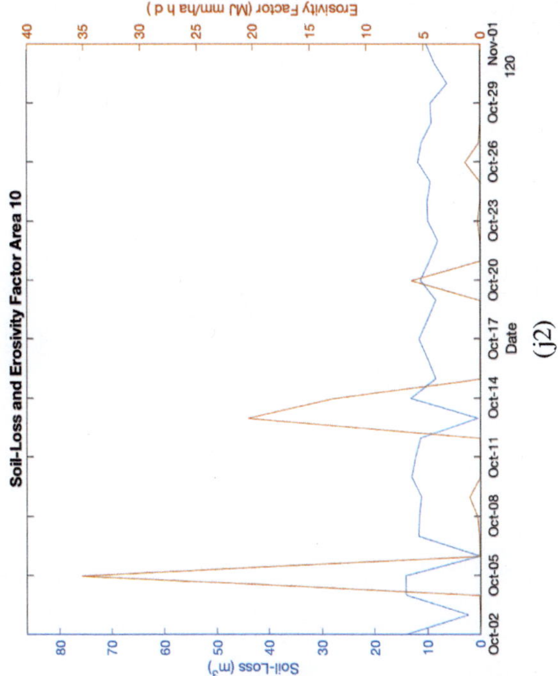

Soil-Loss and Erosivity Factor Area 10

(j2)

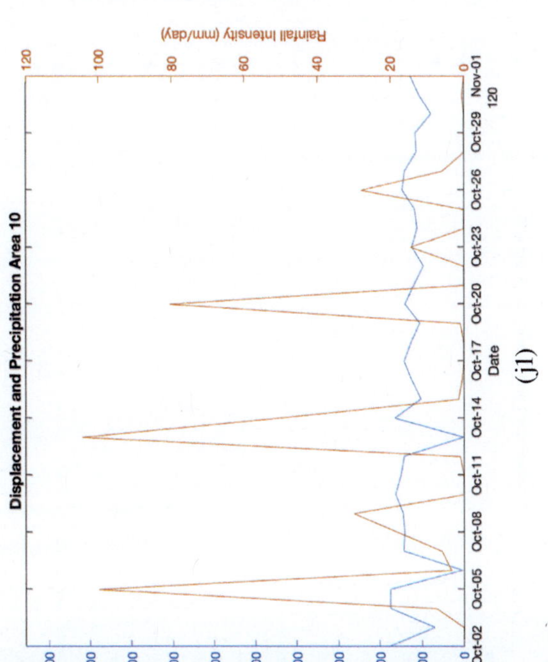

Displacement and Precipitation Area 10

(j1)

FIGURE 5.35 (Continued)

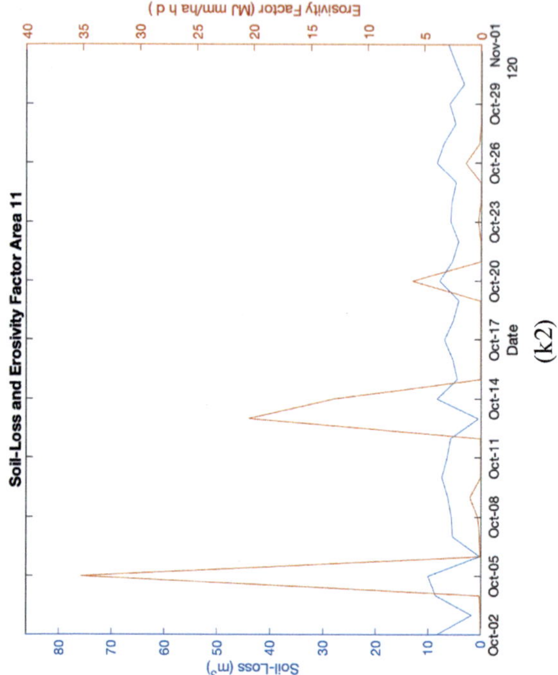

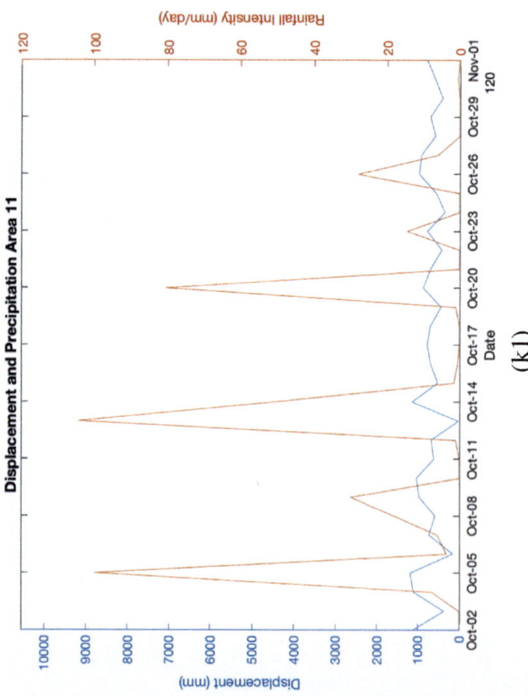

FIGURE 5.35 (Continued)

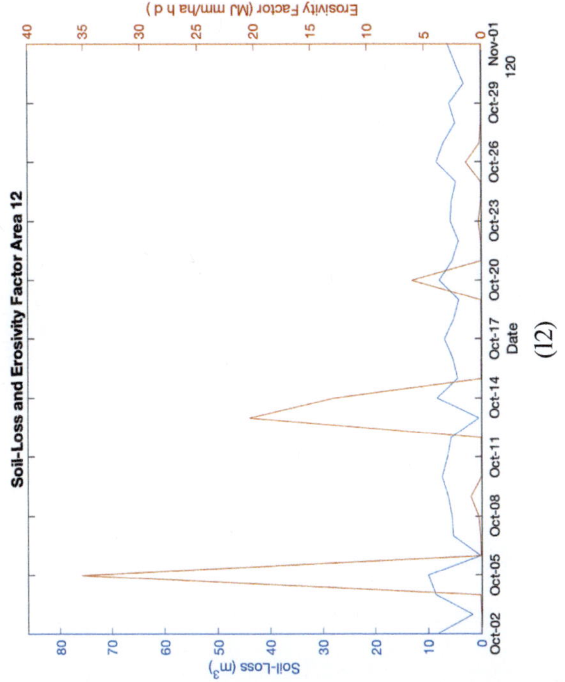

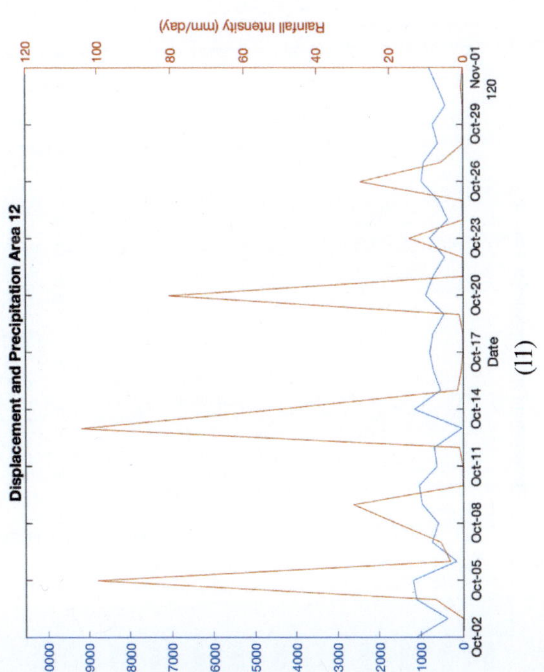

FIGURE 5.35 (Continued)

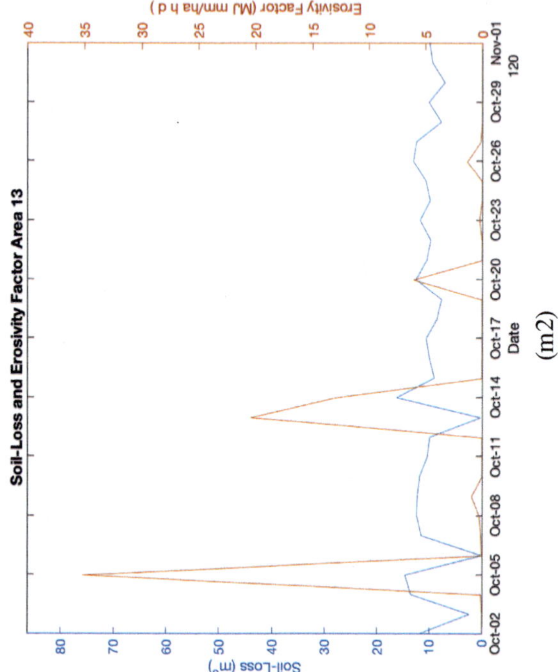

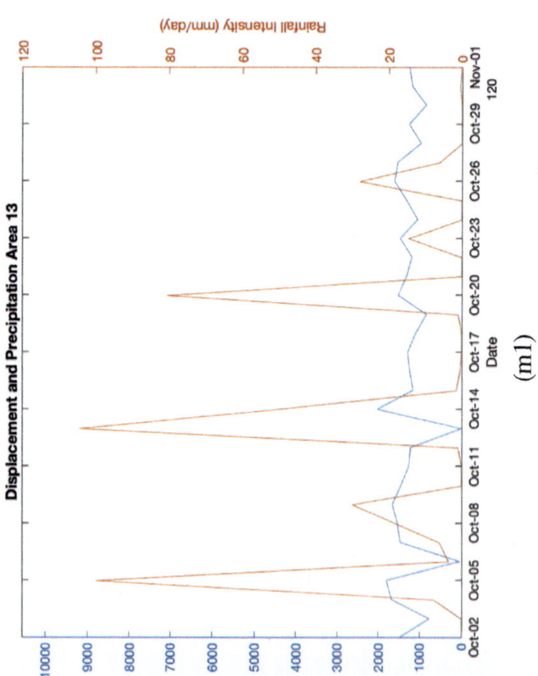

FIGURE 5.35 (Continued)

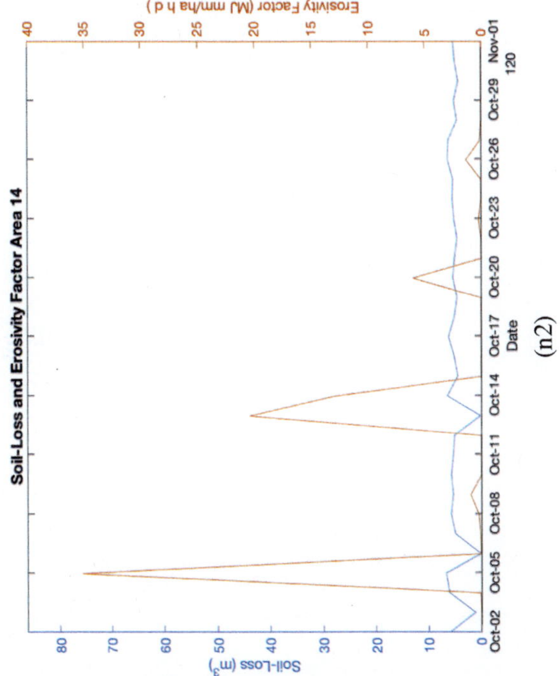

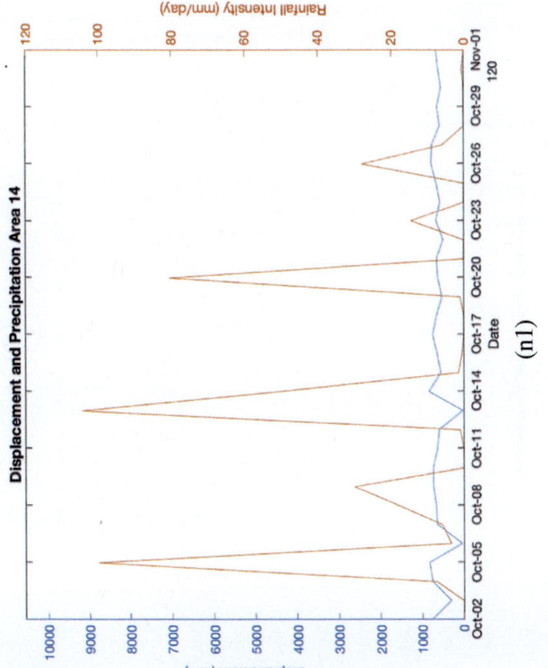

FIGURE 5.35 (Continued)

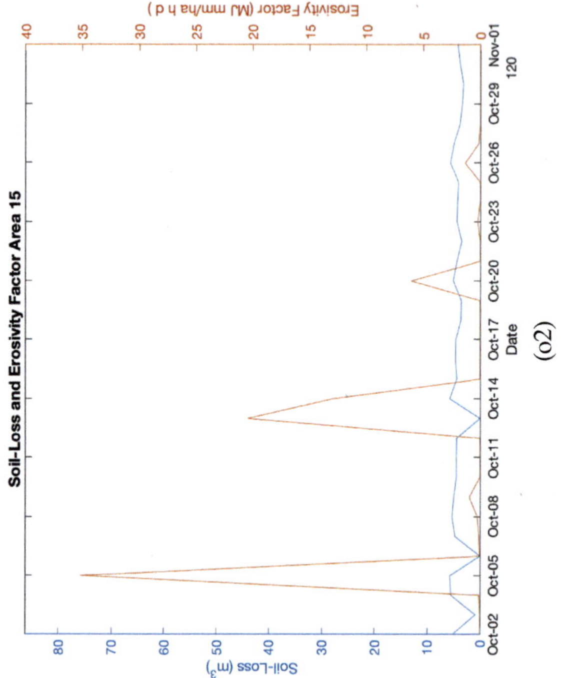

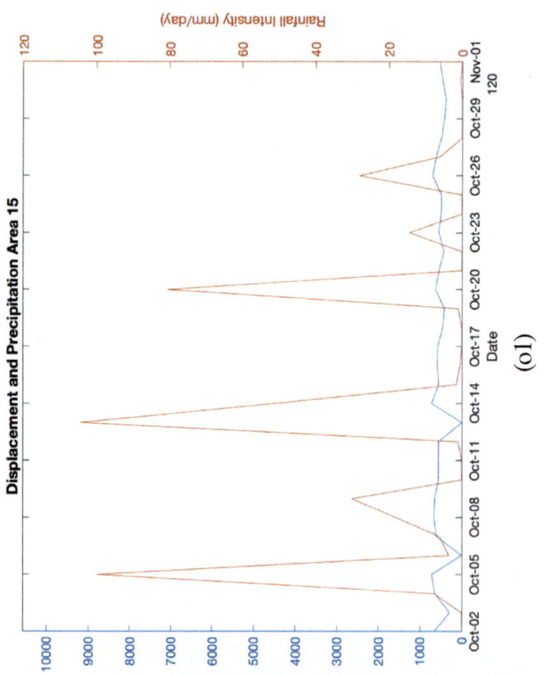

FIGURE 5.35 (Continued)

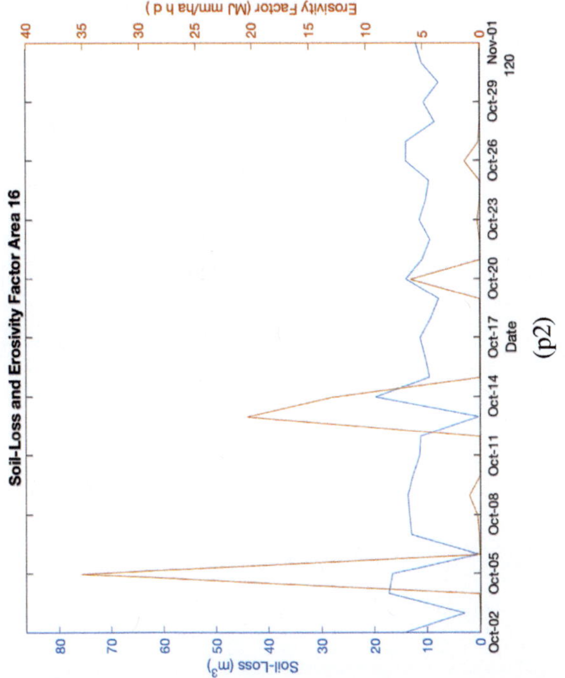

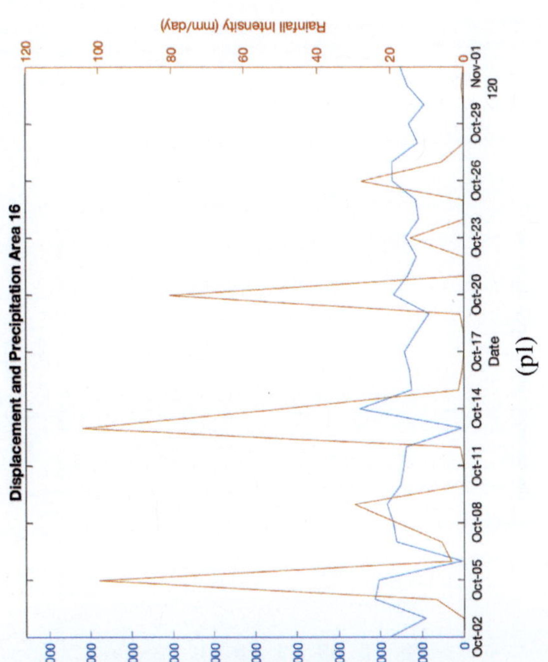

FIGURE 5.35 (Continued)

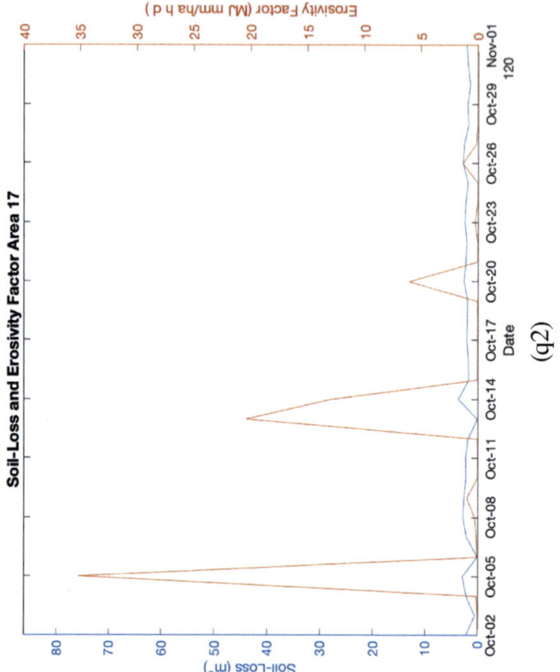

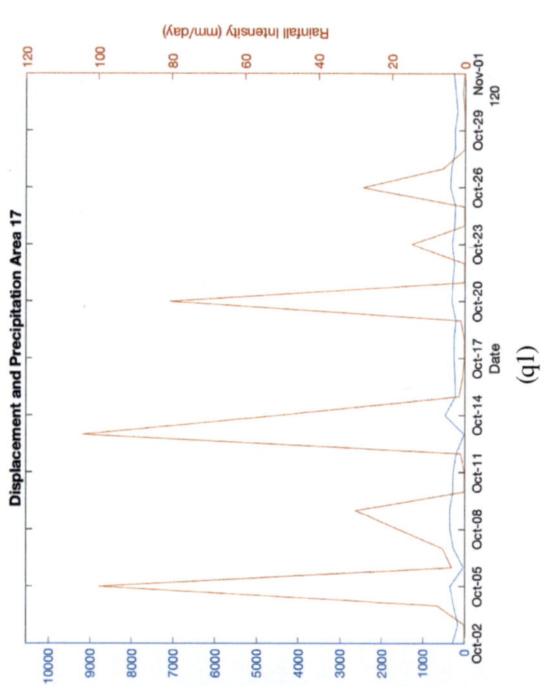

FIGURE 5.35 (Continued)

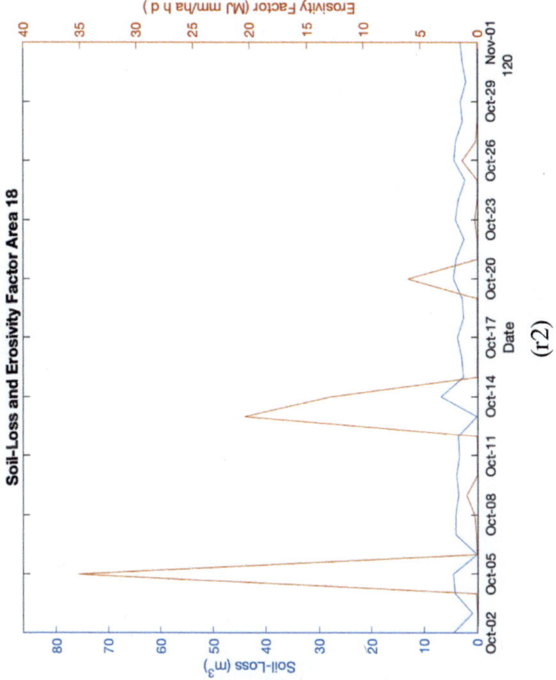

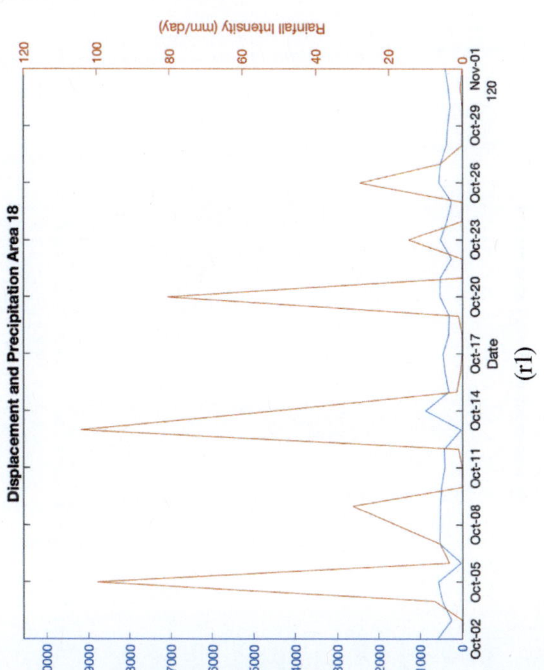

FIGURE 5.35 (Continued)

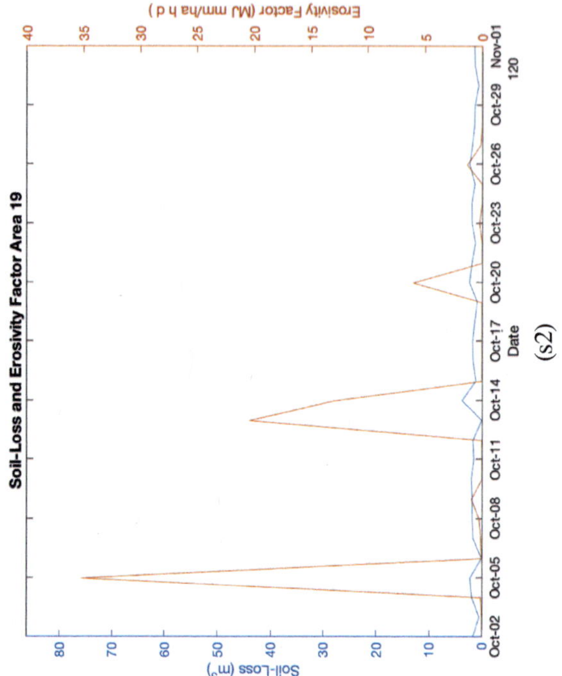

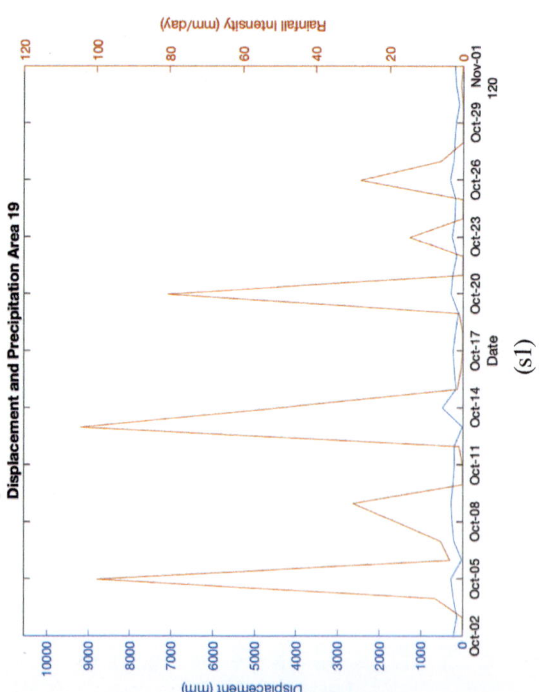

FIGURE 5.35 (Continued)

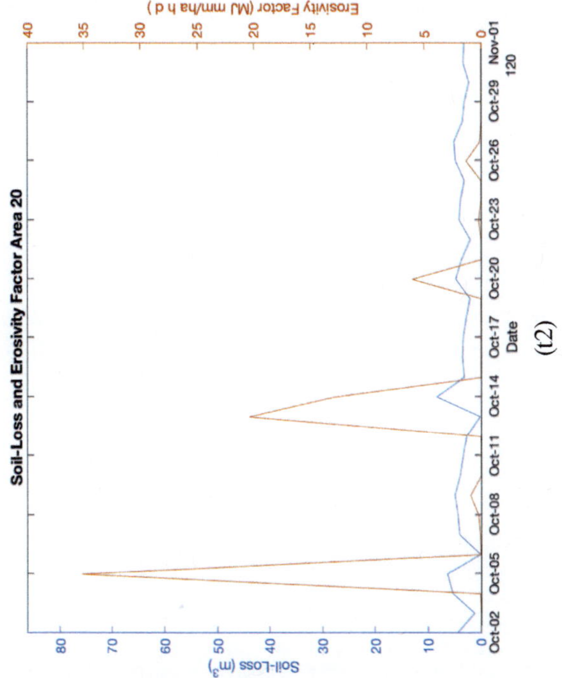

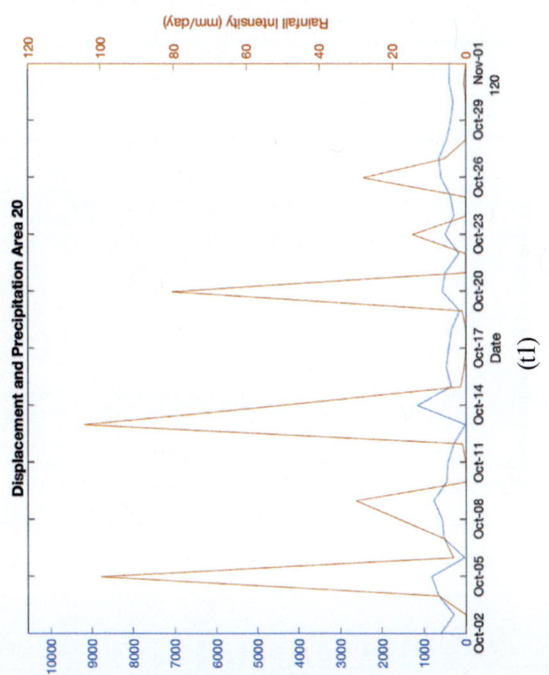

FIGURE 5.35 (Continued)

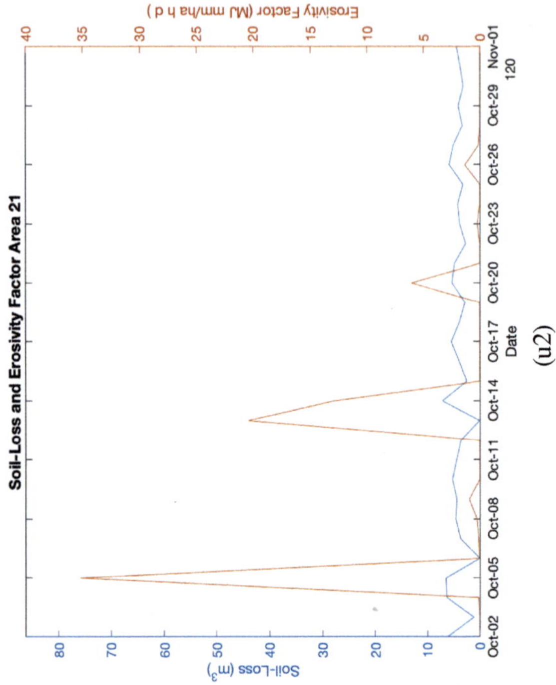

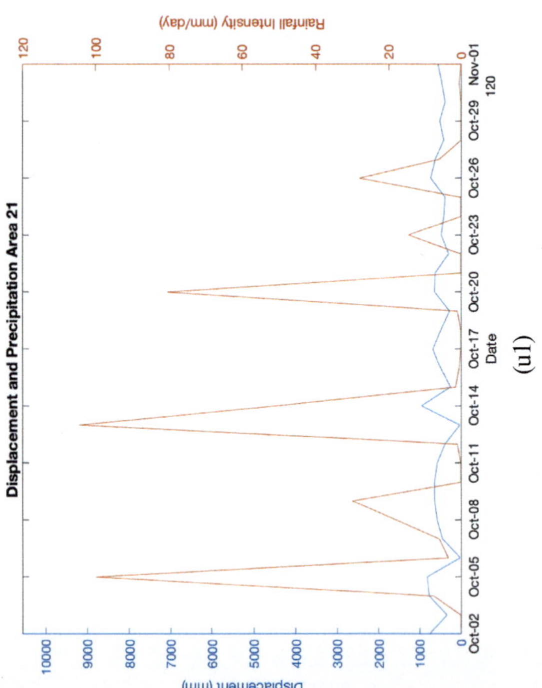

FIGURE 5.35　(Continued)

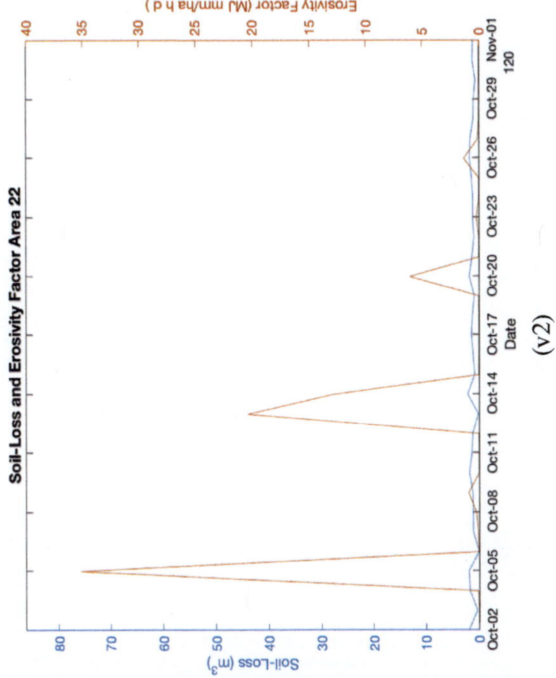

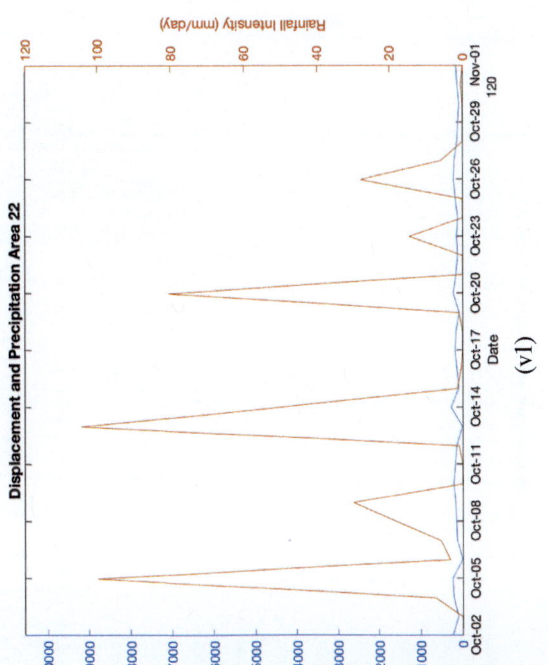

FIGURE 5.35 (Continued)

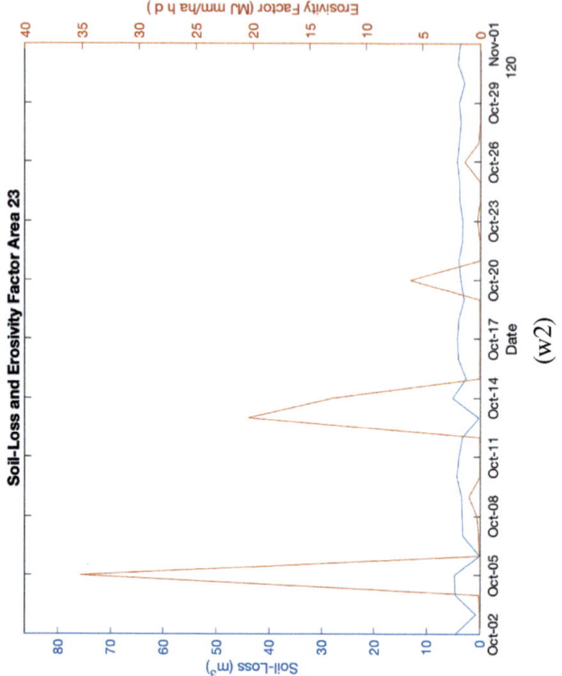

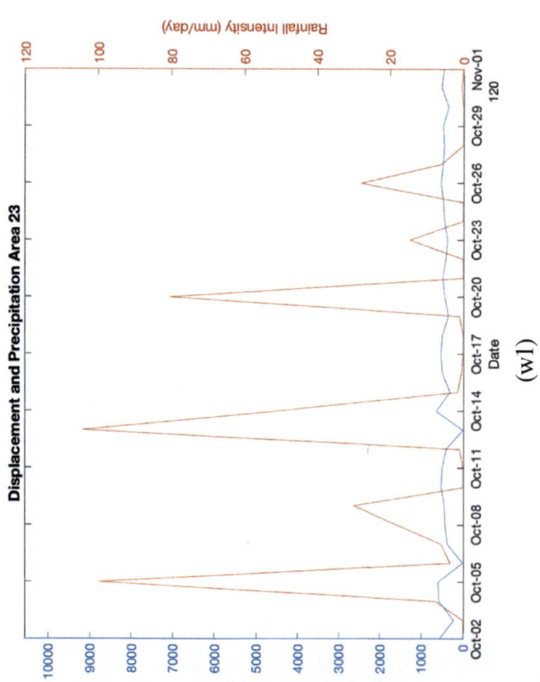

FIGURE 5.35 (Continued)

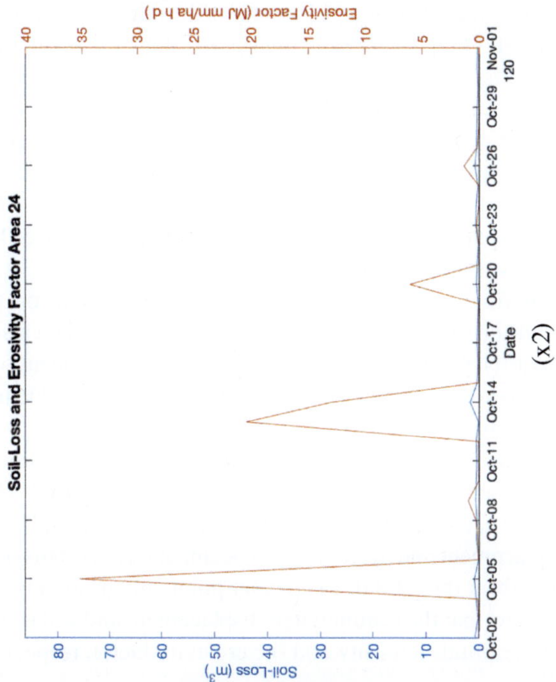

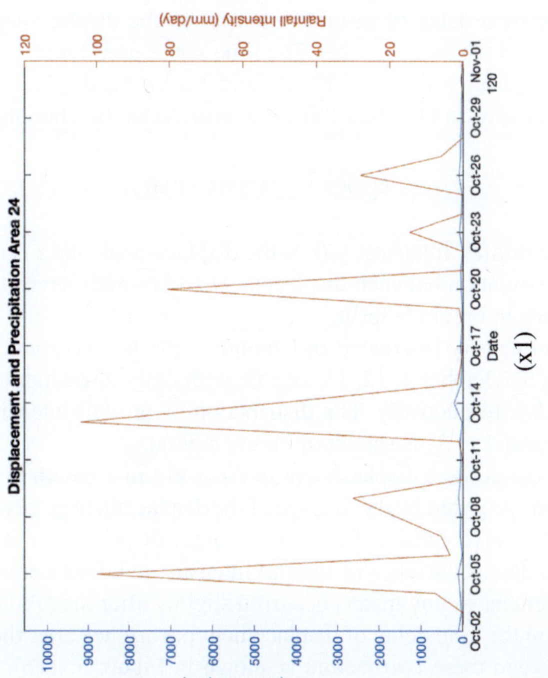

FIGURE 5.35 (Continued)

acquisition time of 30 min. The orange line shows the rainfall intensity on a daily basis from October 1 to 31. On the right side, the figures show the accumulated soil loss in blue for each area. The soil-loss volume is described in m^3. The orange line shows the erosivity factor for each day generated from the rainfall intensity from October 1 to 31.

All the figures show that in Area #5, the area has the largest displacement values. This occurs because of the large size of the area, which covers more displacement points. Accordingly, the soil-loss volume in Area #5 also shows the largest soil loss. While the largest displacement amount in this area approached over 10 m on October 4, the soil loss on the same date approached 80 m^3.

The daily rainfall intensity has a particularly high amount on particular days. Simultaneously, the erosivity factor generated from the same data also gives a higher value on particular days. Due to the diverse size of the area, the amount of displacement is less in some areas, such as from Area #14 to Area #24, due to fewer covered displacement points in each area.

In the accumulated displacement and accumulated soil-loss figures, we cannot see the large increase between October 4 and 12 when the high rainfall intensity occurred, as we can see the obvious increasing displacement and soil-loss values in the averaged displacement and averaged displacement figures. This occurred because the normalization from the total displacement points in one area is diverse. In addition, the figures show that the accumulated displacement and soil loss occurred some days later after the rainfall intensity and the erosivity factor, respectively.

5.3.4 CROSS-CORRELATION BETWEEN DISPLACEMENT AND PRECIPITATION

To estimate the time delay of rainfall intensity to the displacement, we apply the cross-correlation of the two data sets. The time series used for this process is a daily data set, with acquisition times of 30 min and 1 h for the displacement and rainfall intensity, respectively, in October. The cross-correlation function $h(\tau)$ is given by:

$$h(\tau) \cdot = \int_{-\infty}^{\infty} f(t)g(t-\tau)dt \qquad (5.17)$$

where $f(t)$ is the rainfall intensity, $g(t)$ is the displacement, and τ is the time lag.

The cross-correlation between displacement and rainfall intensity in Area #12 in October is shown in Figure 5.36(b).

The rainfall intensity fluctuated in 1 month, with the maximum amount specifically occurring on October 4, 12, 13, and 19, with daily intensities of 100, 104.5, 51, and 80.5 mm/day, respectively. The distribution of rainfall intensity and displacement from October 1 to 31 is shown in Figure 5.36(a).

In addition, the average displacement in Area #12 in 1 month is also shown. The displacement was acquired by the average of the displacement points in Area #12. The displacement is relatively stable, but it shows larger displacement on particular days. From this trend date occurrence of rainfall intensity and displacement, we could see that the displacement by any means occurred slightly after the rainfall intensity date.

To investigate the time delay of displacement occurrence after the rain, the cross-correlation between these two factors is shown in Figure 5.36(b). It clearly shows that after the rainfall intensity date, displacement did not occur on the same date.

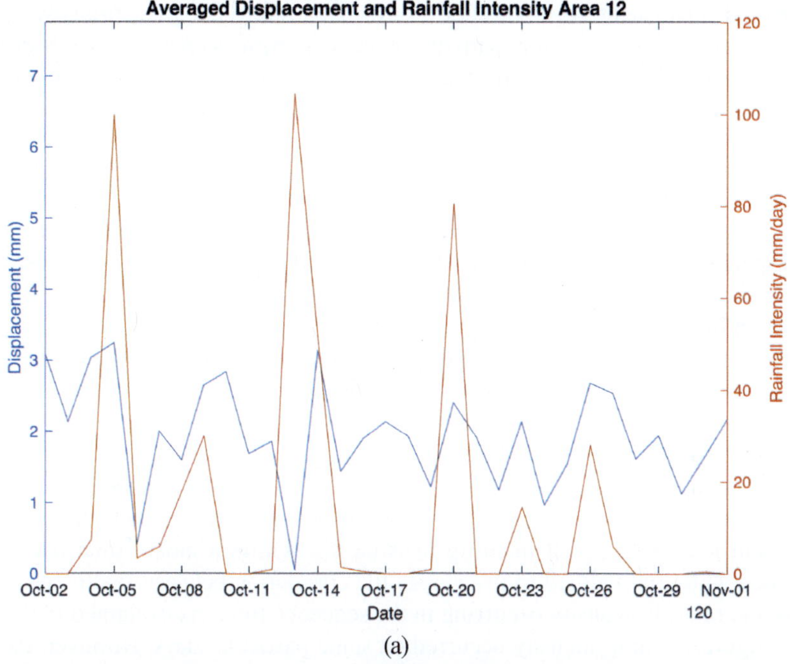

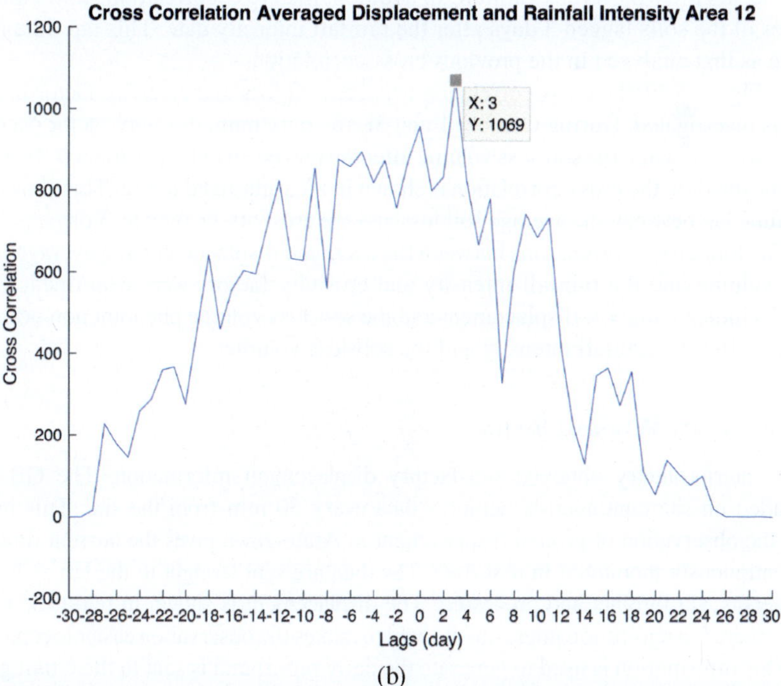

FIGURE 5.36 Distribution and cross-correlation, Area #12, in October. (a) Displacement and rainfall intensity distribution in Area #12. (b) Cross-correlation between displacement and rainfall intensity in Area #12.

Instead, displacement occurred after 3 days. This information is extremely important for soil-loss investigation, particularly in areas with steep slopes and high rainfall intensity.Figure 5.37   Distribution and cross-correlation, Area #12, in October. (a) Displacement and erosivity factor distribution in Area #12. (b) Cross-correlation between displacement and the erosivity factor in Area #12.

The cross-correlation between the average displacement of Area #12 in October is also correlated with the erosivity factor, which is shown in Figure 5.37. The erosivity factor shows the typical peak days only in October 4, 12, and 19 in 1 month, with specific maximum amounts of 35.1, 20.43, and 6 MJ m/ha h d, respectively. The distribution of the erosivity factor and average displacement from October 1 to 31 are shown in Figure 5.37(a).

To investigate the lag of averaged displacement occurrence and the erosivity factor, the cross-correlation between these two factors was processed, as shown in Figure 5.37(b). It clearly shows that the displacement did not happen immediately after the rainwater eroded the soil with the erosivity factor. Instead, through cross-correlation, it is shown that displacement occurred 3 days after the erosivity factor date.

In Figure 5.38, another cross-correlation is also investigated between the averaged soil loss and rainfall intensity in Area #12 within a month time interval in October. This study aimed to investigate the correlation between the losses in soil mass and rainfall intensity occurring in the sequence time. As explained in the previous figure, rainfall intensity occurred on some particular days. However, the soil loss occurs after the strong rainfall. In addition, the cross-correlation shows that the losses of the soils lagged 3 days after the rainfall intensity date. This lag time is the same as that analyzed in the previous cross-correlation.

In Figure 5.39, the cross-correlation between soil loss and erosivity factor in Area #12 is investigated. During October 1 and 31, the maximum erosivity factor occurred on October 4, while the soil-loss volume after the day occurred on October 7. To examine the relation, the cross-correlation is shown in the right-hand figure. This shows that the time lag between the average soil loss and the erosivity factor was 3 days.

The four cross-correlations between the averaged displacement and averaged soil-loss volume and the rainfall intensity and erosivity factors were examined. These results indicate that the displacement and the soil-loss volume phenomenon occurred 3 days after the rainfall intensity and the soil-loss volume.

5.3.5 EARLY WARNING SYSTEM

SAR interferometry obtained satisfactory displacement information. The GB-SAR installed on site continuously acquires data every 30 min from the site. This means that the observation of ground displacement in Arato-zawa gives the current time and is continuously monitored in real time. The data are sent straight to the lab at Tohoku University via Intranet and processed. The displacement is shown in real-time mode. Also, there is a real-time camera on-site, which makes the observation easier to represent.

This information is used to generate the daily report and is sent to the email group consisting of the members of the laboratory and local government. In addition, in terms of early warning, a hazard map is also generated based on the threshold system. The hazard map is created with a 2-h interval of 1 day observation corresponding to displacement map generation. The hazard map is the map of point movement

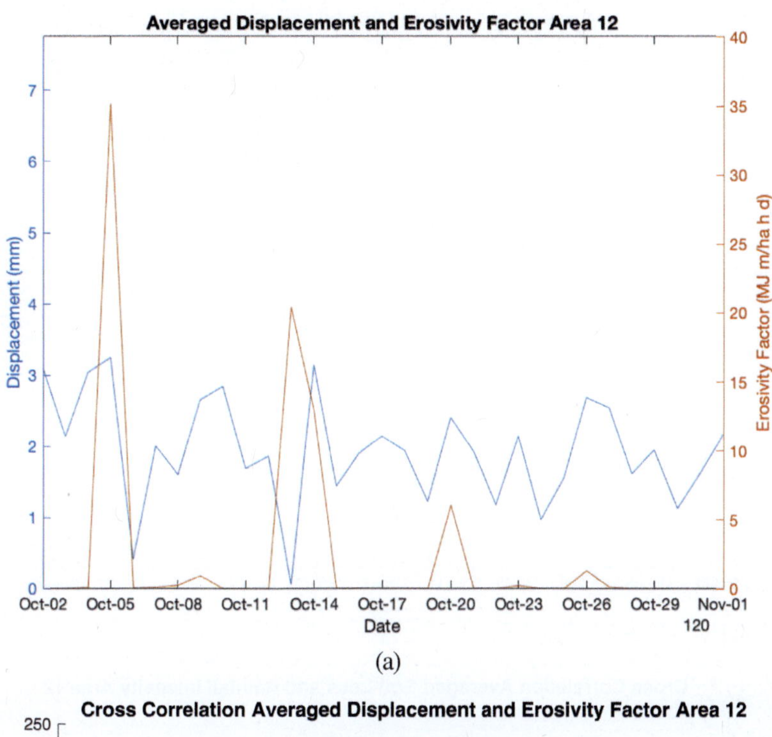

(a)

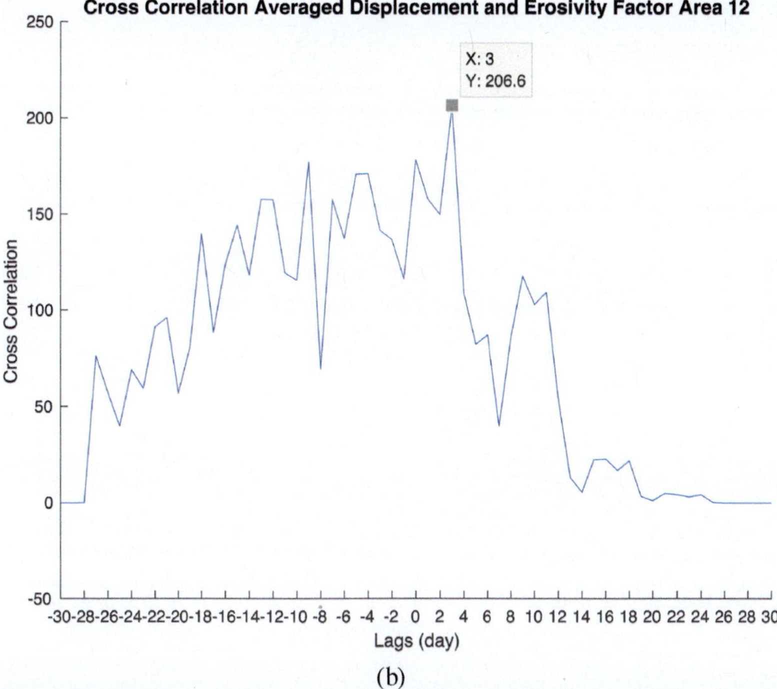

(b)

FIGURE 5.37 Distribution and cross-correlation, Area #12, in October. (a) Displacement and erosivity factor distribution in Area #12. (b) Cross-correlation between displacement and the erosivity factor in Area #12.

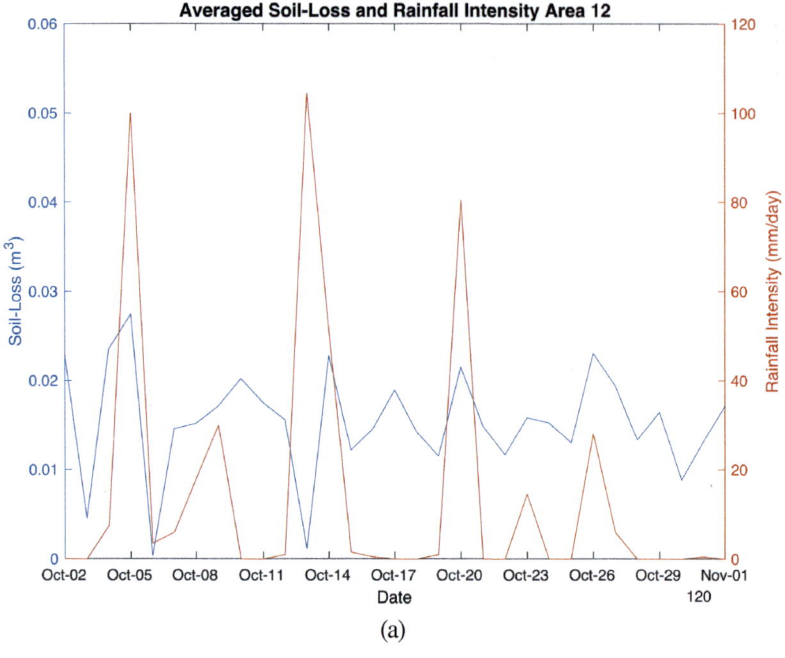

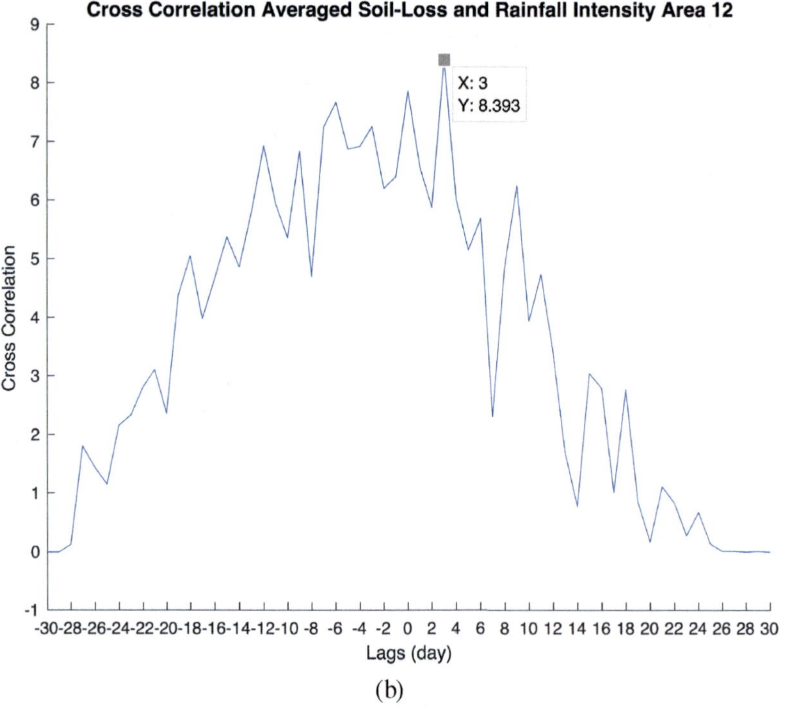

FIGURE 5.38　Distribution and cross-correlation in Area #12 in October. (a) Soil loss and rainfall intensity distribution in Area #12. (b) Cross-correlation between soil loss and rainfall intensity in Area #12.

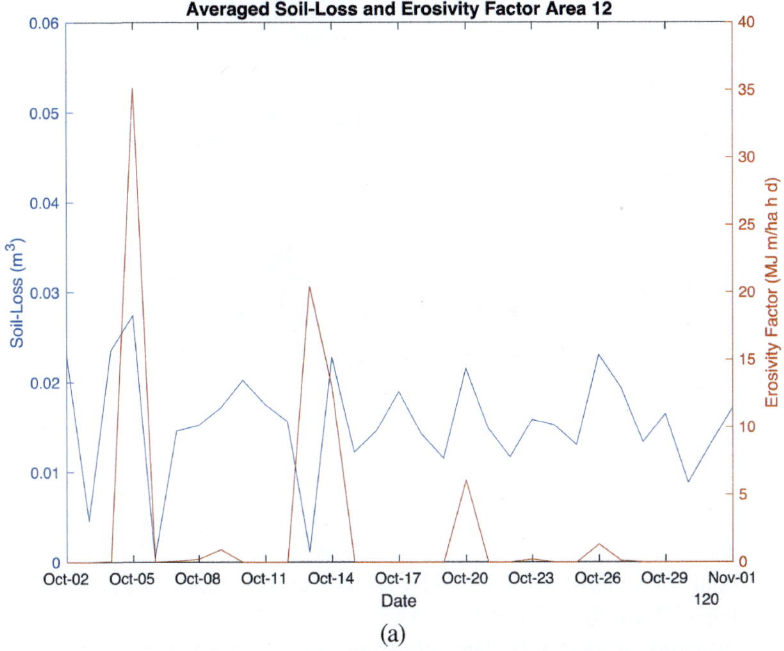

(a)

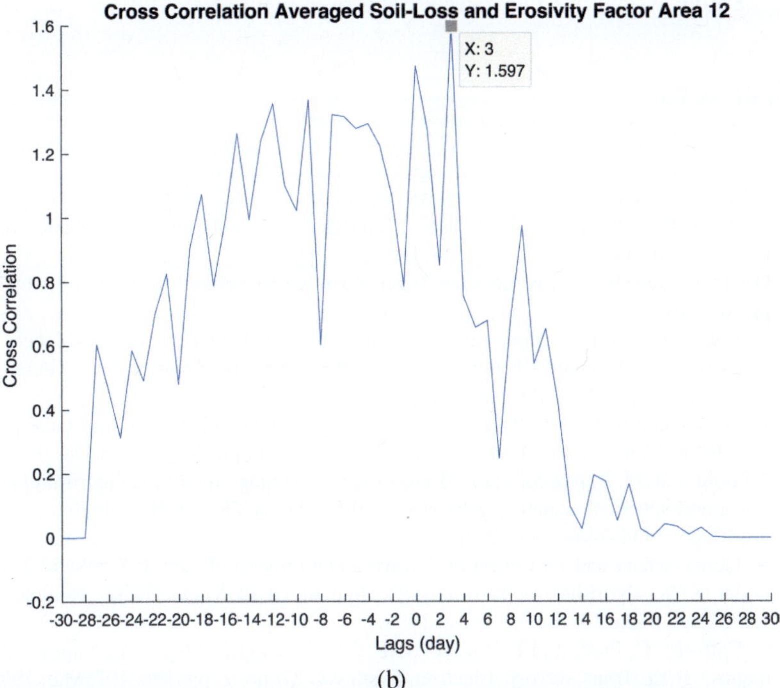

(b)

FIGURE 5.39 Distribution and cross-correlation in Area #12 in October. (a) Soil loss and erosivity factor distribution in Area #12. (b) Cross-correlation between soil loss and erosivity factor in Area #12.

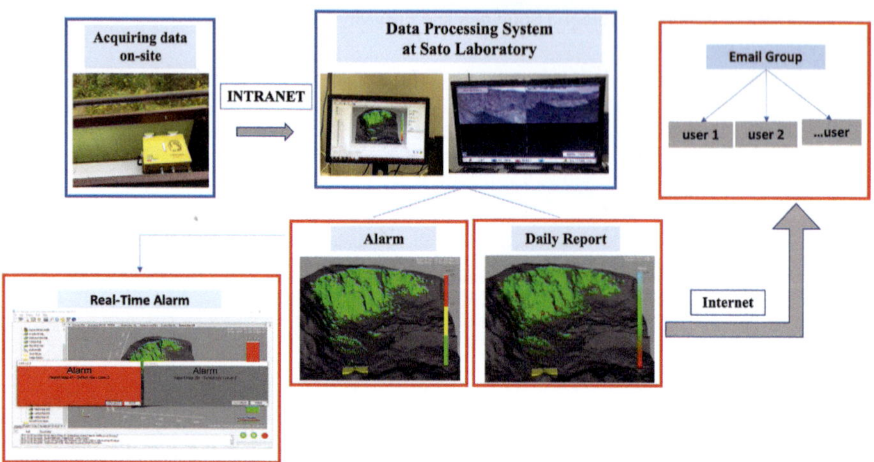

FIGURE 5.40   The early warning system in Arato-zawa.

over an hour. In terms of the threshold level, the hazard map works with three levels of warnings: <3 mm/h, 3–5 mm/h, and >5 mm/h, with the levels of modest, moderate, and warning, respectively. The data flow and the information processing of the early warning system in Arato-zawa are shown in Figure 5.40.

REFERENCES

[1] Z. Zhou, W. Boerner, and M. Sato, "Development of a ground-based polarimetric broadband SAR system for noninvasive ground-truth validation in vegetation monitoring," IEEE Trans. Geosci. Remote Sens., vol. 42, no. 9, pp. 1803–1810, Sept. 2004.

[2] Z. Zhou, K. Takasawa, and M. Sato, "Interferometric polarimetric synthetic aperture radar system," Proc. SPIE, vol. 4548, Oct. 2001, pp. 18–23.

[3] Ő.Yilmaz, Seismic data processing, Tulsa: Society of Exploration Geophysicists, 1987, pp. 403–409.

[4] R. Stolt, "Migration by fourier transform," Geophysics, vol. 43, no. 1, pp. 23–48, Feb. 1978.

[5] J. Chun and C. Jacewitz, "Foundamentals of frequency domain migration," Geophysics, vol. 46, no. 5, pp. 717–733, May 1981.

[6] J. Fortuny and A. Sieber, "Fast algorithm for near-field synthetic aperture radar processor," IEEE Trans. Geosci. Remote Sens., vol. 42, no. 10, pp. 1458–1460, Oct. 1994.

[7] J. Fortuny and J. Lopez-Sanchez, "Extension of 3-D range migration algorithm to cylindrical and spherical scanning geometries," IEEE Trans. Geosci. Remote Sens., vol. 49, no. 10, pp. 1434–1444, May 2001.

[8] A. Gunawarfena and D. Longstaff, "Wave equation formulation of Synthetic Aperture Radar (SAR) algorithms in the time-space domain," IEEE Trans. Geosci. Remote Sens., vol. 36, no. 6, pp. 1995–1999, Nov., 1998.

[9] C. Cafforio, C. Prati, and F. Rocca, "SAR data focusing using seismic migration techniques," IEEE Trans. Aerosp. Electron. Syst., vol. 27, no. 2, pp. 194–207, Mar. 1991.

[10] C. Leuschen and R. Plumb, "A matched-filter-based reverse-time migration algorithm for ground-penetrating radar data," IEEE Trans. Geosci. Remote Sens., vol. 39, no. 5, pp. 929–936, May 1996.

[11] J. Lopez-Sanchez and J. Fortuny-Guasch, "3-D radar imaging using range migration techniques," IEEE Trans. Geosci. Remote Sens., vol. 48, no. 5, pp. 728–737, May 2000.

[12] A. Boag, "A fast multilevel domain decomposition algorithm for radar imaging," IEEE Trans. Antenna Propagat., vol. 49, no. 4, pp. 666–671, Apr. 2001.

[13] Y. Das and W. Boerner, "On radar target estimation using algorithms for reconstruction from projections", IEEE Trans. Antenna Propagat., vol. 26, no. 2, pp. 274–279, 1978.

[14] Z. Zhou and M. Sato, "Development and performance evaluation of a ground-based SAR system," IEICE Tech. Rep., vol. 102, no. 84, pp. 31–36, 2002.

[15] J. M. Lopez-Sanchez, S. R. Cloude, and J. D. Ballester-Berman, "Rice phenology monitoring by means of SAR polarimetry at X-band," IEEE Trans. on Geosci. and Remote Sens., vol. 50, no. 7, pp. 2695–2709, July 2012.

[16] F. Wu, C. Wang, H. Zhang, et al., "Rice crop monitoring in South China with RADARSAT-2 quad-polarization SAR data," IEEE Geosci. and Remote Sens. Lett., vol. 8, no. 2, pp. 196–200, Mar. 2011.

[17] S. Yang, X. Zhao, B. Li, et al., "Interpreting RADARSAT-2 quad-polarization SAR signatures from rice paddy based on experiments," IEEE Geosci. and Remote Sens. Lett., vol. 9, no. 1, pp. 65–69, Jan. 2012.

[18] C. Xu, Y. Chen, L. Tong, et al., "Measuring the microwave backscattering coefficient of paddy rice using FM-CW ground-based scatterometer," Education Technology and Training, 2008. and 2008 International Workshop on Geoscience and Remote Sensing. ETT and GRS 2008,Shanghai, China, 2008, pp. 194–198.

[19] Y. Oh, S. Y. Hong, Y. Kim, et al., "Polarimetric backscattering coefficients of flooded rice fields at L- and C-bands: Measurements, modeling, and data analysis," IEEE Trans. on Geosci. and Remote Sens., vol. 47, no. 8, pp. 2714–2721, Aug. 2009.

[20] J. Susaki and Y. Kawatani, "Decomposition of polarimetirc scattering of paddy rice," Proc. IEEE IGARSS 2008, vol. 4, pp. IV-1117–IV-1120, Boston, MA, 2008.

[21] K. Morrison and M.L. Williams, "High resolution PolInSAR with the Ground-Based SAR (GB-SAR) system: Measurement and modelling," Proc. IEEE IGARSS2005, vol. 2, pp. 1105–1108, Seoul, Korea, July 2005.

[22] S. C. M. Brown, S. Quegan, K. Morrison, et al., "High-resolution measurements of scattering in wheat canopies-implications for crop parameter retrieval," IEEE Trans. on Geosci. and Remote Sens., vol. 41, no. 7, pp. 1602–1610, Jul. 2003.

[23] J. L. Gomez-Dans, S. Quegan, and J.C. Bennett, "Indoor C-band polarimetric interferometry observations of a mature wheat canopy," IEEE Trans. on Geos. and Remote Sens., vol. 44, no. 4, pp. 768- 777, Apr. 2006.

[24] The Iwate-Miyagi Nairiku Earthquake in 2008. https://www.jishin.go.jp/main/chousa/08jun_iwate_miyagi2/index-e.htm

[25] K. Takahashi, M. Matsumoto, and M. Sato, "Continuous observation of natural-disaster-affected areas using ground-based SAR interferometry," IEEE J. Sel. Top. Appl. Earth Obs. Remote Sens., vol. 6, no. 3, 2013.

[26] JMA Precipitation. https://www.jma.go.jp/bosai/en_nowc/#zoom:5/lat:34.016242/lon:135.000000/colordepth:normal/elements:hrpns&slmcs&slmcs_fcst

[27] K. Cooper, Evaluation of the relationship between the Rusle R-factor and mean annual precipitation, 2011. https://www.engr.colostate.edu/~pierre/ce_old/Projects/linkfiles/Cooper%20R-factor-Final.pdf

[28] N. Konz, D. Baenninger, M. Konzet al., "Process identification of soil erosion in steep mountain regions," Hydrol. Earth Syst. Sci, vol. 14, 2010. Doi:10.5194/hess-14-675-2010

6 MIMO Radar Systems

6.1 THEORETICAL BACKGROUND

6.1.1 INTRODUCTION

In order to solve some problems of GB-SAR, such as long data acquisition time and strong sidelobes, advanced multi-input multi-ouput (MIMO) imaging radar techniques are studied in this chapter.

In the first part of this chapter, based on the phase center approximation (PCA) and polynomial factorization (PF) methods [1–2], two linear MIMO radar systems for 2D imaging and a cross-MIMO radar system for 3D imaging are developed. First, linear and planar MIMO array design methods are introduced, and the equivalence between MIMO arrays and virtual arrays is discussed. The angular point spread functions (PSFs) of short-range and far-range targets of the designed arrays are compared with the uniform linear array (ULA) and uniform planar array (UPA). With Vivaldi antennas, spiral antennas, integrated SFCW transceivers, switches, micro-controllers, and vector network analyzers (VNAs), some MIMO radar systems are fabricated. Their parameters, such as working frequency, bandwidth, frequency steps, aperture length, and spatial resolutions, are provided. Then, conventional imaging algorithms for GB-SAR are modified for MIMO radar [3–4]. The imaging and displacement estimation performance of the developed MIMO radar systems is evaluated. Finally, some potential improvement methods of the developed MIMO radar systems are discussed.

In the second part of this chapter, considering the azimuth dependency of target complex amplitudes, a block compressive sensing (CS)–based imaging method is proposed to reduce the data sampling period and improve the imaging quality of the developed linear MIMO radar systems. The target azimuth dependency caused by system errors and/or target reflection properties at short range is analyzed first. Then, a novel sensing matrix is proposed for CS-based imaging of the observation scene with only a few significant targets. In the proposed sensing matrix, the received signal has block sparsity, to exploit which a modified block OMP (BOMP) algorithm [5] is proposed to effectively and efficiently reconstruct the imaging scene with under-sampled data. Finally, a cross-correlation and summation-based method [6] is employed for the fusion of multiple images. The performance and advantages of the proposed imaging algorithm are validated by experiments conducted by the designed linear MIMO radar systems. The practical problems and applications of the proposed imaging algorithm are discussed.

In the third part of this chapter, a tensor CS (T-CS) [7]–based 3D imaging method is proposed for the developed cross-MIMO radar system. With the far-field assumption, the signal model for cross-MIMO imaging is simplified, and a 3D pseudopolar

DOI: 10.1201/9781003316312-6

coordinate system is established, based on which the far-field pseudo-polar image format algorithm (FPFA) [8–9] is extended to its 3D version to significantly reduce the computational cost of 3D high-resolution imaging. Then, the spatially variant apodization (SVA)–based method is adopted to adaptively suppress the sidelobes without losing the spatial resolution. To further reduce the data acquisition time, random spatial under-sampling is implemented, and the T-CS algorithm is applied to exploit the sparsity of the observation scene. Finally, a tensor-based iterative adaptive approach (T-IAA) is proposed to solve the sparse reconstruction problem effectively. The proposed algorithms are assessed by experiments with point and distributed targets. The drawbacks of the proposed algorithms and potential solutions are discussed.

6.1.2 MIMO RADAR DEVELOPMENT

6.1.2.1 MIMO Array Design

6.1.2.1.1 1D MIMO Array Design

In order to obtain an azimuth resolution, a GB-SAR system generates a synthetic aperture by sequentially moving its transceiver along a rail, which is equivalent to a uniform linear array of transceivers, as shown in Figure 6.1. Therefore, a ULA-based radar system could be used to achieve the same functions as GB-SAR. However, to get the same azimuth resolution and ambiguous angle, a large number of transceivers are needed. In order to maintain the aperture length but reduce the number of transceivers, linear MIMO arrays have drawn attention in recent years [10–13].

For a MIMO array with M transmitters and N receivers, an equivalent virtual array with $M \times N$ transceivers can be obtained based on the phase center approximation, as shown in Figure 6.2. According to PCA, the phase of the echoed signal for a transmitter at r_T and a receiver at r_R in a MIMO array can be approximated by the phase of the echoed signal from a transmitter-receiver collocated element at $(r_T + r_R)/2$ in a virtual array [14].

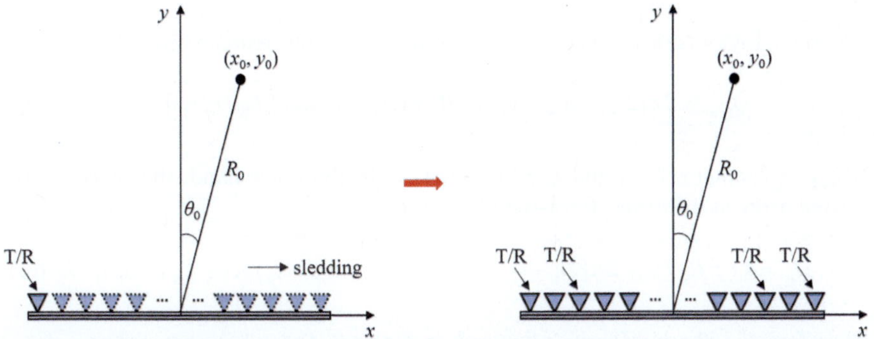

FIGURE 6.1 GB-SAR synthetic aperture and uniform linear array.

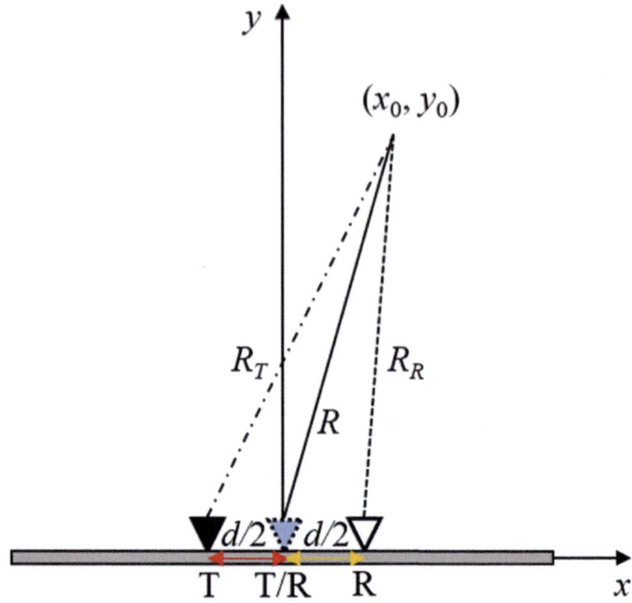

FIGURE 6.2   Virtual element generated based on phase center approximation.

For a linear MIMO array, assuming the m-th transmitter is at $(x_m, 0)$ and the n-th receiver is at $(x_n, 0)$, the p-th element in the virtual array is located at

$$x_n = (x_m + x_n)/2 \mid m = 1,2,\dots,M, n = 1,2,\dots,N, p = 1,2,\dots,MN \quad (6.1)$$

where $p = (m - 1)N + n$. As shown in Figure 6.2, there is indeed a phase difference between the virtual array and the MIMO array. Given a target at (x_0, y_0), the phase of the MIMO array at the center frequency f_c is given by

$$\varphi_{m,n} = 2\pi f_c [R_T^m + R_R^n]/c = 2\pi f_c [\sqrt{(x_m - x_0)^2 + y_0^2} + \sqrt{(x_n - x_0)^2 + y_0^2}]/c \quad (6.2)$$

Based on Taylor series expansion, this phase can be approximated by

$$\varphi_{m,n} \simeq 2\pi[2r_0 - (x_m + x_n)\sin\theta_0 + (x_m^2 + x_n^2)\cos^2\theta_0 / (2r_0)]/\lambda_c \quad (6.3)$$

where $\theta_0 = \mathrm{atan}(x_0 / y_0)$ and $r_0 = (x_0^2 + y_0^2)^{1/2}$. On the other hand, the phase of the virtual array at the center frequency f_c is given by

$$\varphi_p = 4\pi f_c R_p / c \simeq 2\pi[2r_0 - (x_m + x_n)\sin\theta_0 + (x_m + x_n)^2 \cos^2\theta_0 / (4r_0)]/\lambda_c \quad (6.4)$$

Therefore, the phase difference between (6.3) and (6.4) can be expressed as

$$\Delta\varphi_{m,n} = 2\pi[d_{m,n}^2 \cos^2\theta_0 / 4r_0]/\lambda_c \quad (6.5)$$

where $d_{m,n}$ is the distance between the m-th transmitter and the n-th receiver. Assuming the defocusing effect caused by a phase error of $\pi/4$ or less is negligible [15], it can be derived that if the following condition can be satisfied, then the phase difference in (6.5) can be neglected.

$$\left|\Delta\varphi_{m,n}\right|_{\max} = 2\pi\left[\left|d^2_{m,n}\right|_{\max}\Big/4r_0\right]\Big/\lambda_c \leq \pi/4 \Rightarrow r_0 \geq 2\left|d^2_{m,n}\right|_{\max}\Big/\lambda_c \tag{6.6}$$

It can be found from (6.6) that, if the target is at far field, the MIMO array can be considered equivalent to the virtual array. Therefore, given a ULA as the virtual array, a MIMO array can be designed accordingly. For ULA design, according to the angular resolution given in (2.28) and the unambiguous condition given in (2.25), the aperture length L and the inter-element spacing Δx can be determined. Then, assuming the ULA has P elements, the transmitting and receiving arrays of the MIMO array can be designed by the polynomial factorization method as

$$P(z) = \sum_{p=1}^{P} z^{p-1} = P_T(z)P_R(z) \tag{6.7}$$

where z^{p-1} corresponds to the p-th element at $(x_p, 0)$. For instance, given a ULA with 64 elements, the factorization of $P(z)$ can be expressed as

$$P(z) = (z+1)(z^2+1)(z^4+1)(z^8+1)(z^{16}+1)(z^{32}+1) \tag{6.8}$$

To minimize the total number of transmitters and receivers, $M + N$, the transmitter number M and the receiver number N that satisfy $MN = P$ should be close (if they cannot be equivalent) to each other. Furthermore, in order to minimize the length of the MIMO array, that is, to have the best spatial efficiency, the following transmitting and receiving arrays should be selected.

$$\begin{aligned} P_T(z) &= (z^4+1)(z^8+1)(z^{16}+1) \\ P_R(z) &= (z+1)(z^2+1)(z^{32}+1) \end{aligned} \tag{6.9}$$

where $P_T(z)$ and $P_R(z)$ can be switched. Then, to design a linear MIMO array with non-overlapping and symmetrical topology, the transmitting/receiving arrays should be adjusted to

$$\begin{aligned} P_T(z) &= (z^4+1)(z^8+1)(z^{16}+1)z^r \\ P_R(z) &= (z+1)(z^2+1)(z^{32}+1)z^{-r} \end{aligned} \tag{6.10}$$

where r is a constant to adjust the positions of the transmitting/receiving arrays.

Based on this approach, the MIMO array can be designed. For instance, given a central frequency of 5 GHz, inter-element spacing of 1.5 cm, and number of elements of 64 for a ULA, the MIMO array designed based on (6.9) is shown in Figure 6.3(a), where a height difference is introduced to clearly show the MIMO array and its equivalent ULA. Figure 6.3(b) shows the MIMO array obtained by (6.10), where the MIMO array and ULA are symmetrical and non-overlapped. To evaluate the performance of MIMO radar compared with GB-SAR (or ULA-based radar), the point spread function can be calculated [12], which is normally divided into the range and angular directions. Since only the approach used to get a high angular direction is changed by the MIMO array, and the range direction parameters (central frequency and frequency bandwidth, numbers, and step) are the same as GB-SAR, only the angular PSF, which is defined as the profile at the same range as the target, is derived for comparison in the following.

For SFCW-based GB-SAR (or ULA-based radar), the angular PSF can be expressed as

$$\chi_{ULA}(\theta \,/\, r_0, \theta_0) = \sum_{q=1}^{Q}\sum_{p=1}^{P} s_0(p,q)\exp\{j4\pi f_q R_p(r_0,\theta)/c\} = \sum_{q=1}^{Q} F(\theta, f_q \,/\, r_0, \theta_0) \quad (6.11)$$

where $s_0(p,q) = \exp(-j4\pi f_q R_p(r_0,\theta_0)/c)$ is the GB-SAR received signal from the target (r_0, θ_0), and $F(\theta, f_q \,/\, r_0, \theta_0)$ can be viewed as the radiation pattern of GB-SAR

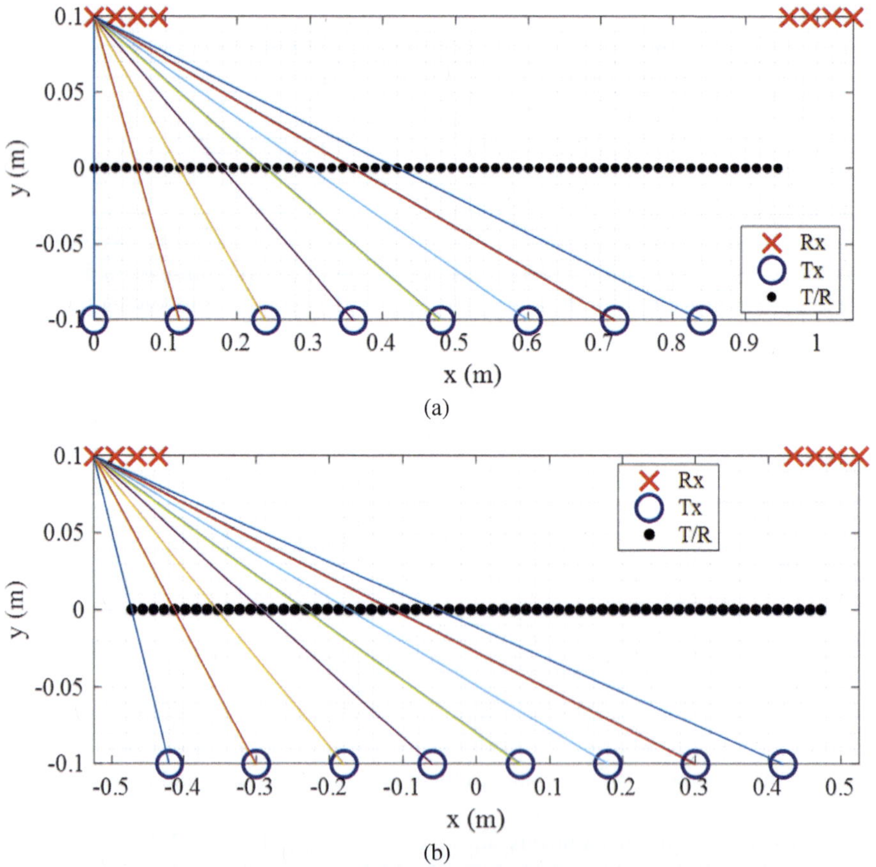

FIGURE 6.3 MIMO arrays and their equivalent ULAs: (a) asymmetric and (b) symmetric.

at the frequency f_q. For SFCW-based linear MIMO radar, the angular PSF can be expressed as

$$\chi_{MIMO}(\theta \,/\, r_0, \theta_0) = \sum_{q=1}^{Q} \left\{ \sum_{m=1}^{M} s_0(m,q) \exp\{j2\pi f_q R_m(r_0,\theta)\,/\,c\} \right.$$

$$\left. \sum_{n=1}^{N} s_0(n,q) \exp\{j2\pi f_q R_n(r_0,\theta)\,/\,c\} \right\}$$

$$= \sum_{q=1}^{Q} F_T(\theta, f_q \,/\, r_0, \theta_0) F_R(\theta, f_q \,/\, r_0, \theta_0)$$

$$= \sum_{q=1}^{Q} F_{TR}(\theta, f_q \,/\, r_0, \theta_0)$$

(6.12)

where $s_0(m,q) = \exp(-j2\pi f_q R_m(r_0,\theta_0)\,/\,c)$ and $s_0(n,q) = \exp(-j2\pi f_q R_n(r_0,\theta_0)\,/\,c)$ are the received signals for m-th transmitter and n-th receiver, respectively, and $F_T(\theta, f_q \,/\, r_0, \theta_0)$ and $F_R(\theta, f_q \,/\, r_0, \theta_0)$ can be viewed as the transmitting and receiving radiation patterns of MIMO radar at the frequency f_q. The radiation pattern $F_{TR}(\theta, f_q \,/\, r_0, \theta_0)$ of the MIMO array is expressed as the product of transmitting and receiving radiation patterns.

For a central frequency of 5 GHz and the MIMO array shown in Figure 6.3(b), given a target at $(100, 0°)$, the radiation patterns of the MIMO array and its equivalent ULA are presented in Figure 6.4. It can be seen that, for the MIMO array and the target at the far field ($\geq$ about 30 m), the grating lobes of the receiving pattern are canceled well by the deep nulls of the transmitting pattern. Therefore, the radiation patterns of the MIMO array and the ULA are equivalent. It also happens for targets at different positions, such as $(100, 30°)$, as shown in Figure 6.5.

However, since the phase difference between the virtual array and the MIMO array can only be neglected for targets at the far field, as indicated by (6.6), for targets at the near field, the phase difference will introduce a mismatch between the transmitting and receiving arrays. Therefore, high-level grating lobes of the receiving pattern cannot be exactly canceled by the nulls of the transmitting pattern, as shown in Figure 6.6. Therefore, the MIMO array is no longer equivalent to the ULA. It can be seen from Figure 6.6(b) that the sidelobe level of the MIMO array is slightly higher than the ULA. Therefore, to get the same near-range imaging performance with GB-SAR, phase compensation should be conducted for the MIMO radar. However, for practical displacement estimation, the far-field assumption can usually be used since the slightly higher-level sidelobes are considered acceptable and can also be suppressed by other methods.

Also, it is worth noting that previous simulations only show the radiation patterns of the MIMO array and the ULA at the central frequency. When full bandwidth is considered, the radiation patterns at all frequencies should be summed together to obtain the angular PSF of the MIMO array and ULA, as indicated by (6.12) and shown in Figure 6.7, where the bandwidth is 300 MHz and the number of frequencies is 501. Moreover, it should be pointed out that all the simulation results are obtained with a height difference (0.2 m) between the transmitting and receiving arrays. In

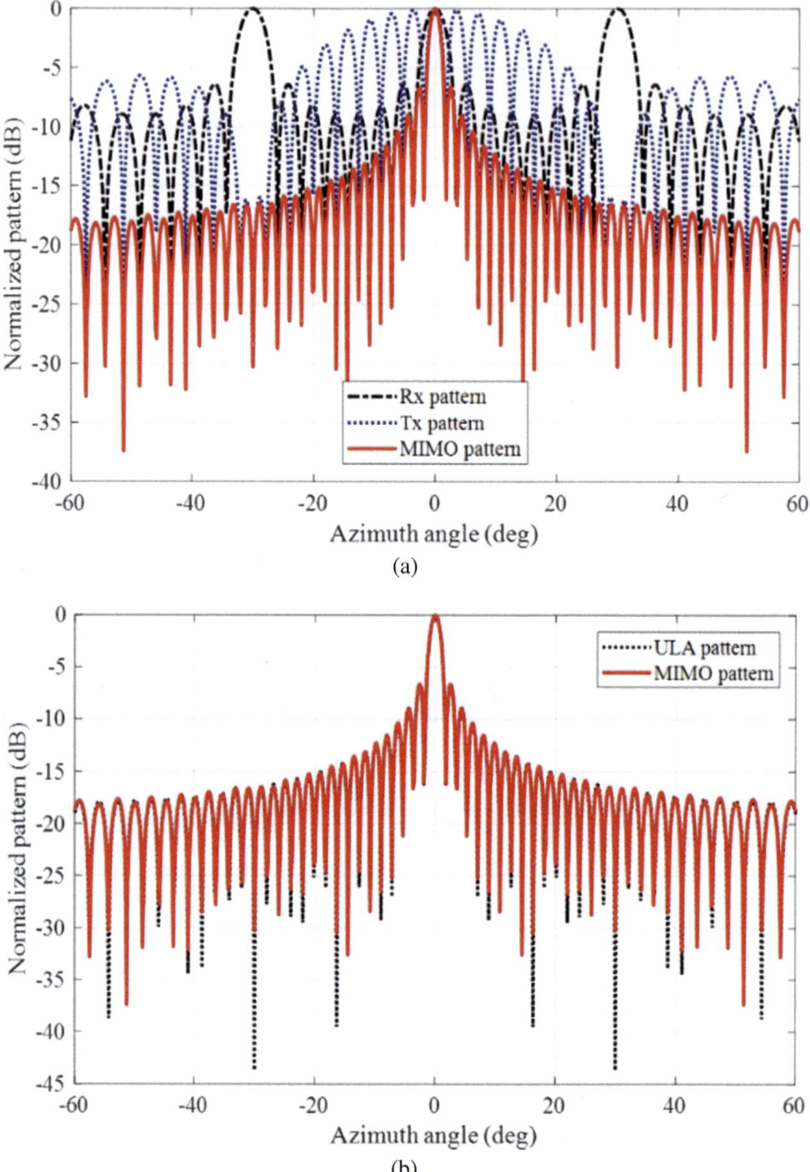

FIGURE 6.4   Radiation patterns of (a) the MIMO array and (b) the ULA for a target at (100, 0°) at f_c = 5GHz.

practice, considering the physical size of the antennas (e.g., broadband Vivaldi and spiral antennas), a certain height difference is unavoidable and can be used to reduce the mutual coupling between antennas. From the simulation results, it can be concluded that the MIMO array can indeed provide a similar PSFs with the ULA for the target at the far field.

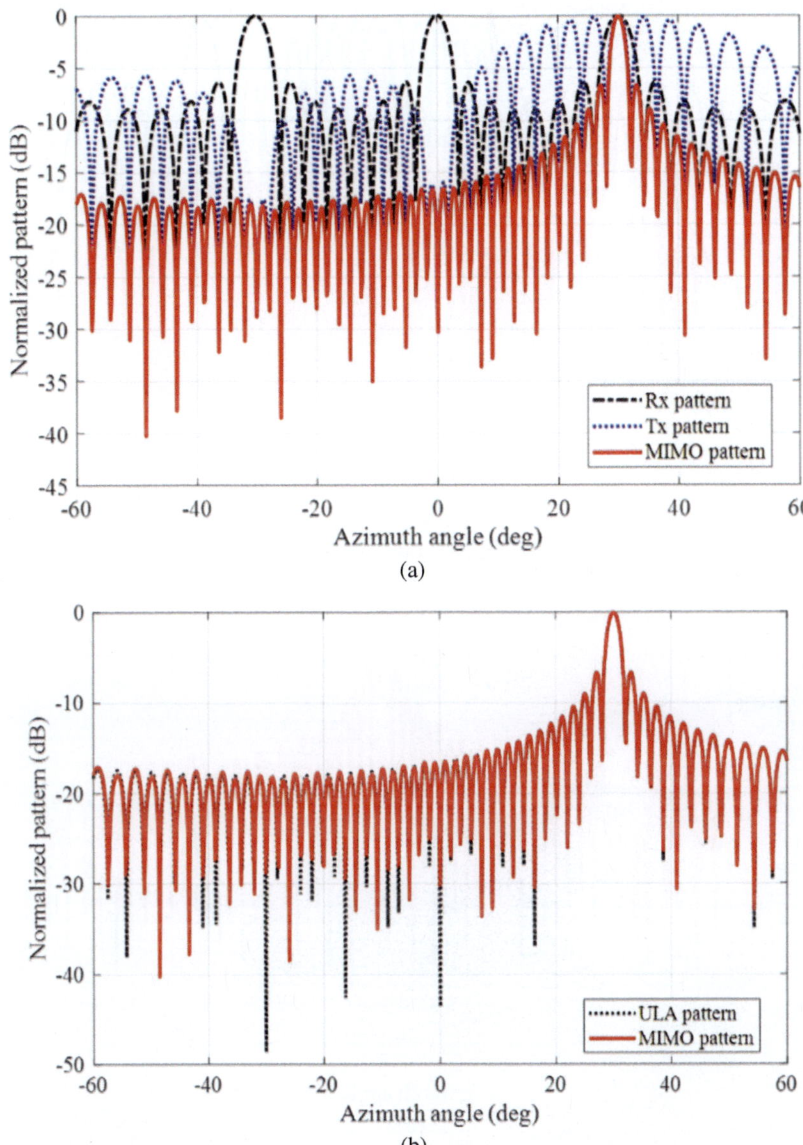

FIGURE 6.5 Radiation patterns of (a) the MIMO array and (b) the ULA for a target at (100, 30°) at f_c = 5GHz.

6.1.2.1.2 2D MIMO Array Design

Since only 2D images can be obtained by GB-SAR and the linear MIMO array-based radar system, a further step is considered to get the 3D high-resolution image based on a 2D planar MIMO array. Similarly, first, a uniform planar array can be designed according to the desired ambiguities and resolutions in the azimuth and

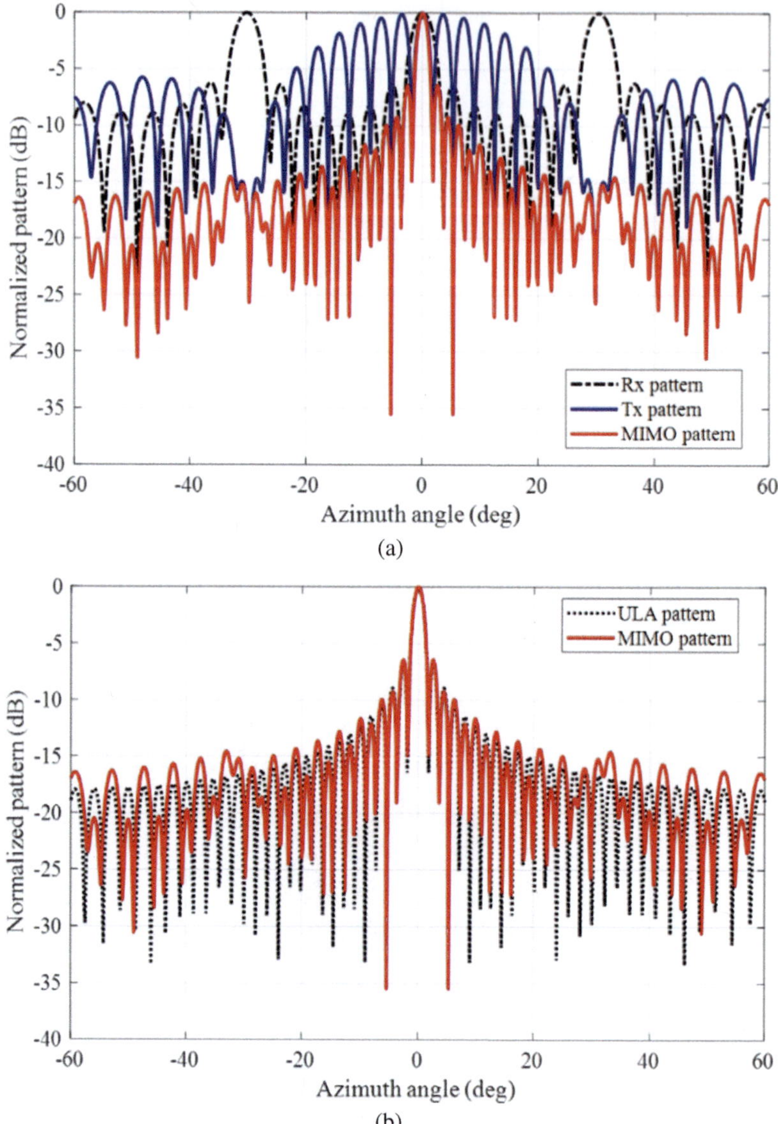

FIGURE 6.6 Radiation patterns of (a) the MIMO array and (b) the ULA for a target at (10, 0°) at f_c = 5GHz.

elevation directions. The aperture length L_X (L_Z), the inter-element spacing Δx (Δz), and number of transceivers can be determined. For a 2D MIMO array, assuming the m-th transmitter is at (x_m, z_m) and the n-th receiver is at (x_n, z_n), then, according to PCA, as shown in Figure 6.8, a virtual element can be generated at

$$[x_p, z_p] = [(x_m + x_n)/2, (z_m + z_n)/2] \mid m = 1, 2, ..., M,$$

$$n = 1, 2, ..., N, p = 1, 2, ..., MN \qquad (6.13)$$

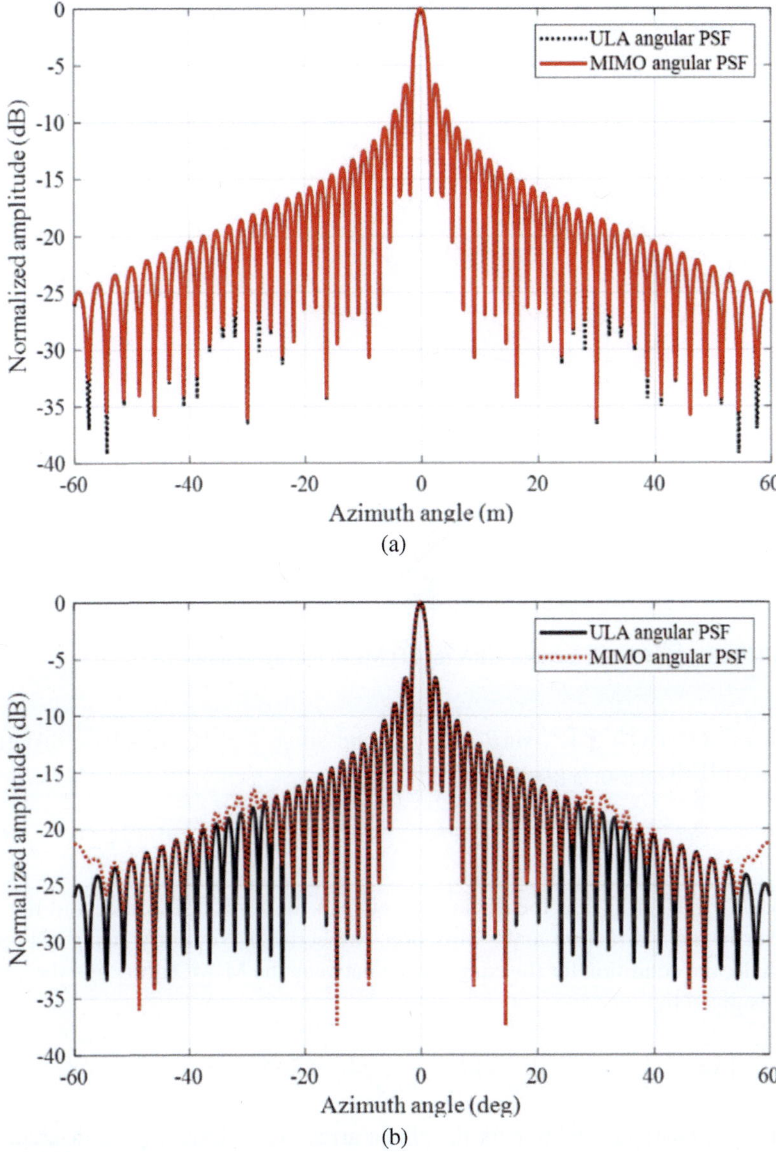

FIGURE 6.7 Angular PSFs for targets at (a) $(100, 0°)$ and (b) $(10, 0°)$.

Then, given a target at (x_0, y_0, z_0), the phase of the virtual array can be approximated by

$$\phi_p \simeq 4\pi f_c [r_0 - x_p \sin\theta_0 - z_p \sin\varphi_0 + \frac{x_p^2 + z_p^2 - (x_p \sin\theta_0 + z_p \sin\varphi_0)^2}{2r_0}]/c \qquad (6.14)$$

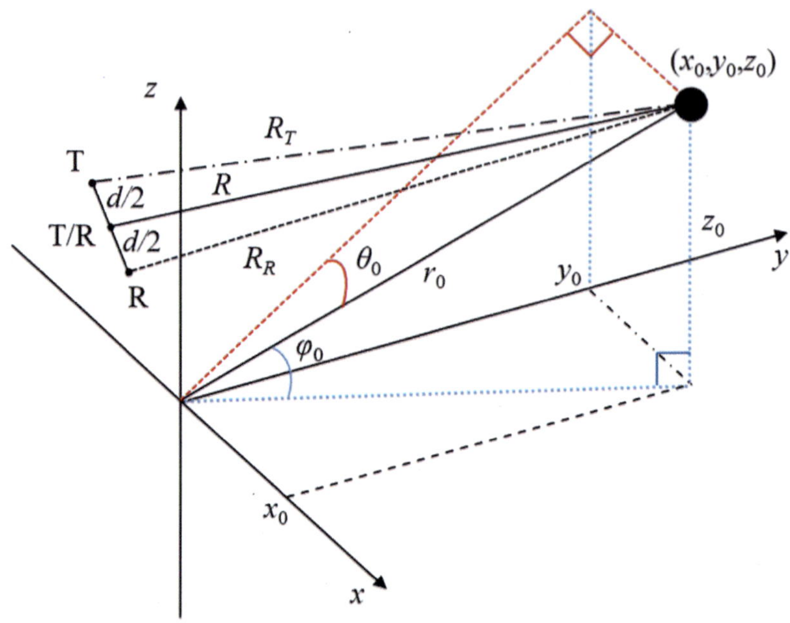

FIGURE 6.8   Virtual element of UPA for 2D MIMO array design.

where $r_0 = (x_0^2 + y_0^2 + z_0^2)^{1/2}$, $\sin\theta_0 = x_0 / r_0$, and $\sin\varphi_0 = z_0 / r_0$. Its phase difference from the MIMO array is given by

$$\Delta\phi_{m,n} \simeq \pi[d_{x,m,n}^2 + d_{z,m,n}^2 - (d_{x,m,n}\sin\theta_0 + d_{z,m,n}\sin\varphi_0)^2]/(2r_0\lambda) \qquad (6.15)$$

where $d_{x,m,n}$ and $d_{z,m,n}$ are the distances between the m-th transmitter and the n-th receiver in the x direction and z direction, respectively. Similarly, using $\pi/4$ as the threshold, the condition for the equivalence between the MIMO array and the virtual array is given by

$$\left|\Delta\phi_{m,n}\right|_{\max} \simeq \left[\left|d_{x,m,n}^2\right|_{\max} + \left|d_{z,m,n}^2\right|_{\max}\right]/4r_0 \leq \lambda/8 \Rightarrow r_0 \geq 2\left|d_{m,n}^2\right|_{\max}/\lambda \qquad (6.16)$$

which is the far-field condition for the planar array. After designing the desired UPA and assuming it has $P = P_x P_z$ elements, where $P_{x/z}$ denotes the element number in the x/z direction, the transmitting and receiving arrays of the 2D MIMO array can be designed by the 2D PF method as

$$P(h,l) = \sum_{p_x=1}^{P_x} \sum_{p_z=1}^{P_z} h^{p_x-1} l^{p_z-1} = P_T(h,l)P_R(h,l) \qquad (6.17)$$

Then, similarly to the design of the linear MIMO array, the transmitting and receiving arrays should be adjusted to make the planar MIMO array symmetrical and non-overlapped.

With a central frequency of 5 GHz and a UPA of 16×16 elements that are uniformly distributed in the x-z domain with an inter-element spacing of $\Delta x = \Delta z = 3$ cm, the MIMO array that has the highest efficiency (i.e., has the smallest size) can be obtained, as shown in Figure 6.9(a). On the contrary, the one that has the lowest efficiency is shown in Figure 6.9(b). However, the cross-MIMO array (i.e., the one

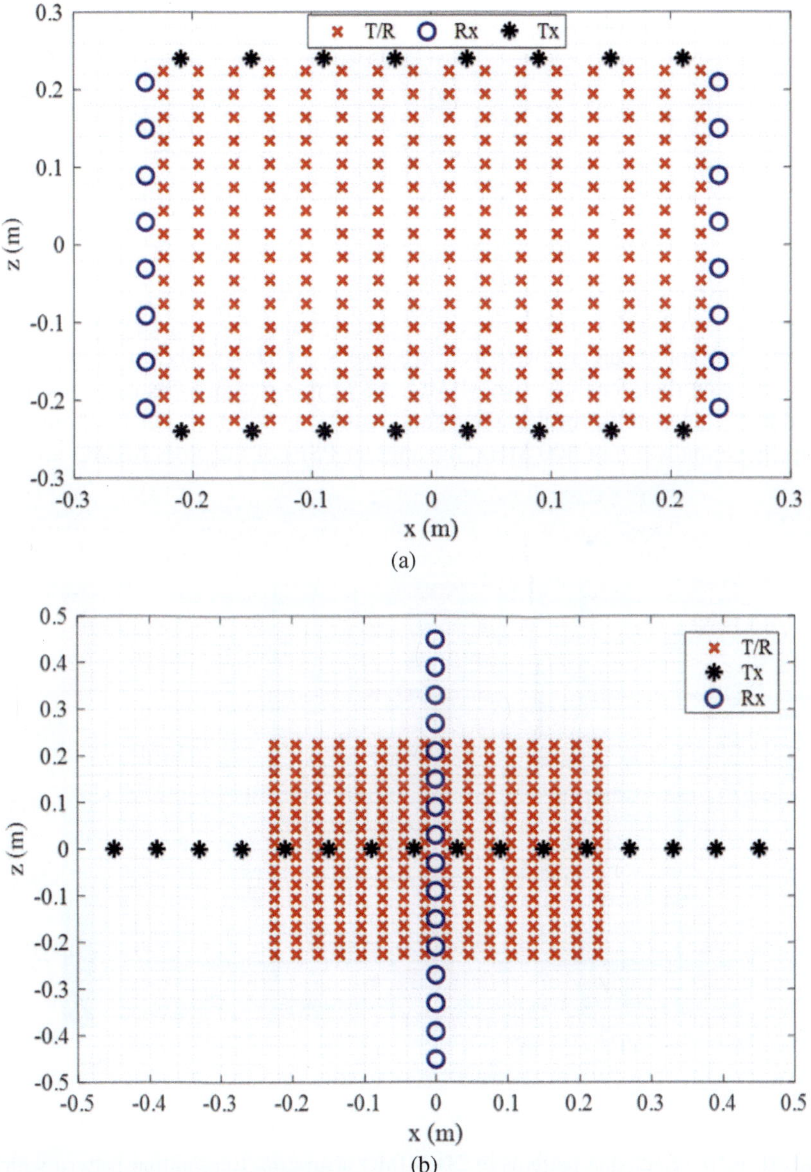

FIGURE 6.9 2D MIMO arrays and its equivalent virtual arrays that have (a) the highest efficiency and (b) the lowest efficiency.

with the lowest efficiency) is easier to develop in practice, and its mutual coupling is smaller. Also, the imaging process of the cross-MIMO array can be simplified.

For SFCW-based 2D MIMO radar, the angular PSF can be expressed as

$$\chi_{MIMO}(\theta,\varphi\,/\,r_0,\theta_0,\varphi_0) = \sum_{q=1}^{Q} F_T(\theta,\varphi,f_q\,/\,r_0,\theta_0,\varphi_0)F_R(\theta,\varphi,f_q\,/\,r_0,\theta_0,\varphi_0) \quad (6.18)$$

where

$$\begin{cases} F_T(\theta,\varphi,f_q\,/\,r_0,\theta_0,\varphi_0) = \sum_{m=1}^{M} s_0(m,q)\exp\{j2\pi f_q R_m(r_0,\varphi,\theta)/c\} \\ F_R(\theta,\varphi,f_q\,/\,r_0,\theta_0,\varphi_0) = \sum_{n=1}^{N} s_0(n,q)\exp\{j2\pi f_q R_n(r_0,\varphi,\theta)/c\} \end{cases} \quad (6.19)$$

are the transmitting and receiving radiation patterns at frequency f_q.

For the two MIMO arrays given in Figure 6.9 and a target at (100, 0°, 0°), the transmitting and receiving patterns at frequency 5 GHz are shown in Figure 6.10. It can be seen that, for these two MIMO arrays, the grating lobes can be well canceled by the deep nulls, resulting in a good radiation main lobe. Furthermore, with a frequency bandwidth of 300 MHz, the angular PSFs of the cross-MIMO array and

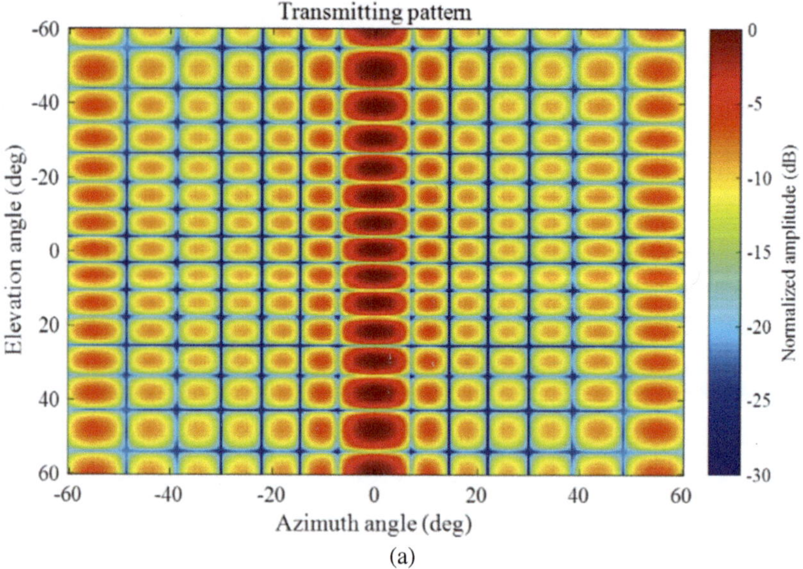

FIGURE 6.10 Radiation patterns of 2D MIMO arrays: (a) transmitting pattern with the highest efficiency, (b) transmitting pattern with the lowest efficiency, (c) receiving pattern with the highest efficiency, (d) receiving pattern with the lowest least efficiency, (e) MIMO pattern with the highest efficiency, (f) MIMO pattern with the lowest efficiency.

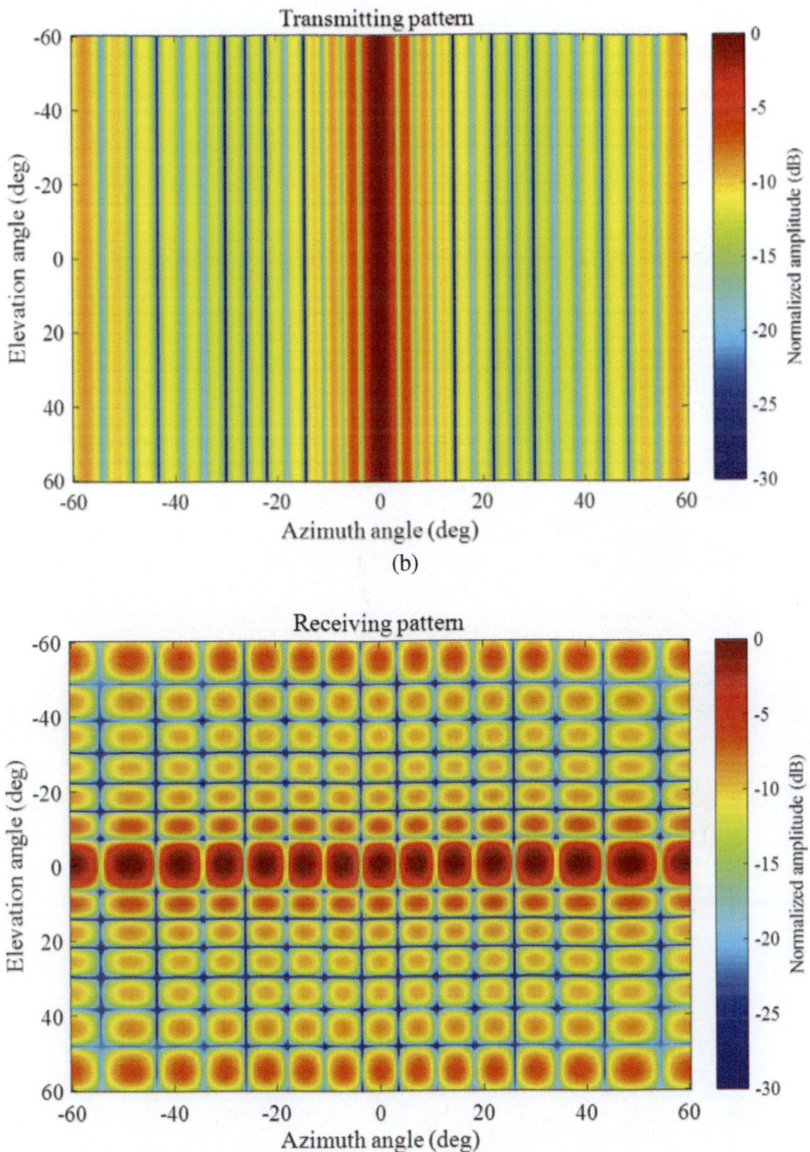

(b)

(c)

FIGURE 6.10   (Continued)

its equivalent UPA are presented in Figure 6.11, from which it can be found that the cross-MIMO array can achieve a similar PSF with the UPA.

However, it should be noted that, for the designed UPA, due to the consideration of the antenna size, the ambiguous angles in the azimuth and elevation directions are about 60°. This is because the width of the used Vivaldi antenna is about 7.5 cm,

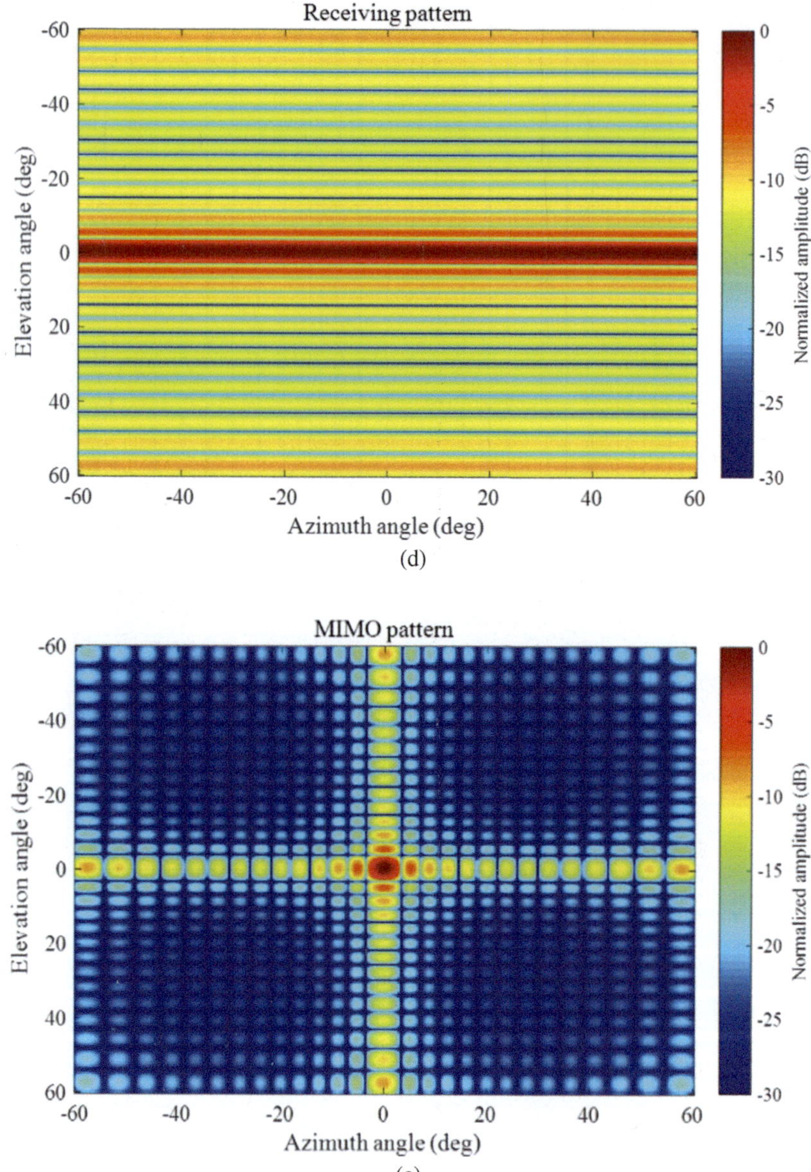

FIGURE 6.10 (Continued)

and the size of the used spiral antenna is about 5.5×5.5 cm. For targets located at different positions, different radiation patterns will be generated. For example, given a target at $(100, 30°, 30°)$, the radiation patterns of the cross-MIMO array and its equivalent UPA are shown in Figure 6.12, where some grating lobes are generated.

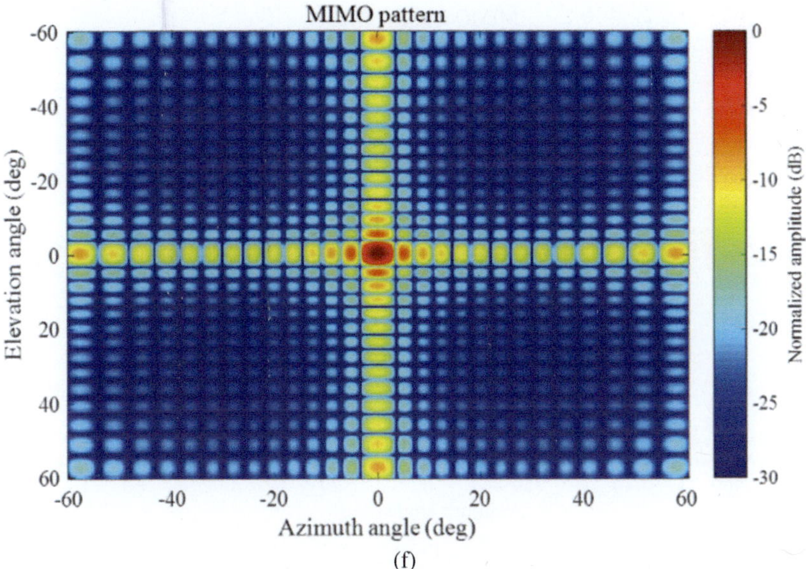

(f)

FIGURE 6.10   (Continued)

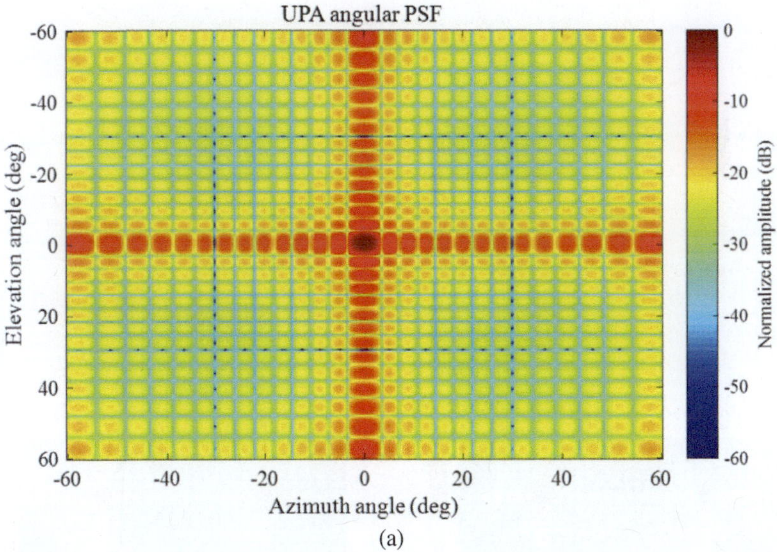

(a)

FIGURE 6.11   Angular PSFs of (a) the UPA and (b) the cross-MIMO array (i.e., the array with the lowest efficiency) for a target at (100, 0°, 0°).

Therefore, in practical applications, the looking angle of the designed MIMO radar system should be adjusted to make the targets inside the main lobe of the radiation patterns. Otherwise, angular ambiguities may have a negative influence on the following process.

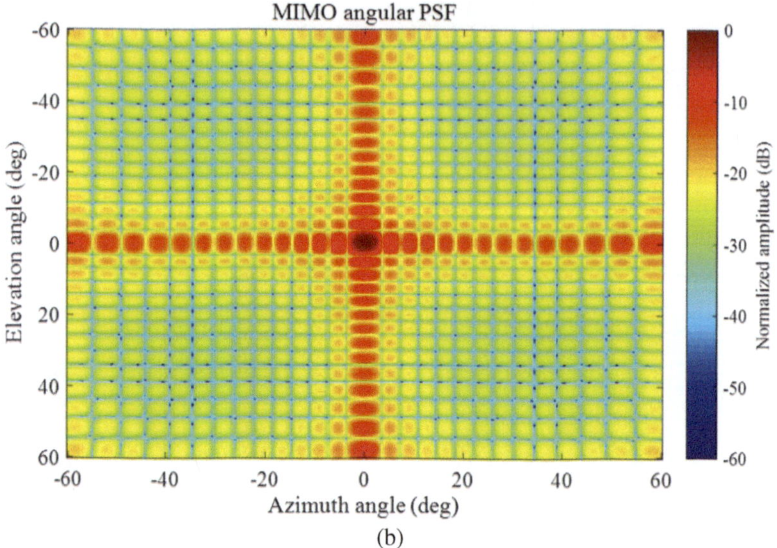

(b)

FIGURE 6.11   (Continued)

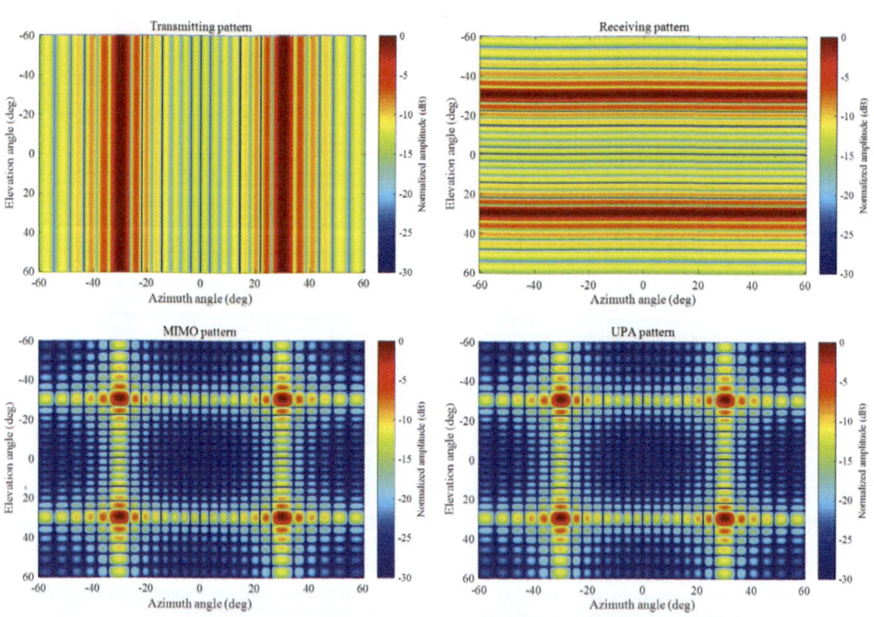

FIGURE 6.12   Radiation patterns of the UPA and the cross-MIMO array for a target at (100, 30°, 30°). The grating lobes are generated by the large inter-element spacing.

6.1.2.2 MIMO Radar Systems

The first linear MIMO radar system, which is only used to evaluate the performance of the designed MIMO arrays and the basic processing algorithms, is developed based on an integrated SFCW transceiver, as shown in Figure 6.13. For the integrated transceiver, as shown in the right of Figure 6.13, a maximum of eight transmitting channels and eight receiving channels can be used. Its working frequency is from 3.81 to 8.01 GHz (unchangeable), with 1024 frequency steps (adjustable, the minimal number is 8), resulting in a range resolution of 3.53 cm and an unambiguous range of about 18 m. To achieve the transmitting orthogonality, the time division multiplexing (TDM) approach is used: each transmitter emits a signal sequentially, while each receiver records the echoed signal sequentially. The length of the equivalent ULA of the designed MIMO array is about 0.89 m, giving an angle resolution of about 0.028 rad. The data sampling rate of this system is about 8 Hz, faster than most commercial GB-SAR systems. However, the main problem of this system is its short ambiguous range caused by the ultra-wide bandwidth and the limited frequency numbers, making it unsuitable for practical applications. Also, since only the in-phase signal component is obtained, the unambiguous range of this system is halved ($c/4\Delta f$). Therefore, after testing the designed MIMO arrays, some imaging algorithms, and several calibration methods and measuring displacements and vibrations of short-range targets, this system is abandoned and modified to the second linear MIMO radar system, as shown in Figure 6.14.

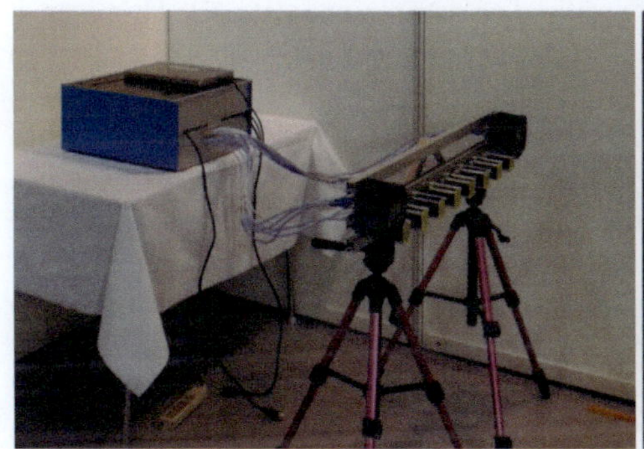
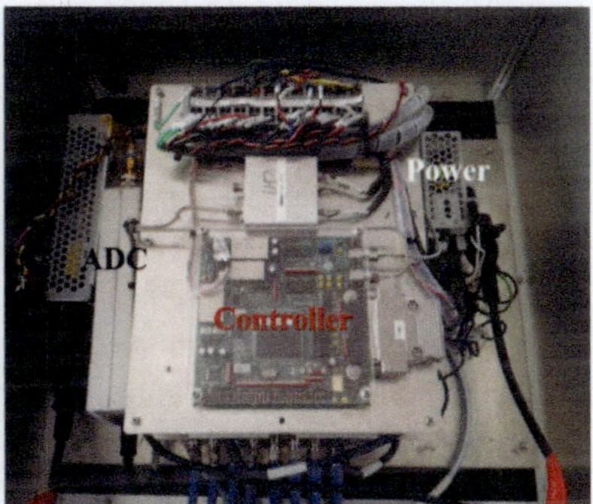

FIGURE 6.13 First linear MIMO radar system.

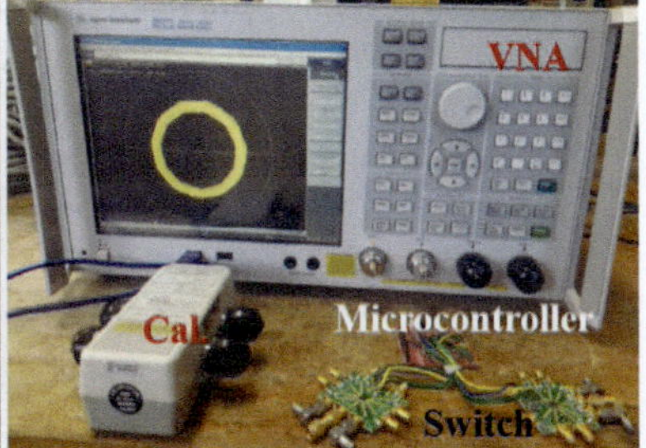

FIGURE 6.14 Second linear MIMO radar system.

Apart from the MIMO array, which will be designed according to the working frequency, the second linear MIMO radar system mainly includes a vector network analyzer (VNA E5071C) to generate the SFCW signal, two 1 × 8 switches (HMC321ALP4E, working frequency from 0 to 8 GHz, isolation higher than 30 dB, and insertion loss 2.4 dB at 5 GHz) to realize the time-divided signal transmitting and receiving, a micro-controller (OLIMEXINO-32U4) that is used to control the switches, a low noise amplifier (LNA, OLIMEXINO-32U4), and a PC used to save and post-process the VNA data and remotely control the micro-controller and the VNA. Compared to the first linear MIMO radar system, the second one can randomly change the working frequency within the broadband of the VNA, LNA, switches, and antennas. The maximum number of frequencies of the VNA is 1601, and the ambiguous detection range can be adjusted according to the used bandwidth. Also, both the in-phase and quadrature (I/Q) signal components can be recorded. The key problem of this system is its data acquisition period. Since only four ports are available and ports 1–2 are used for the MIMO radar, each pair of transceivers should be turned on to measure S21 sequentially; that is, the receivers cannot sample the data at the same time, which decreases the data sampling rate. Moreover, to compensate for the time delays caused by the switches and the coaxial cables, the calibration state is saved in the VNA and should be recalled for each measurement. Therefore, a long data acquisition time is needed for the second linear MIMO radar system. For example, with a central frequency of 5 GHz, a bandwidth of 500 MHz, and 201 steps, about 15 seconds are required to sample one data set. However, it should be noted that this data acquisition rate is sufficient for most applications in practice.

For VNA-based linear MIMO radar, another potential modification is to use multiple systems to estimate the 2D or 3D displacement vector [16]. For example, since the used VNA (E5071C) has four ports, two MIMO radar systems can measure S21 and S43, respectively, as shown in Figure 6.15. For this configuration, the positions of MIMO radar systems can be changed with different spatial baselines with the extension of coaxial cables. The displacement estimation accuracies in different directions are dependent on the spatial baseline between two systems, which should be adjusted in practical applications. Two micro-controllers can be interleaved: one micro-controller switch on a specific pair of transceiver of the first MIMO array, then the same pair of transceiver of the second MIMO array will be switched on by another micro-controller. In such a case, it is assumed that two MIMO radar systems can sample the data near-simultaneously. Thus, two images of the observation scene can be obtained. After conducting high-resolution and high-accuracy image co-registration, the 2D displacement vector of the scene can be derived.

Vivaldi antennas and spiral antennas [17], as shown in Figure 6.16, are used to build the MIMO arrays. Their radiation patterns at 5 GHz, maximum gains, and S11s are shown in Figures. 6.17, 6.18, and 6.19, respectively. The width and thickness of the Vivaldi antenna are about 7.5 cm and 1.5 mm, respectively. The size of the spiral antenna is about 5.5 × 5.5 cm. Both antennas are wideband, which was selected to benefit the arbitrary frequency selection of the designed MIMO radar systems. Compared with the spiral antenna, the Vivaldi antenna has a broader beam-width

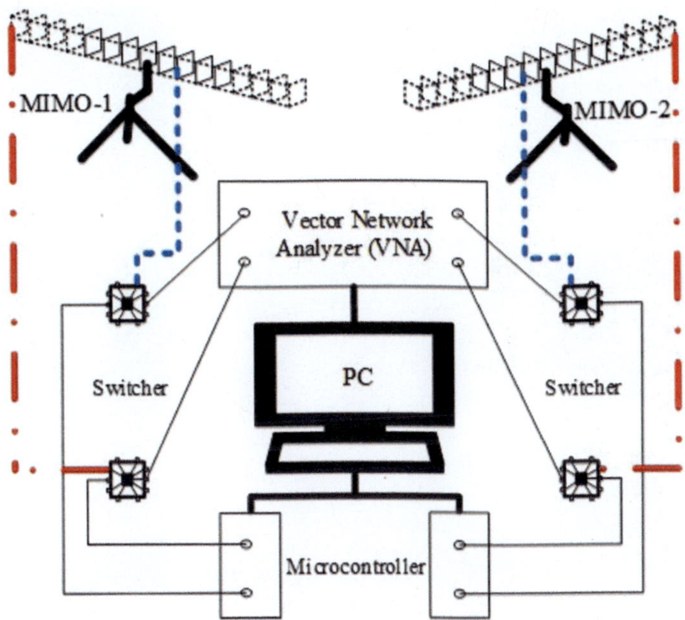

FIGURE 6.15 Radar system with two linear MIMO arrays for 2D displacement vector estimation.

and a higher gain and can be easily spaced at half of the wavelength, which is important in the real implementation of MIMO radar systems. On the other hand, because the transmitting or receiving array can be non-Nyquist spaced (i.e., with spacing larger than half of the wavelength), the spiral antenna can be used for polarized transmission or reception (the combination of dual-polarization components of the target reflection can thus be received).

The configuration of the developed VNA-based cross-MIMO radar system for 3D imaging is shown in Figure 6.20. The typology of the cross-MIMO array is similar to that in Figure 6.9(b), while the transmitting array and the receiving array are switched. Four 1 × 8 switches and two micro-controllers are used to realize the transmitting orthogonality. In such a case, all four ports of the VNA have been used: the sampling of S21 and S43 is interleaved. The uniform transmitting array includes 16 right-hand circularly polarized spiral antennas, and the uniform receiving array includes 16 Vivaldi antennas.

With a central frequency of 5 GHz, due to the antenna size limitation, as mentioned in the previous part, the inter-element spacing of the transmitting array and the receiving array are designed to be 5 and 6 cm, respectively, giving an ambiguous angles of the equivalent UPA about 60° in the azimuth direction and 73.7° in the elevation direction. Considering the narrow radiation pattern of the spiral antenna, as shown in Figure 6.17, if the looking angle of the cross-MIMO radar system is

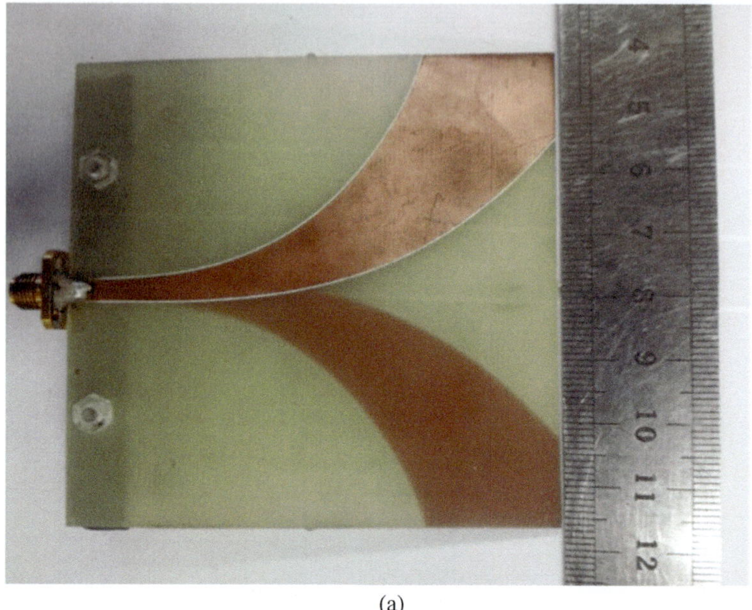

(a)

(b)

FIGURE 6.16 (a) Vivaldi antenna and (b) spiral antenna.

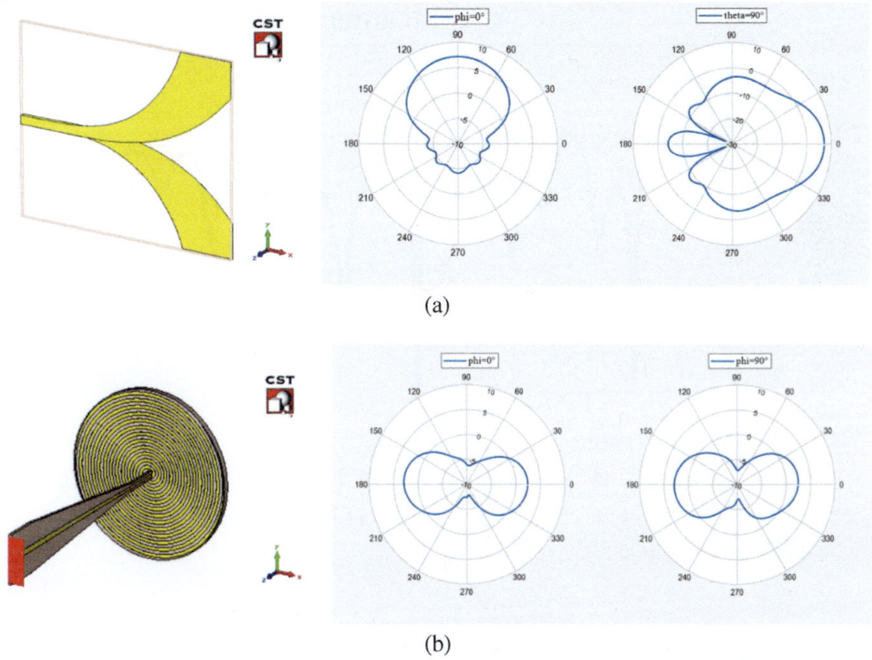

FIGURE 6.17   Simulated radiation patterns of (a) Vivaldi antenna and (b) spiral antenna at the frequency of 5 GHz.

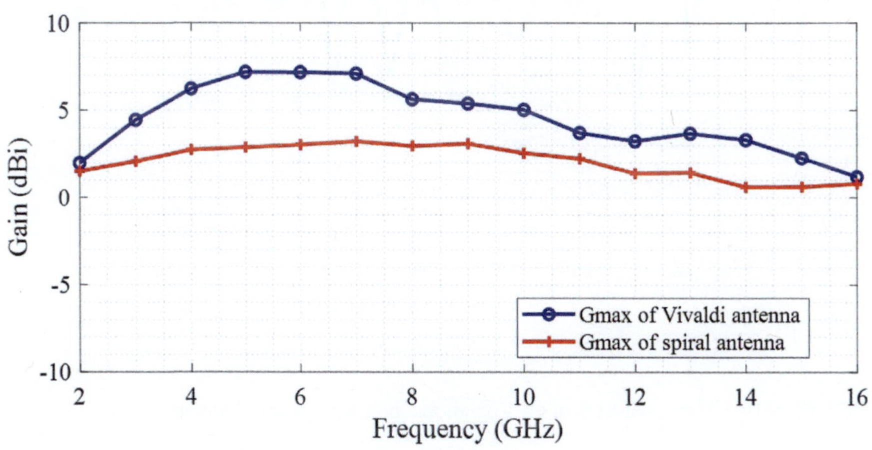

FIGURE 6.18   Simulated maximum gains of Vivaldi antenna and spiral antenna.

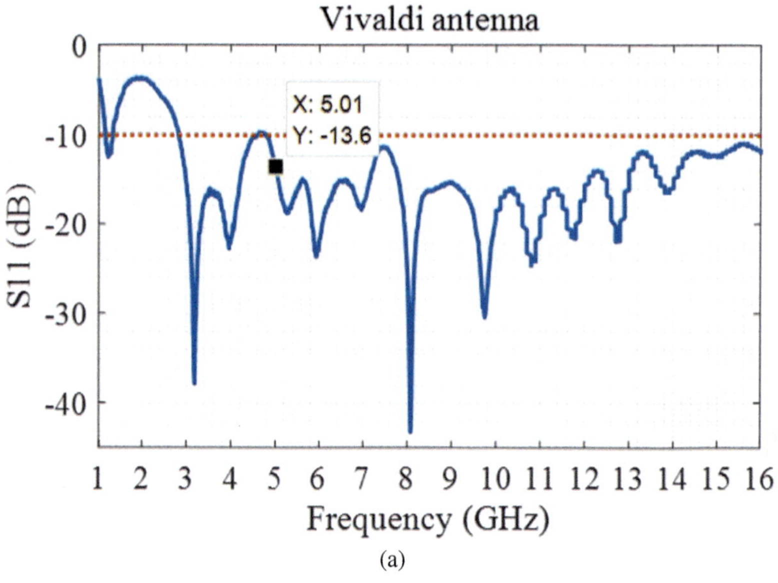

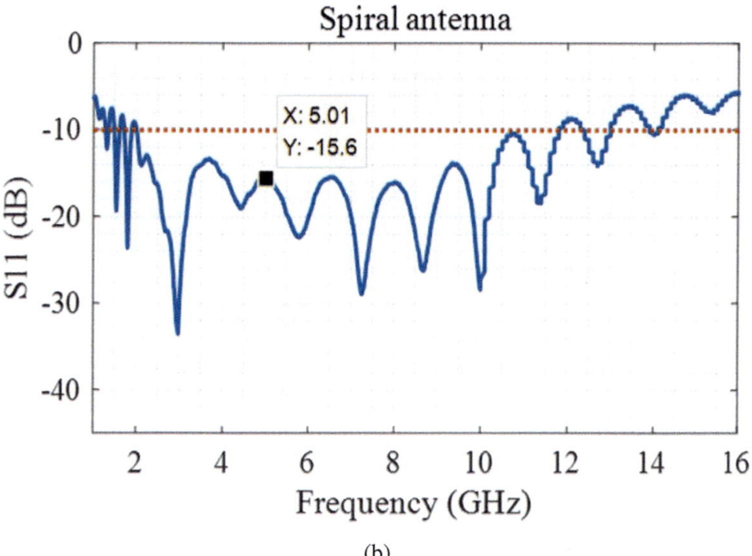

FIGURE 6.19   Measured S11s of (a) Vivaldi antenna and (b) spiral antenna.

adjusted to make the observation scene located at the main lobe of the antennas, the influences of the ambiguities can be avoided. For this system, the resolution in the azimuth direction is about 0.067 rad, and the resolution in the elevation direction is about 0.080 rad.

FIGURE 6.20 The developed cross-MIMO radar system.

6.1.2.3 Experiments and Results

6.1.2.3.1 1D Linear MIMO Radar

Consider an SFCW radar system with a linear MIMO array, shown in Figure 6.21, where the m-th transmitter is at $(x_m, 0)$ and the n-th receiver is at $(x_n, 0)$, $m = 1, 2, \ldots, M$ and $n = 1, 2, \ldots, N$. Supposing a point target is located at (x_0, y_0), the two-way distance of the target is given by

$$R_{m,n} = \sqrt{(x_m - x_0)^2 + y_0^2} + \sqrt{(x_n - x_0)^2 + y_0^2} \tag{6.20}$$

For the q-th frequency $f_q = f_0 + (q-1)\Delta f$ with $q = 1, 2, \ldots, Q$, the demodulated received signal for the m-th transmitter and the n-th receiver can be expressed as

$$S_0(m,n,q) = \sigma(x_0, y_0)\exp(-j2\pi f_q R_{m,n}/c) + N_0(m,n,q) \tag{6.21}$$

where f_0 is the start frequency, Δf is the frequency step, $\sigma(x_0, y_0)$ is the reflection coefficient of the target, c is the speed of light, and $N(m, n, q)$ is the additive thermal noise. Therefore, the received signal from all targets in the whole observation area is given by

$$S(m,n,q) = \iint_\Omega \sigma(x,y)\exp(-j2\pi f_q R_{m,n}(x,y)/c)\,dxdy + N(m,n,q) \tag{6.22}$$

where $R_{m,n}(x, y)$ is the two-way distance of the target at (x, y), $\sigma(x, y)$ is the reflection coefficient, and Ω denotes the observation area. Assuming that the illuminated scene

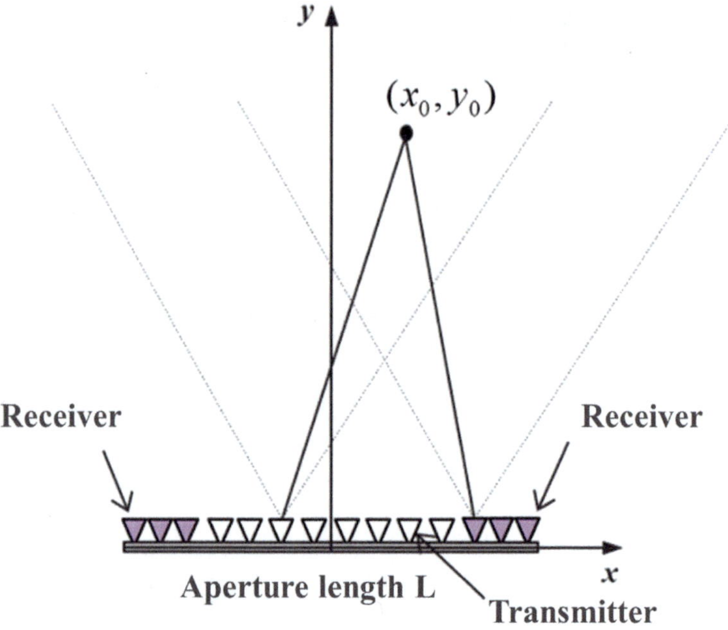

FIGURE 6.21   Imaging geometry of the designed linear MIMO radar system.

is discretized by I grids in the x axis and J grids in the y axis, the discrete expression of the received signal is given by

$$S(m,n,q) = \sum_{i=1}^{I}\sum_{j=1}^{J}\sigma(x_i,y_j)\exp(-j2\pi f_q R_{m,n}(x_i,y_j)/c) + N(m,n,q) \qquad (6.23)$$

According to the signal model in (6.23), the classical matched filtering-based imaging methods for GB-SAR can be modified for MIMO radar imaging. For example, the frequency-domain and time-domain BP algorithms can be modified to

$$\tilde{\sigma}_{FDBP}(x,y) = \sum_{m=1}^{M}\sum_{n=1}^{N}\sum_{q=1}^{Q}S(m,n,q)\exp\{+j2\pi f_q R_{m,n}(x,y)/c\} \qquad (6.24)$$

and

$$\tilde{\sigma}_{TDBP}(x,y) = \sum_{m=1}^{M}\sum_{n=1}^{N}S_t(m,n,R_{m,n}(x,y)/c) \qquad (6.25)$$

Also, if the transmitting and receiving arrays are both uniform, then RMA can be adopted by

$$S(k_x,k_y) = \iint_{\Omega}\sigma(x,y)\exp\{-j[k_x x + k_y y]\}dxdy \qquad (6.26)$$

with

$$k_x = k_x^T + k_x^R, \quad k_y = \sqrt{k^2 - (k_x^T)^2} + \sqrt{k^2 - (k_x^R)^2} \tag{6.27}$$

where k_x^T and k_x^R are the spatial wavenumbers obtained by the Fourier transform with respect to x_m and x_n, respectively [4]. Then, the interpolation from the 3D domain (k_x^T, k_x^R, k) to the 2D (k_x, k_y) domain should be conducted to achieve uniformly distributed k_x and k_y. Finally, the image can be obtained by the 2D FFT of the interpolated signal as

$$\tilde{\sigma}_{RMA}(x, y) = \boldsymbol{F}_x^H \left\lceil \tilde{S}(k_x, k_v) \right\rceil \boldsymbol{F}_v^* \tag{6.28}$$

Moreover, for targets located at the far field, the MIMO array can be viewed as equivalent to ULA, as indicated by (6.6). Therefore, similar approximations as those for GB-SAR imaging can be used to employ the FPFA-based method for MIMO radar imaging, which gives

$$\tilde{\sigma}_{FPFA}(\alpha, \beta) = \sum_{o=1}^{O-1} \frac{1}{o!} [-j2\pi\beta / f_c]^o \{ \boldsymbol{F}_\alpha^H [S(p,q) \cdot (x_p \hat{f}_q)^o] \boldsymbol{F}_\beta \} \tag{6.29}$$

where $S(p, q)$ corresponds to the p-th ($p = 1, 2, \ldots, P$ and $P = MN$) spatial sampling point of MIMO radar, that is, the p-th pair of transmitter and receiver.

Based on the signal model in (6.23) and the modified imaging algorithms, the performance of the designed two linear MIMO radar systems are evaluated. First, for the first linear MIMO radar system, three trihedral corner reflectors (CRs) are used as the targets, and the imaging result is obtained by the time domain BP algorithm, as shown in Figure 6.22. To avoid the negative influence of antenna mutual coupling, a Hanning window is used in the range direction, which is applied to all the following experiments. It can be seen that, although high-level sidelobes in the azimuth direction are generated, three CRs can be well focused and distinguished.

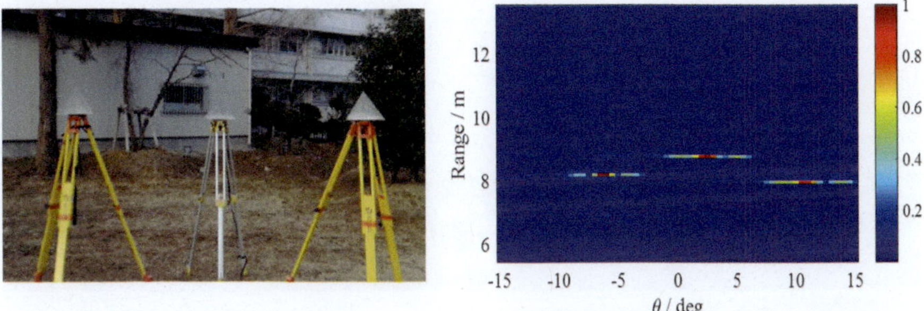

FIGURE 6.22 Three trihedral corner reflectors are imaged by the first linear MIMO radar system.

One of the CRs is then mounted on a linear scanner that can be controlled with an accuracy of 1 μm, as shown in Figure 6.23(a). Along the line of sight (LOS) direction, this CR is moved away from the radar system from 2 to 22 mm with a 2-mm step. The displacement estimation result is shown in Figure 6.23(b), where the root mean square error (RMSE) of the estimated displacement is about 0.37 mm, verifying the displacement estimation capacity of the first MIMO radar system.

Since the first linear MIMO radar system has a high data acquisition rate (about 8 Hz), some vibrations or fast displacements can be measured. Therefore, another experiment is conducted to estimate the displacement of a bridge caused by the passing of a truck in a period of 40 seconds. The structure of the bridge under measurement is shown in Figure 6.24(a), from which it can be seen that the bridge has four horizontal bars and five vertical bars, which are the major monitoring parts during this experiment. The radar image of the bridge is shown in Figure 6.24(b). It can be seen that the structure of the bridge can be well focused. Four different points that have the highest amplitudes are selected to estimate the displacements, and the results are shown in Figure 6.25, from which it can be observed that these four points enjoy different displacements. Different from real aperture radar measurements commonly used for bridge monitoring [18], which can only provide a high-range resolution and thus lose the information in the azimuth direction, the 2D imaging capacity and the fast data sampling rate of the designed MIMO radar system make the separation of targets in both the range and azimuth directions possible. Therefore, a more accurate displacement estimation is achieved.

(a)

FIGURE 6.23 Displacement estimation of a trihedral CR by the first linear MIMO radar system.

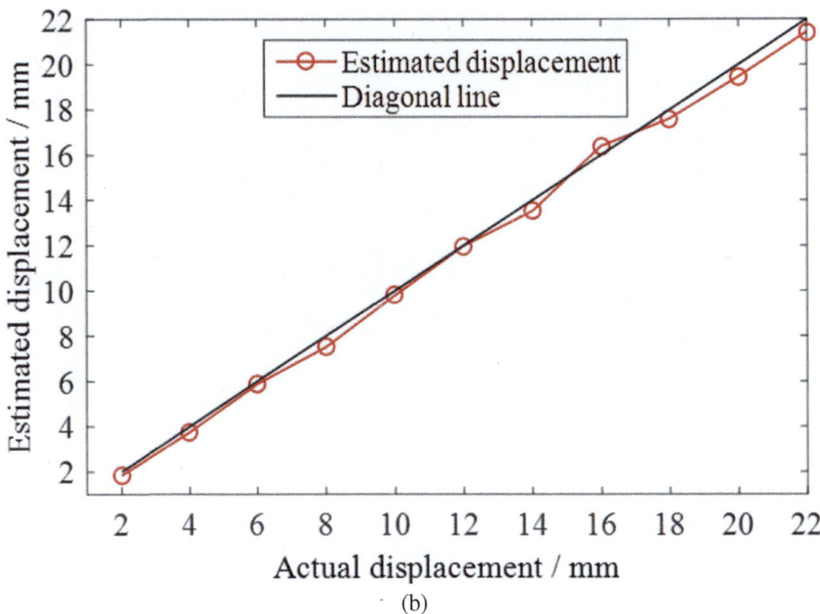

(b)

FIGURE 6.23 (Continued)

(a)

FIGURE 6.24 A bridge is measured by the first linear MIMO radar system.

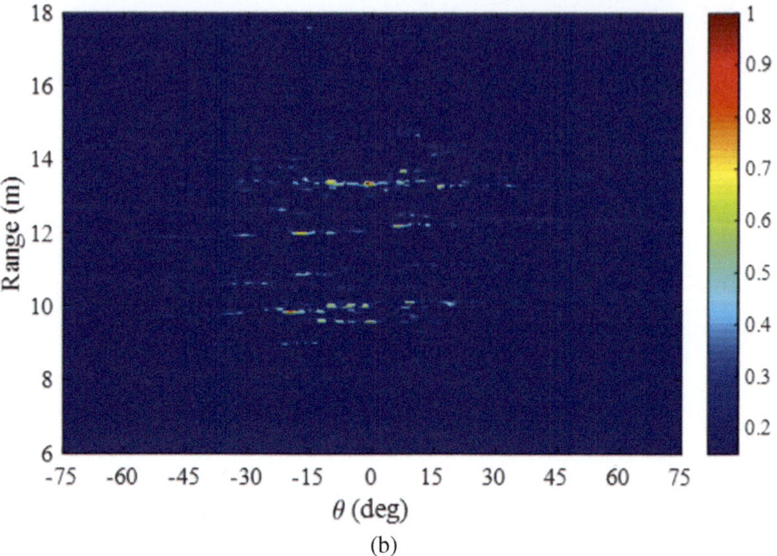

FIGURE 6.24   (Continued)

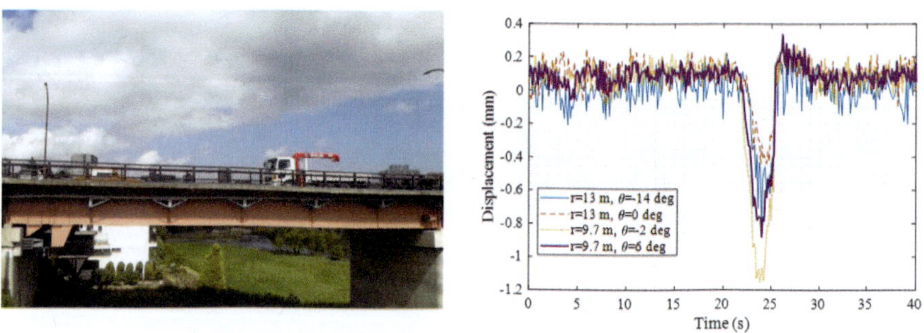

FIGURE 6.25   Displacement estimation of four points of the bridge during the passing of a track.

The main problem of the first linear MIMO radar system is its short ambiguous range, due to which the MIMO array cannot well approximate its equivalent ULA. Thus, high-level sidelobes in the azimuth direction are generated, which cannot be suppressed by the conventional methods. Therefore, although it has a longer data acquisition time, the second linear MIMO radar system is developed and validated, as shown in Figure 6.26(a), where a natural scene including some trees, cars, light poles, and buildings is imaged. The working frequency for this experiment is from 4.75 to 5.25 GHz with 1601 frequency steps, which are determined according to the hardware parameters and the desired performance, giving a range resolution of 30 cm and an ambiguous range of 480 m. The topology of this linear MIMO array is shown in Figure 6.3(b), while no height difference is introduced to the transmitting and receiving arrays since only Vivaldi antennas are used in this experiment, resulting in an angular resolution of about 0.034 rad. The imaging result obtained by the time domain BP algorithm is shown in Figure 6.26(b). As indicated by the rectangles and circles, the building, small tree fence, light pole, and cars can be clearly detected, validating the imaging performance of the second linear MIMO radar system.

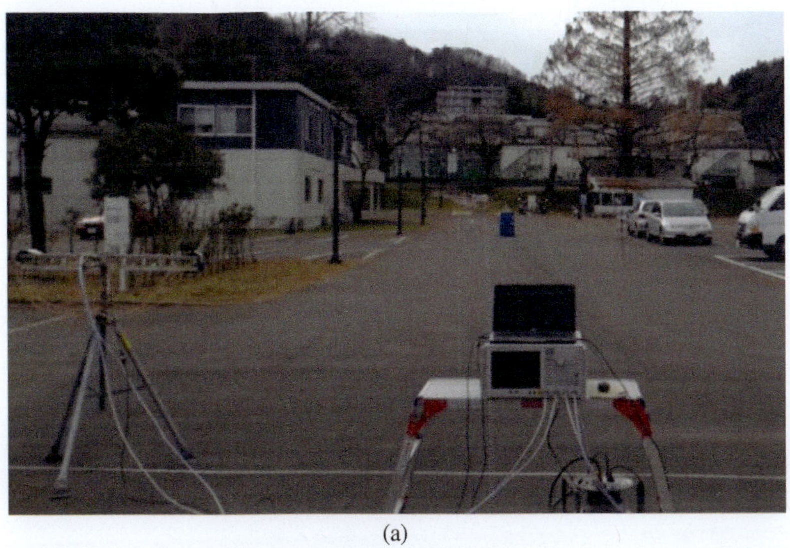

(a)

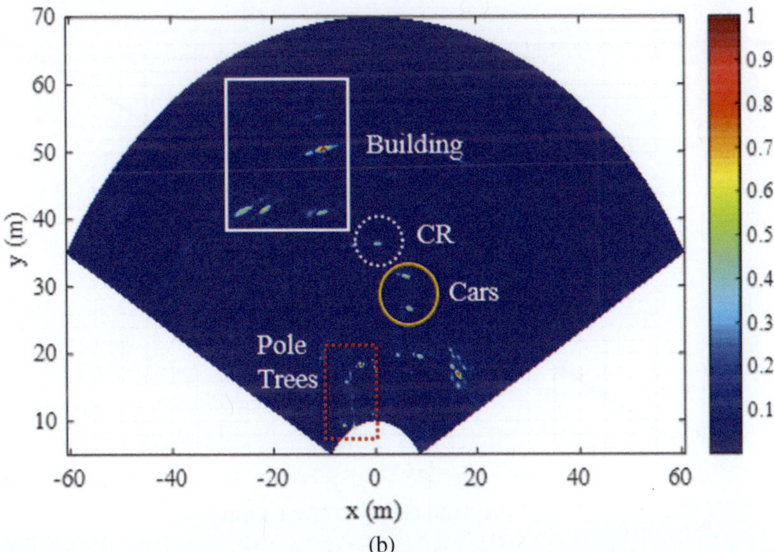

(b)

FIGURE 6.26 The second linear MIMO radar system is used for natural scene measurement.

Then, a trihedral CR, as indicated by the dotted circle in the radar image shown in Figure 6.26(b), is moved on the positioner, as shown in Figure 6.27(a). The displacements of this CR are measured, and the results are shown in Figure 6.27(b). It can be seen that the displacements can be accurately estimated. The RMSE of the displacement estimation is about 0.14 mm, which is better than the first linear MIMO radar system, as the VNA can provide a more stable phase and a deeper dynamic range than the integrated SFCW transceiver.

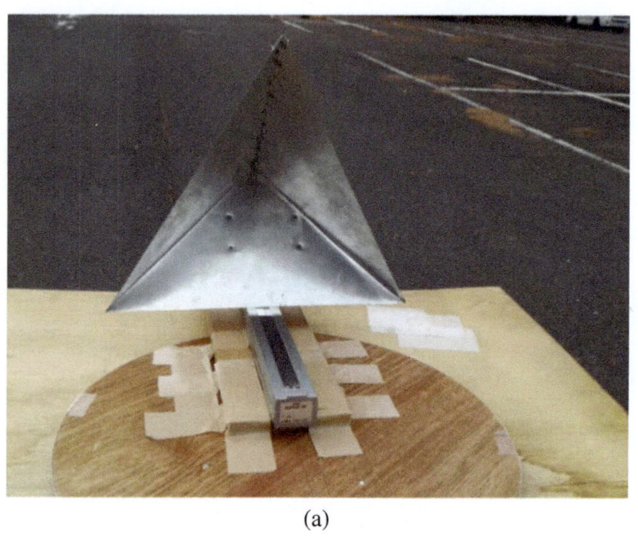

(a)

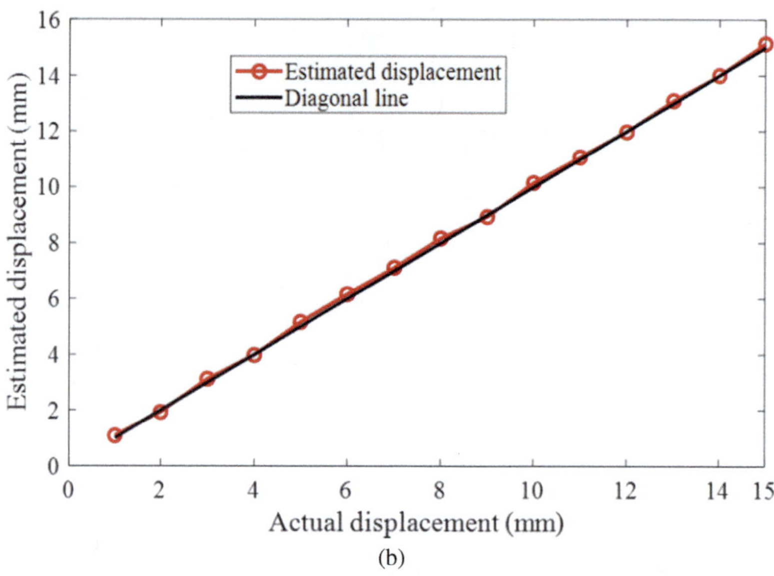

(b)

FIGURE 6.27   Displacement estimation by the second linear MIMO radar system of a CR.

6.1.2.3.2   2D Cross-MIMO Radar

The imaging geometry of the designed cross-MIMO radar system is shown in Figure 6.28, where the m-th receiver is at $(x_m, 0, 0)$, the n-th transmitter is at $(0, 0, z_n)$, $m = 1, 2, \ldots, 16$, and $n = 1, 2, \ldots, 16$. With a target located at (x_0, y_0, z_0), the distance between the m-n-th spatial sampling point and the target is given by

$$R^0_{m,n} = \sqrt{(x_m - x_0)^2 + y_0^2 + z_0^2} + \sqrt{x_0^2 + y_0^2 + (z_n - z_0)^2} \tag{6.30}$$

With the SFCW transmitted signal, the received signal at the m-n-th spatial sampling point and the q-th frequency can be expressed as

$$S_0(m,n,q) = \sigma_0 \exp(-\mathrm{j}2\pi f_q R^0_{m,n} / c) + N_0(m,n,q) \tag{6.31}$$

where $f_q = f_0 + (q-1)\Delta f$, $q = 1, 2, \ldots, Q$, f_0 is the start frequency, Δf is the frequency step, σ_0 is the reflection coefficient of the target, c is the velocity of light, and $N_0(m, n, q)$ is the additive noise.

Therefore, the received signal from the whole observation 3D scene is given by

$$S(m,n,q) = \iiint_\Omega \sigma(x,y,z) \exp\{-\mathrm{j}2\pi f_q R_{m,n}(x,y,z) / c\}\,\mathrm{d}x\mathrm{d}y\mathrm{d}z + N(m,n,q) \tag{6.32}$$

where $R_{m,n}(x, y, z)$ is the distance between the target at (x, y, z) and the m-n-th sampling point, $\sigma(x, y, z)$ is the target reflection coefficient, and Ω denotes the observation area.

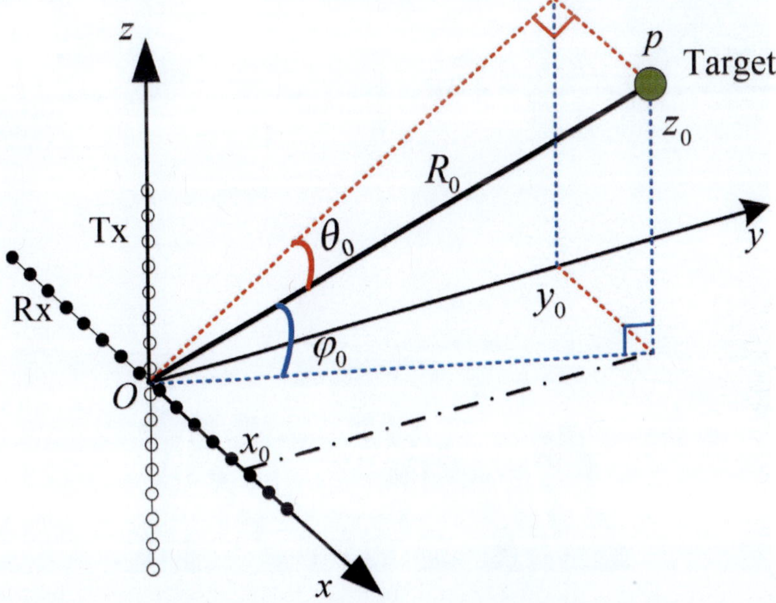

FIGURE 6.28   Imaging geometry of the designed cross-MIMO radar system.

The reflection coefficient of the target at (x, y, z) can be estimated by the frequency domain BP algorithm as follows

$$\tilde{\sigma}_{FDBP}(x, y, z) = \sum_{m=1}^{M}\sum_{n=1}^{N}\sum_{q=1}^{Q} s(m, n, q)\exp\{+j2\pi f_q R_{m,n}(x, y, z)/c\} \tag{6.33}$$

Imaging of the whole observation scene using (6.33) has an extremely high computational complexity, which prevents it from practical applications. Similarly to linear MIMO radar, other imaging algorithms for GB-SAR 2D imaging can be modified to reduce the computing time. One of the modified algorithms, 3D FPFA, will be introduced in Section 6.3.

To test the imaging performance of the designed cross-MIMO radar system, two experiments are conducted. The working frequency of the designed system is from 4.75 to 5.25 GHz with 201 frequency steps, giving a range resolution of 30 cm and an ambiguous range of 60 m. The first experiment, as shown in Figure 6.29(a), is to image two trihedral CRs located at different positions. The radar image obtained by the time-domain BP algorithm is shown in Figure 6.29(b), where two CRs can be well focused in the 3D spatial domain with a dynamic range of 20 dB. Therefore, a better understanding of the observation scene can be achieved than GB-SAR and linear MIMO radar. However, it should be noted that, caused by the limited aperture size and frequency bandwidth, conventional matched filtering based methods

(a)

FIGURE 6.29 Two trihedral CRs are imaged by the cross-MIMO radar system.

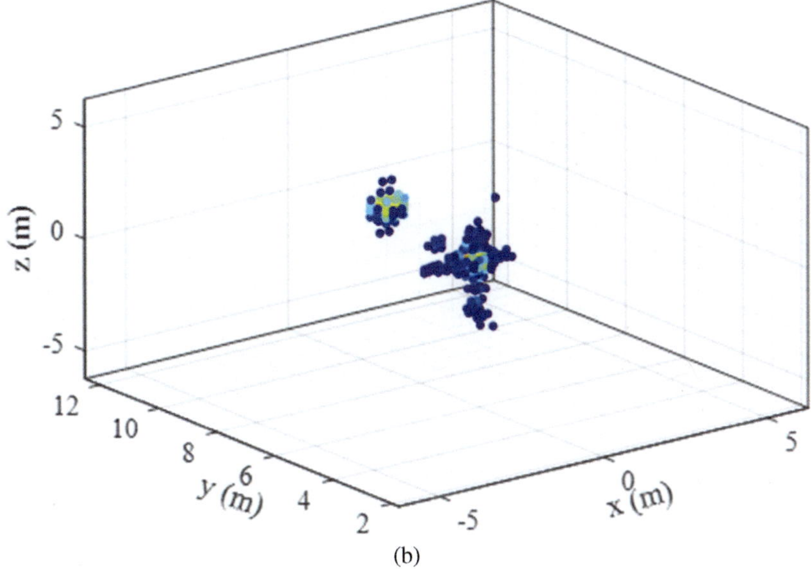

FIGURE 6.29   (Continued)

generate high-level sidelobes in the azimuth and elevation directions (as mentioned previously, sidelobes in the range direction have been suppressed by a Hanning window). Some advanced methods, such as the SVA-based method and the CS-based method, should be used to improve the imaging quality, which will be introduced in Section 6.3. Then, the bigger CR with a size of 40 cm is sledded on a positioner from 2 to 20 mm. Figure 6.30(a) shows the estimation result of a 2-mm displacement in the 3D spatial domain, while the estimation results for all ten displacements are shown in Figure 6.30(b), where the performance of some other imaging algorithms is also displayed, which will be detailed in Section 6.3.

In the second experiment, a four-layer building is measured by the cross-MIMO radar system. As shown in Figure 6.31, where the dynamic range is 25 dB, although the 3D image can provide some structures of the building, strong sidelobes and artifacts make the result difficult to interpret, which will be overcome in Section 6.3.

6.1.3 DISCUSSION

The linear MIMO array and cross-MIMO array have been designed based on the approximation of the phase center. In the far field, these MIMO arrays can be viewed to equivalent to ULA or UPA. However, for short-range applications, the equivalence will be destroyed and the imaging quality will be degraded. Since the design methods are simple and easy to implement in practice, the developed MIMO radar systems can be used, but some post-processing algorithms should be used to improve the imaging performance. Also, when an ultra-wide bandwidth is considered [19–20], which could be adopted for high-range resolution imaging with the considerations

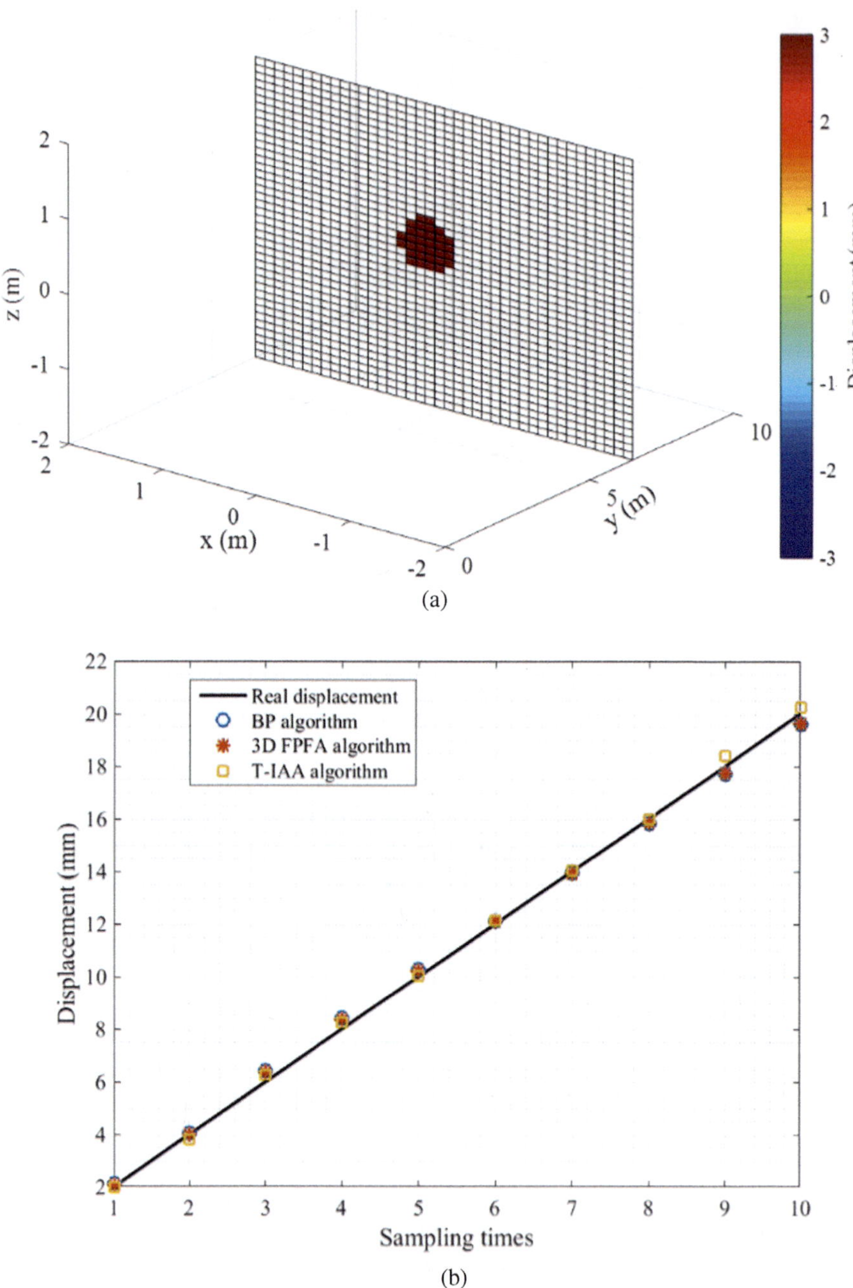

FIGURE 6.30 Displacement estimation by the cross-MIMO radar system of a CR.

of different scattering parameters of the targets and different wave propagation properties in the atmosphere, the ULA/UPA and their equivalent MIMO arrays will provide different angular PSFs from those shown previously. Therefore, different array design methods should be used to benefit from the advantages brought by

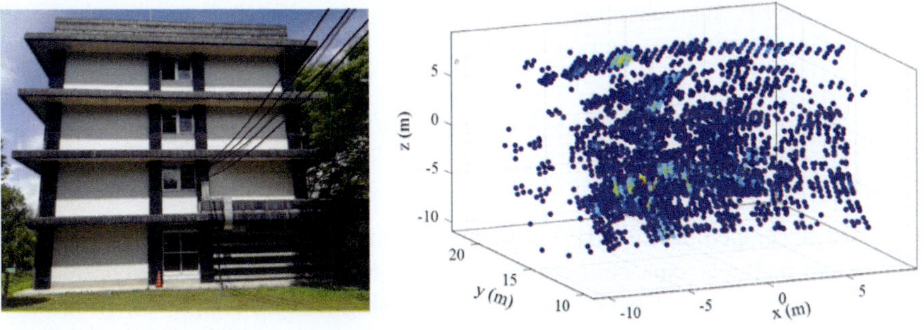

FIGURE 6.31   Imaging of a four-layer building by the cross-MIMO radar system.

the ultra-wide bandwidth. For the linear MIMO radar system, the ULA acts as the virtual array, and then the PF method is used to design the MIMO array. Some other types of virtual arrays can also be considered to reduce the minimum spacing requirement of its equivalent MIMO arrays, such as the MIMO radar systems designed in [13] and [21]. For the 2D MIMO radar system, the cross-MIMO array that has the lowest efficiency has been adopted, which can be modified if more precise and compact hardware is available. However, it should be pointed out that the advantage of the cross-MIMO array is its simple signal model, and thus, with some approximations, faster and advanced imaging methods can be easily applied. The problem of the cross-MIMO radar system is its ambiguous angles; thus, some smaller-size antennas should be fabricated to replace the used Vivaldi antennas and spiral antennas.

The developed radar systems, no matter if they are VNA based or integrated transceiver based, have some system errors, such as time delay error, antenna position error, and phase/gain error, caused by antennas, switches, and coaxial cables. These errors will degrade the imaging quality, especially the antenna position error, with which the grating lobes of the transmitting array cannot be effectively canceled by the deep nulls of the receiving array. Another problem is the long data acquisition period of VNA-based radar systems. Although it can provide a higher data sampling rate than most GB-SAR systems, the potential of MIMO radar has not been fully exploited. For example, with the TDM approach, an integrated system with multiple channels should be used to receive the echoed signal at the same time. Also, the code division multiplexing (CDM) method should be considered to transmit and record all the channels simultaneously. As GB-SAR can use only four antennas to realize full-polarimetric high-resolution imaging, another promising development direction may be a full-polarimetric MIMO radar system if dual-polarized antennas are available. Finally, similar to GB-SAR, MIMO radar imaging based on conventional matched filtering method suffers from high-level sidelobes in the range and azimuth directions. The spatial resolution is also limited by the frequency bandwidth and synthetic aperture size. The computing complexities of the conventional methods are high, which may prevent them from practical applications, especially for the 3D high-resolution imaging. Adaptive amplitude windowing methods, such as the spatially variant apodization–based method, and compressive sensing–based method, can be applied to overcome these limitations. For the CS-based method, the observation model of the MIMO radar should be considered for specific applications, which will be introduced in the next section.

6.2 BLOCK SPARSITY-BASED IMAGING METHOD

6.2.1 INTRODUCTION

For the developed linear MIMO radar systems, similar to GB-SAR imaging, CS-based methods can be used for resolution-improved and sidelobe-reduced imaging. For CS-based radar imaging, the sensing matrix, that is, the product of the dictionary where the received signal is sparse or compressible and the measurement matrix which is used to sample (or under-sample) the reflections of targets, is extremely important. Commonly, the dictionary is assumed to be a unity matrix, and the measurement matrix is constructed based on the mathematical model of the radar observation process. Inaccuracies of the observation model may cause the defocusing problem of the reconstructed images. Several methods have been proposed to mitigate the influences of these inaccuracies, such as the methods in [22] proposed to solve the off-grid problem and the methods in [23–24] proposed to compensate the system phase/gain errors.

In the research on CS-based MIMO radar imaging, it is found that the observation model applied in existing methods is not suitable for imaging targets at the near range or for the cases with system errors. Specifically, the assumptions of far range, limited imaging region, and no channel mismatch are usually used in the existing CS-based imaging methods; range attenuation and antenna phase/gain effects are ignored; and thus the complex amplitudes of targets are assumed to be azimuth independent [25]. However, in the case of near-range targets or system errors, these effects cannot be ignored because of the difference of target complex amplitude for different spatial sampling points. Otherwise, the imaging quality will be degraded. Although some compensation methods can be employed, the azimuth dependency of targets cannot be accurately achieved in advance. In this section, a novel sensing matrix is proposed for CS-based linear MIMO radar imaging, which takes the azimuth dependency of target complex amplitudes into consideration. When projected into the proposed sensing matrix, the received signal vector has the block sparse property. Based on this property and the advanced block CS algorithm, a 2D image of targets can be obtained at each spatial sampling point. The final 2D radar image is then obtained by the cross-correlation and summation process, which can further suppress the strong artifacts and improve the imaging quality. Experimental results are presented to show the effectiveness of the proposed imaging method and its advantages over conventional methods.

6.2.2 TARGET AZIMUTH DEPENDENCY

First, to facilitate the following formulations, the signal model in (6.23) is rewritten as

$$s(p,q) = \sum_{i=1}^{I}\sum_{j=1}^{J} \sigma(x_i,y_j)\exp(-j4\pi f_q R_p(x_i,y_j)/c) + n(p,q) \qquad (6.30)$$

where $R_p(x_i, y_j) = R_{m,n}(x_i, y_j)/2$ is the average range from the target to the p-th pair of transmitter and receiver (i.e., the p-th spatial sampling point), $p = 1, 2, \ldots, P$, and $P = MN$. Then, the received signal vector can be expressed as

$$s = \Theta\sigma + n \tag{6.31}$$

where

$$\sigma = [\sigma(x_1, y_1), \ldots, \sigma(x_i, y_j), \ldots, \sigma(x_I, y_J)] \in C^{IJ \times 1} \tag{6.32}$$

is the vectorized complex amplitudes of all targets,

$$s = [s(1,1), \ldots, s(1,Q), \ldots, s(P,1), \ldots, s(P,Q)] \in C^{PQ \times 1} \tag{6.33}$$

is the vectorized received signal at all frequencies and spatial sampling points, and

$$\Theta = \begin{bmatrix} e^{-j4\pi f_1 R_1(x_1,y_1)/c} & e^{-j4\pi f_1 R_1(x_2,y_1)/c} & \cdots & e^{-j4\pi f_1 R_1(x_I,y_J)/c} \\ e^{-j4\pi f_2 R_1(x_1,y_1)/c} & e^{-j4\pi f_2 R_1(x_2,y_1)/c} & \cdots & e^{-j4\pi f_2 R_1(x_I,y_J)/c} \\ \vdots & \vdots & \ddots & \vdots \\ e^{-j4\pi f_Q R_P(x_1,y_1)/c} & e^{-j4\pi f_Q R_P(x_2,y_1)/c} & \cdots & e^{-j4\pi f_Q R_P(x_I,y_J)/c} \end{bmatrix} \tag{6.34}$$

is the $PQ \times IJ$ sensing matrix designed according to the radar observation model.

Based on (6.31), the BP-based imaging method can be expressed as

$$\sigma_{FDBP} = \Theta^H s \tag{6.35}$$

and the CS-based imaging method can be expressed as

$$\sigma_{CS} \leftarrow \min\|\sigma\|_0 \quad \text{s.t. } \|s - \Theta\sigma\|_2 < \varepsilon \tag{6.36}$$

where $\|\cdot\|_0$ denotes the number of nonzero elements of a vector, $\|\cdot\|_2$ is the Euclidean norm of a vector, and ε is the noise level.

It can be seen from (6.30) that the complex amplitude of target is assumed to be independent of the spatial sampling points. This is to say, at each spatial sampling point, the complex amplitude of a specific target is supposed to be a constant. This assumption can be satisfied if the targets are at far range, the reconstruction area is small, and all transceiving channels are matched, where the effects of range attenuation and antenna phase/gain error can be ignored.

When this assumption is not satisfied, that is the complex amplitude of the target depends on the spatial sampling points, which is commonly the case of near-range targets or system phase/gain errors, (6.30) should be modified to

$$s(p,q) = \sum_{i=1}^{I} \sum_{j=1}^{J} \sigma(p, x_i, y_j) \exp(-j4\pi f_q R_p(x_i, y_j)/c) + n(p,q) \tag{6.37}$$

This is called target azimuth dependency in the following, which is shown in Figure 6.32. In this sketch, the noise is ignored and we simply set $P = 3$, $Q = 2$, and $IJ = 7$. The empty boxes correspond to zeros, and the shape-filled boxes denote nonzero values, while the value is not determined by the shape. In such a case, some compensation methods should be applied to the BP imaging method and the CS-based imaging method. Otherwise, the imaging performance will be degraded. As to the range attenuation effect, we can add the range-dependent path loss factor into the sensing matrix Θ for the CS-based imaging method, or we can introduce the correction after we obtain the 2D radar image by the BP method. As to the antenna phase/gain effect, the radiation pattern and phases of the used antennas can be measured or calculated and then added into the sensing matrix Θ. However, these methods are too complicated to implement. At each imaging point, the range attenuation should be calculated, which is also dependent on many practical factors, for all spatial sampling points. Also, the phase/gain of all transceivers should be measured for each imaging point, and the phases of the antennas may change over time. Moreover, the complex amplitude of the target is also related to the target itself and the incident angle of radar wave, especially at the near range, which cannot be measured or calculated in advance. Therefore, more advanced methods without pre-measurements or pre-calculations are required.

To solve these problems, a sensing matrix for CS-based imaging is proposed. First, the vectorized received signal for all frequencies at the p-th spatial sampling point is expressed as

$$\boldsymbol{s}_p = [s(p,1), s(p,2), ..., s(p,Q)]^{\mathrm{T}} \in C^{Q \times 1} \tag{6.38}$$

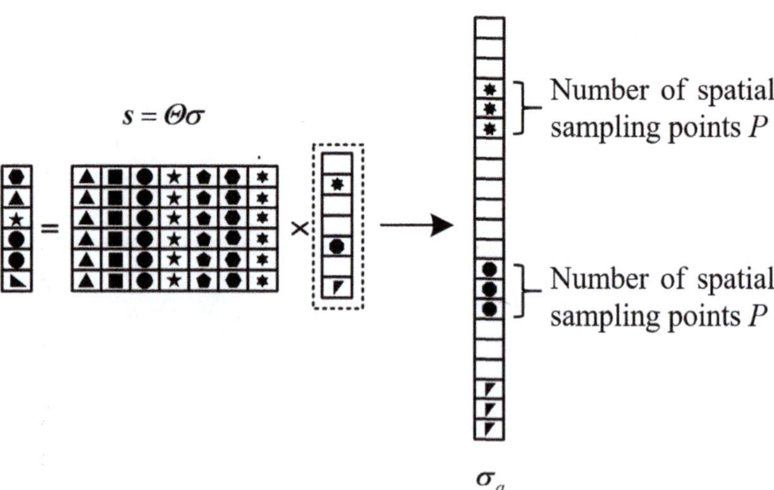

FIGURE 6.32 Sketch showing the azimuth dependency of target complex amplitude.

where $[\cdot]^{\mathrm{T}}$ denotes the transpose operator. Based on (6.38), the received signal for all frequencies and spatial sampling points can be rewritten as

$$s = [s_1^{\mathrm{T}}, s_2^{\mathrm{T}}, ..., s_P^{\mathrm{T}}]^{\mathrm{T}} = \sum_{i=1}^{I} \sum_{j=1}^{J} \boldsymbol{\Theta}(x_i, y_j) \boldsymbol{\sigma}(x_i, y_j) + n \tag{6.39}$$

where

$$\boldsymbol{\sigma}(x_i, y_j) = [\sigma(1, x_i, y_j), \sigma(2, x_i, y_j), ..., \sigma(P, x_i, y_j)]^{\mathrm{T}} \in C^{P \times 1} \tag{6.40}$$

is the complex amplitude of the target at (x_i, y_j) with respect to all spatial sampling points, and $\boldsymbol{\Theta}(x_i, y_j)$ is a $PQ \times P$ matrix given by

$$\boldsymbol{\Theta}(x_i, y_j) = diag[\boldsymbol{\Theta}(1, x_i, y_j), \boldsymbol{\Theta}(2, x_i, y_j), ..., \boldsymbol{\Theta}(P, x_i, y_j)] \in C^{PQ \times P} \tag{6.41}$$

where $diag[\cdot]$ denotes the diagonalization operator and $\boldsymbol{\Theta}(p, x_i, y_j)$ with $p = 1, 2, \ldots,$ P is a $Q \times 1$ vector defined as

$$\boldsymbol{\Theta}(p, x_i, y_j) = [e^{-j4\pi f_1 R_p(x_i, y_j)/c}, e^{-j4\pi f_2 R_p(x_i, y_j)/c}, ..., e^{-j4\pi f_Q R_p(x_i, y_j)/c}]^{\mathrm{T}} \tag{6.42}$$

Then, written in matrix form, the received signal can be expressed as

$$s = \boldsymbol{\Theta}_a \boldsymbol{\sigma}_a + n \tag{6.43}$$

where

$$\boldsymbol{\Theta}_a = [\boldsymbol{\Theta}(x_1, y_1), \boldsymbol{\Theta}(x_2, y_1), ..., \boldsymbol{\Theta}(x_I, y_J)] \in C^{PQ \times PIJ} \tag{6.44}$$

is the proposed sensing matrix and

$$\boldsymbol{\sigma}_a = [\sigma^{\mathrm{T}}(x_1, y_1), \sigma^{\mathrm{T}}(x_2, y_1), ..., \sigma^{\mathrm{T}}(x_I, y_J)]^{\mathrm{T}} \in C^{PIJ \times 1} \tag{6.45}$$

is the vectorized complex amplitudes of the targets corresponding to all spatial sampling points.

Compared to (6.31), it can be seen that (6.43) takes the target azimuth dependency into account. In such a model, the target complex amplitudes of different spatial sampling points are separated. $\boldsymbol{\sigma}_a$ contains the factors that may influence the target amplitude, including range attenuation factor, antenna phase/gain difference, and incident angle–related target scattering property. The relationship between the proposed model in (6.43) and the conventional model in (6.31) is shown in Figure 6.33. The goal is to solve (6.43), that is, to estimate the amplitudes of targets with respect to different spatial sampling points. Similar to the conventional FDBP method, the complex conjugate process can be used to get

$$\boldsymbol{\sigma}_a = \boldsymbol{\Theta}_a^{\mathrm{H}} s \tag{6.46}$$

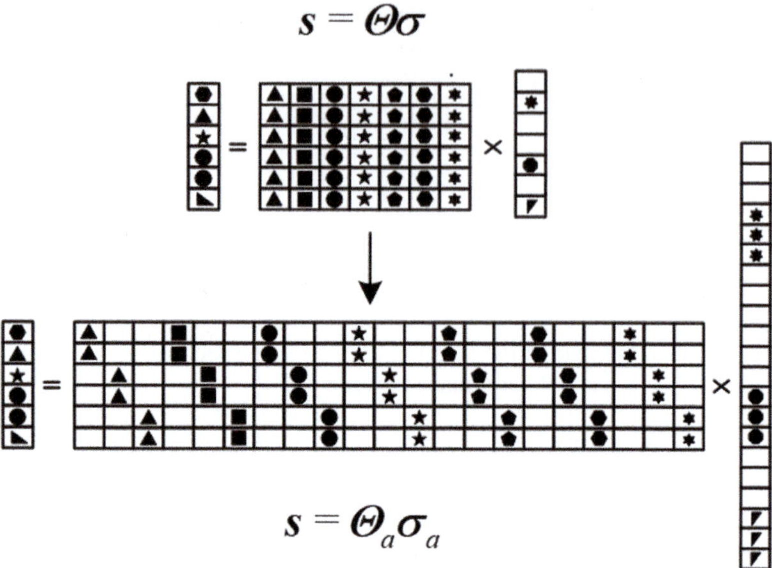

FIGURE 6.33   Sketch showing the relationship between (6.31) and (6.43).

Then, if the summation of $\sigma(x_i, y_j)$ is taken, we have

$$\sigma_a^\lozenge = \left[\sum \sigma(x_1, y_1), \sum \sigma(x_2, y_1), ..., \sum \sigma(x_I, y_J) \right] \in C^{IJ \times 1} \tag{6.47}$$

It can be derived that $\sigma_a^\lozenge$ is equal to $\sigma = \Theta^H s$, which is the result of (6.31) obtained by the FDBP method. The physical meaning of this equivalence lies on the equivalence of the FDBP method and the TDBP method. If there are only a few significant targets in the imaging scene, the CS-based method can still be applied to solve (6.43). In this case, if there are K nonzero elements in σ, the number of nonzero elements in σ_a will be PK. Since the dimension of Θ_a is $PQ \times PIJ$, the sparsity of σ_a is the same as σ. Therefore, σ_a can be estimated by solving

$$\sigma_a \leftarrow \min \|\sigma_a\|_0 \quad \text{s.t.} \quad \|s - \Theta_a \sigma_a\|_2 < \varepsilon \tag{6.48}$$

However, it should be noted that, because of the large size of Θ_a, the computational complexity of solving (6.48) is much larger than solving (6.36). Moreover, the reconstruction accuracy should be improved in the noisy conditions. Therefore, advanced sparse recovery algorithms and additional processing methods are required to estimate σ_a more accurately and effectively.

6.2.3   BLOCK SPARSITY AND UNDER-SAMPLING

It can be seen from (6.45) and Figure 6.33 that, in comparison with σ, σ_a exhibits additional sparse structure in the form of nonzero elements occurring in clusters.

This type of sparsity is referred to as group sparsity or block sparsity, and the corresponding CS algorithm is called block CS in the literature. Exploiting block sparsity will help to achieve better reconstruction results than just treating the signal as being arbitrarily sparse. For block sparsity [26–27], the mixed l_2/l_0 norm is normally utilized to count the number of nonzero blocks, defined as

$$\left\| \sigma_1 \right\|_{2,0} = \sum_{i=1}^{I} \sum_{j=1}^{J} I\left(\left\| \sigma(x_i, y_j) \right\|_2 \right) \tag{6.49}$$

where $I(\mathrm{g})$ denotes the indicator function, which is zero when its argument is zero and is one otherwise. Thus, (6.43) can be solved by the following minimization problem, in which the mixed l_2/l_0 norm is used to restrict the set of possible solutions.

$$\sigma_a \leftarrow \min \left\| \sigma_a \right\|_{2,0} \quad \text{s.t.} \quad \left\| s - \Theta_a \sigma_a \right\|_2 < \varepsilon \tag{6.50}$$

To solve (6.50), the block OMP algorithm [5] can be used. Since the block sparsity of σ_a is exploited by the mixed l_2/l_0 norm, the estimation accuracy of (6.50) is improved compared with (6.48). However, although the memory usages to save Θ and Θ_a are equivalent, the computing complexity of solving (6.50) using BOMP is much higher than solving (6.36) using OMP. In the OMP/BOMP algorithm, the most time-consuming part is the index selection step (i.e., selecting the atom/block that is best matched to the residual of the previous iteration from the sensing matrix), compared with which the complexities of other steps are negligible. In the conventional CS-based method using OMP, the size of Θ is $PQ \times IJ$, giving the total complexity of index selection as $O(PQIJT_{omp})$, where T_{omp} is the iteration number of OMP. For solving (6.50) via BOMP, due to the l_2 norm, the computational complexity of the index selection is increased to $O[(P+1)QIJT_{bomp}]$, where T_{bomp} is the iteration number of BOMP. Therefore, a more efficient method is required to estimate σ_a from (6.43). Instead of the mixed l_2/l_0 norm, in order to exploit the specific block sparsity of σ_a in Θ_a, similar to the TDBP algorithm, a mixed norm function is defined as

$$\left\| \sigma_a \right\|_{\mathrm{sum},0} = \sum_{i=1}^{I} \sum_{j=1}^{J} I\left(\left| \sum \sigma(x_i, y_j) \right| \right) \tag{6.51}$$

In such a case, rather than calculating the l_2 norm of $\sigma(x_i, y_j)$ in (6.50), the elements of $\sigma(x_i, y_j)$, that is, the complex amplitudes of target at (x_i, y_j) for different spatial sampling points, are coherently summed to help to count the number of nonzero blocks of σ_a. Then σ_a can be estimated from the following minimization problem

$$\sigma_a \leftarrow \min \left\| \sigma_a \right\|_{\mathrm{sum},0} \quad \text{s.t.} \quad \left\| s - \Theta_a \sigma_a \right\|_2 < \varepsilon \tag{6.52}$$

To effectively solve (6.52), a modified BOMP algorithm that can use the same index selection procedure as OMP is proposed, which is summarized in Table 6.1. Given the same iteration number, the computing complexity of solving (6.52) via the modified BOMP algorithm is slightly higher than solving (6.36) via OMP algorithm.

TABLE 6.1

Modified BOMP Algorithm

Input: s, Θ, Θ_a, ε, and the maximum iteration number T.

Procedure:

1) Initialization: $r_0 = s$, $L = \varnothing$, $A_0 = \varnothing$, $t = 1$;

2) Index selection: $\lambda_t \leftarrow \arg \max\limits_{o=1, 2,...,O} \left| \left\langle r_{t-1}, \Theta^o \right\rangle \right|$;

3) Index set expansion: $L_t = L_{t-1} \cup \{\lambda_t\}$, $A_t = A_{t-1} \cup \Theta_a^{\lambda_t}$;

4) Estimation: $\sigma_t = \left(A_t^{\mathrm{H}} A_t \right)^{-1} A_t^{\mathrm{H}} s$;

5) Residual updating: $r_t = s - A_t \sigma_t$;

6) Iteration: $t \leftarrow t + 1$, if $t \le T$, $\left\| r_{t-1} \right\|_2 \ge \varepsilon$, go back to 2), otherwise, stop.
Output: $\sigma_a \{L_T\} = \sigma_T$.

Also, Θ_a is not necessary to save in practice. In the proposed algorithm, the λ_t-th sub-matrix $\Theta_a^{\lambda_t}$ can be easily generated from Θ. Therefore, the memory usage of (6.52) is also similar to (6.36). Moreover, if a proper noise level ε is adopted, the iteration number of the proposed method (T_{mbomp}) will be smaller than that of the conventional CS-based method (T_{omp}), resulting in reduced running time and memory usage.

As mentioned previously, long data acquisition time is still a problem for the developed VNA-based linear MIMO radar system. Also, large P and Q make (6.36) and (6.52) difficult to solve. The memory usage needed to hold Θ/Θ_a is also large when a large observation area is imaged and fine resolution is preferred. One of the attractive characteristics of the CS-based imaging method is that it can still work well with under-sampled data. Therefore, to reduce the data acquisition time, the data size, and the computational complexity, random selections of frequencies and spatial sampling points are conducted. Q_u numbers, which indicate the indices of frequencies, are randomly selected from $[1, 2, \ldots, Q]$. Then, P_u numbers that indicate the indices of spatial sampling points are randomly selected from $[1, 2, \ldots, P]$. Based on these selections, the under-sampled received signal can be expressed as

$$s_u = \Theta_u \sigma_u + n_u \tag{6.53}$$

where s_u and n_u are the under-sampled signal vector and noise vector, $\Theta_u \in C^{P_u Q_u \times P_u IJ}$ is the sensing matrix with a reduced size, and $\sigma_u \in C^{P_u IJ \times 1}$ is the complex amplitudes of targets that correspond to the selected spatial sampling points. Θ_u and σ_u have the following forms

$$\Theta_u = [\underbrace{\phi_1, \cdots, \phi_{P_u}}_{\Theta_u[1,1]}, \underbrace{\phi_{P_u+1}, \cdots, \phi_{2P_u}}_{\Theta_u[2,1]}, \cdots, \underbrace{\phi_{(IJ-1)P_u+1}, \cdots, \phi_{P_u IJ}}_{\Theta_u[I,J]}] \tag{6.54}$$

$$\sigma_u = [\underbrace{\sigma_1, \cdots, \sigma_{P_u}}_{\sigma_u[1,1]}, \underbrace{\sigma_{P_u+1}, \cdots, \sigma_{2P_u}}_{\sigma_u[2,1]}, \cdots, \underbrace{\sigma_{(IJ-1)P_u+1}, \cdots, \sigma_{P_u IJ}}_{\sigma_u[I,J]}]^{\mathrm{T}} \tag{6.55}$$

In (6.54) and (6.55), Θ_u can be viewed as a concatenation of column-blocks $\Theta_u[i, j]$, and σ_u can be viewed as a concatenation of blocks with $\sigma_u[i, j]$ denoting the i-j-th block. Then, (6.53) can be solved by

$$\sigma_u \leftarrow \min \|\sigma_u\|_{\text{sum},0} \quad \text{s.t.} \quad \|s_u - \Theta_u \sigma_u\|_2 < \varepsilon_u \tag{6.56}$$

It should be noted that the number of entries in $\sigma_u[i, j]$ is changed to the number of selected spatial sampling points P_u.

6.2.4 Cross-Correlation–Based Fusion

After solving (6.56) by the modified BOMP algorithm, at each selected spatial sampling point, the complex amplitudes of targets can be obtained. For the z-th ($z = 1, 2, \ldots, P_u$) selected spatial sampling point, we have

$$\sigma_z = [\alpha_z, \alpha_{z+P_u} \cdots, \alpha_{z+(IJ-1)P_u}]^{\text{T}} \in \sigma_u \tag{6.57}$$

Therefore, by metricizing σ_z, a focused 2D image of targets can be obtained, which means there are P_u radar images in all that can be produced by the proposed imaging method. With $P_u = 3$, $Q_u = 2$, and $IJ = 7$, the proposed imaging method and its results are shown in Figure 6.34.

Conventionally, using only one spatial sampling point, it is impossible to get the 2D-focused radar image. However, the block sparsity property of the received signal in the proposed sensing matrix together with the block CS algorithm makes it possible. The proposed mixed norm in (6.51) restricts the possible solutions of the under-determined problem (6.53), resulting in different 2D radar images for different spatial sampling points. Complex amplitudes of targets influenced by different range attenuation, antenna phases/gains, and different incident angles of radar wave are characterized by these images. Although these images can be interpreted separately, to facilitate post-processing, a final fused 2D image is required.

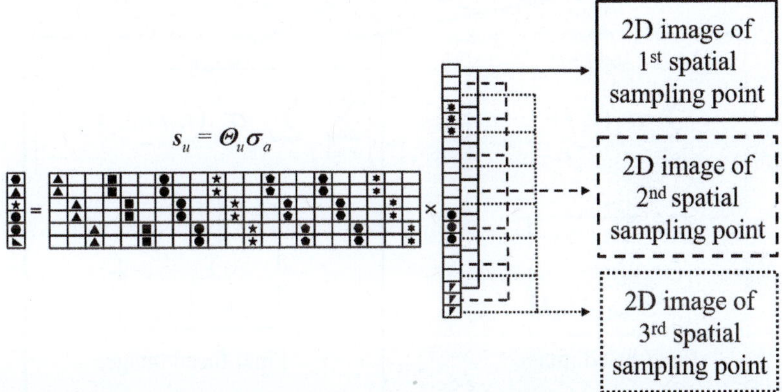

FIGURE 6.34 All P_u radar images of the targets can be obtained by the proposed method.

Similar to the process in (6.47), a simple method is taking summation of these images, which gives $\sigma_{sum} = \sum_{z=1}^{P_u} \sigma_z$. However, if more advanced methods can be employed, it can be expected that the artifacts caused by the estimation errors in solving (6.53) can be further suppressed. For the TDBP method, a cross-correlation–based method is proposed in [28] to suppress azimuth artifacts, which is similar to the delay-multiply-sum method used in medical imaging [6]. Different from the formulation of TDBP in (6.47), the cross-correlation–based method first calculates the cross-correlation of the time domain values of all spatial sampling points, and then a summation is made to get the final result. Inspired by this process, we employ the cross-correlation method to fuse all the radar images of the selected sampled sampling points. The simple summation fusion method and the cross-correlation–based fusion method are shown in Figure 6.35.

Compared with the simple summation method, the artifacts will be further suppressed by the cross-correlation process. Moreover, the cross-correlation and summation operations are easier to implement for our case than the cross-correlation–based TDBP method, as only P_u selected sampling points are used here. Mathematically, the final fused 2D image of targets obtained by the cross-correlation–based fusion method is given by

$$\sigma_{cc} = \begin{bmatrix} \sum_{z_1=1}^{P_u-1} \sum_{z_2=z_1+1}^{P_u} \sigma_{z_1}(1,1)\sigma_{z_2}(1,1) & \cdots & \sum_{z_1=1}^{P_u-1} \sum_{z_2=z_1+1}^{P_u} \sigma_{z_1}(1,J)\sigma_{z_2}(1,J) \\ \vdots & \ddots & \vdots \\ \sum_{z_1=1}^{P_u-1} \sum_{z_2=z_1+1}^{P_u} \sigma_{z_1}(I,1)\sigma_{z_2}(I,1) & \cdots & \sum_{z_1=1}^{P_u-1} \sum_{z_2=z_1+1}^{P_u} \sigma_{z_1}(I,J)\sigma_{z_2}(I,J) \end{bmatrix} \tag{6.58}$$

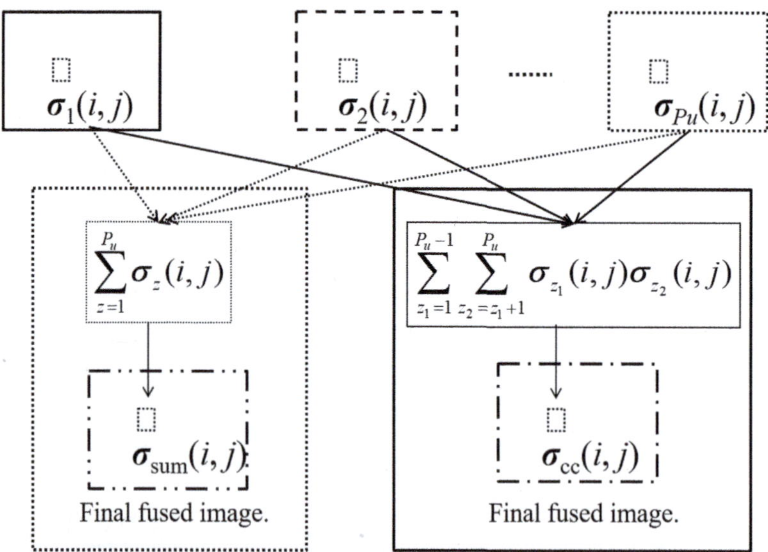

FIGURE 6.35 Summation-based and cross-correlation–based fusion methods to get the final image.

6.2.5 Experiments and Results

Some experiment results are presented to show the effectiveness and advantages of the proposed CS-based imaging method. Experiments are conducted with the designed two linear MIMO radar systems. It should be pointed out that, in order to improve the sparsity of the observation scene, the background subtraction is carried out to reduce the influence of the surroundings and antenna mutual coupling. For the first linear MIMO radar system, which has a limited detection range, three targets with different shapes, a metallic sphere, a landmine model, and a reversed-placed trihedral corner reflector, as shown in Figure 6.36, where the radar system is also

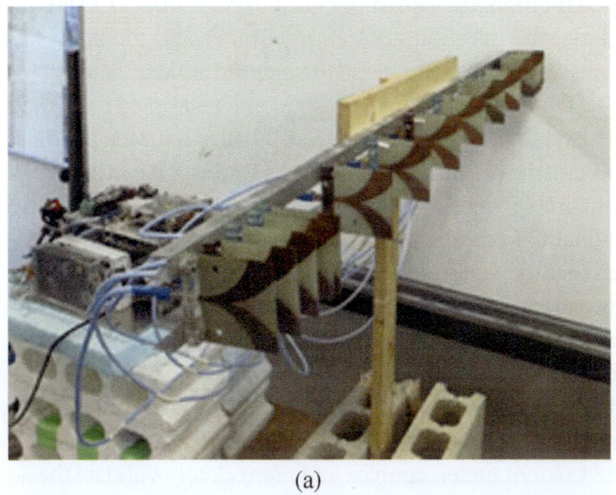

(a)

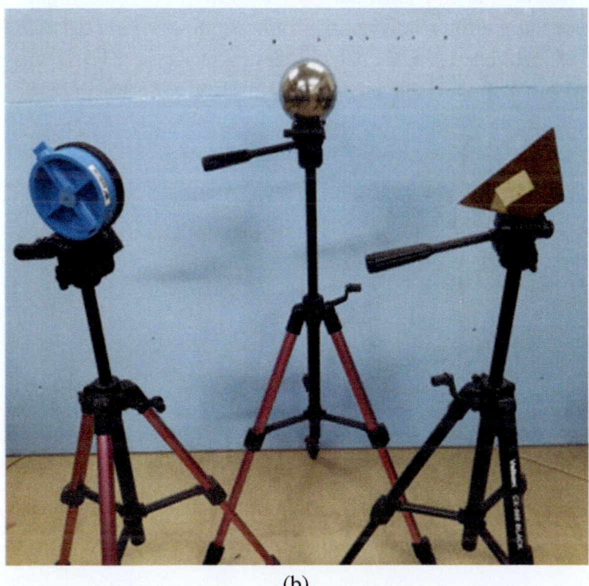

(b)

FIGURE 6.36 (a) The developed linear MIMO radar system and (b) three different-shaped targets used in the experiment.

shown, are used for imaging. These three targets have different reflection properties with the radar incident angles, which is suitable to demonstrate the azimuth dependence of target complex amplitude.

The imaging scene is 2×2 m and discretized into 41×41 grids, and the reconstructed image by the classical time domain BP method with full data is shown in Figure 6.37(a). It can be seen that there are many high-level artifacts in the focused image, which has negative influences on the following process. In order to assess the imaging performances of different CS-based methods with under-sampled data, 32 frequencies are randomly selected from 1024 frequencies. This means that only 1/32 data will be used. The reconstructed images by the conventional CS-based method and the proposed method are shown in Figure 6.37(b–d). For OMP and the modified BOMP algorithms, the maximum iteration number and the noise level are set to be 20 and $0.5\|s_u\|_2$, respectively, which is sufficient to image the targets in the scene.

As we can see from Figure 6.37, even with under-sampled data, the imaging performances of CS-based imaging methods are still better than the BP method. Fewer artifacts are produced in the reconstructed images. However, it can also be seen from Figure 6.37(b) that comparable values around the actual positions of targets will be produced by the conventional CS-based method. That is to say, the signal model of the conventional CS-based method is not perfectly suitable in this case, that is, for the near range targets. On the contrary, since the developed observation model in the proposed method is more suitable for the practical received signal in such a case, the imaging results of the proposed method have fewer strong artifacts than the conventional CS-based method, especially in the imaging result obtained by the proposed cross-correlation–based fusion method. In order to verify the robustness of the proposed imaging method against different random under-samples and more clearly validate the advantages of the proposed method over the conventional CS-based method, the estimation residuals, the ratio of target power, and the running time for both methods are calculated as a function of the number of frequencies. For each frequency number, 50 runs are performed and averaged. To simplify, the number of spatial sampling points is unchanged.

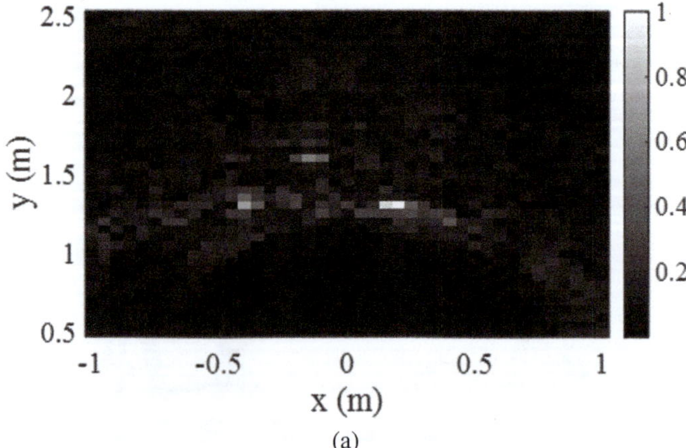

(a)

FIGURE 6.37 Imaging results obtained by: (a) the BP method with full data, (b) the conventional CS-based method with 1/32 data, (c) the proposed method with 1/32 data and summation based fusion, and (d) the proposed method with 1/32 data and cross-correlation–based fusion.

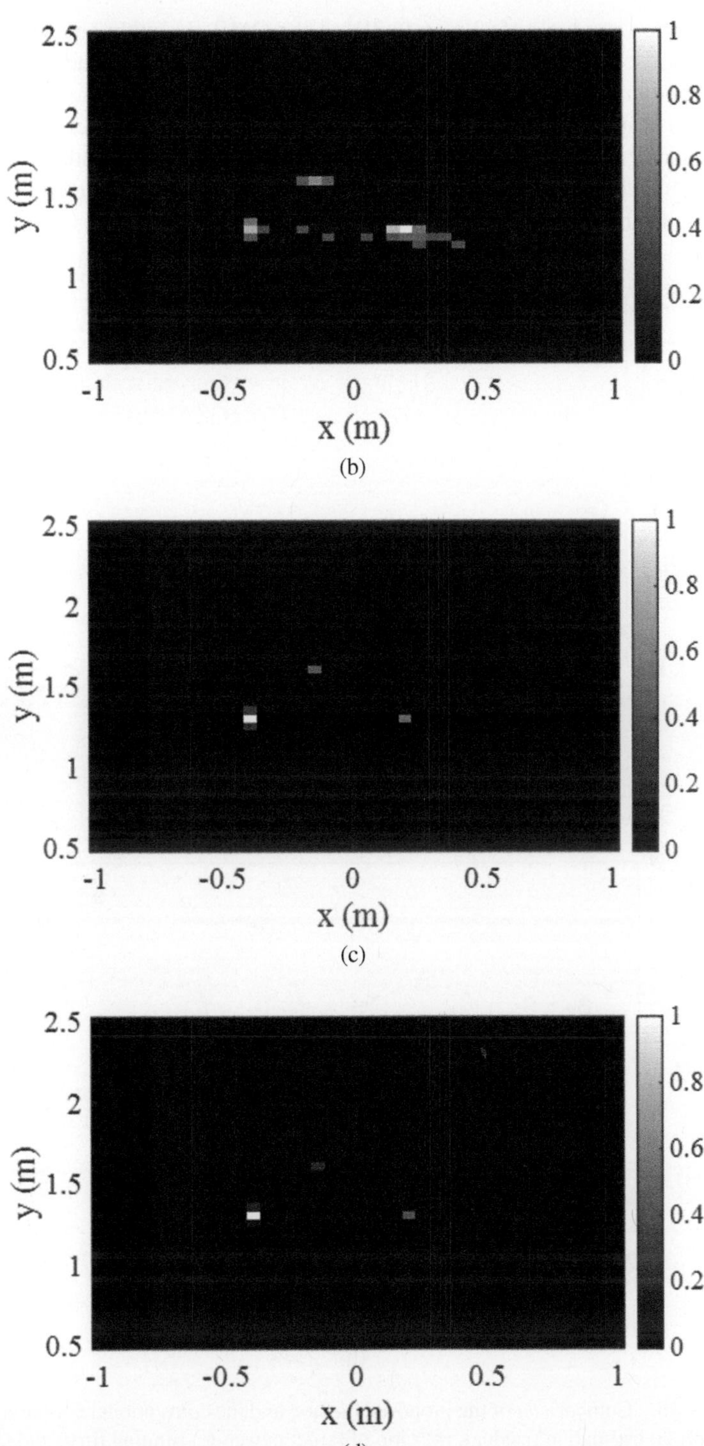

(b)

(c)

(d)

FIGURE 6.37 (Continued)

The estimation residuals of solving (6.31) via the OMP algorithm and solving (6.43) via the modified BOMP algorithm are shown in Figure 6.38(a). The assumption is that the smaller the reconstruction residual, the more suitable the processing model. It is shown that, for all selections of the under-sampled frequencies, the proposed method enjoys a much smaller estimation residual than the conventional CS-based method, which indicates the superiority of the proposed method. Figure 6.38(b) gives the ratios of target powers, which is defined as the power of targets divided by the

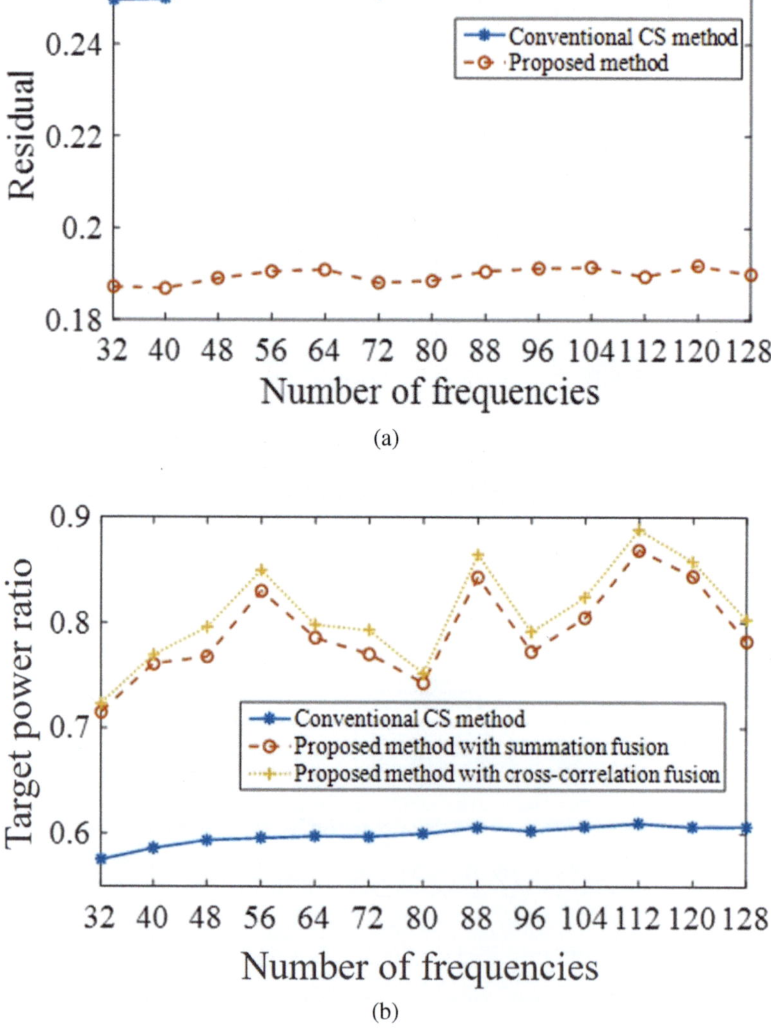

FIGURE 6.38 Comparison of the proposed method and the conventional CS-based method in terms of: (a) estimation residual, (b) ratio of target power, (c) running time, and (d) reconstructed amplitude of the metallic sphere.

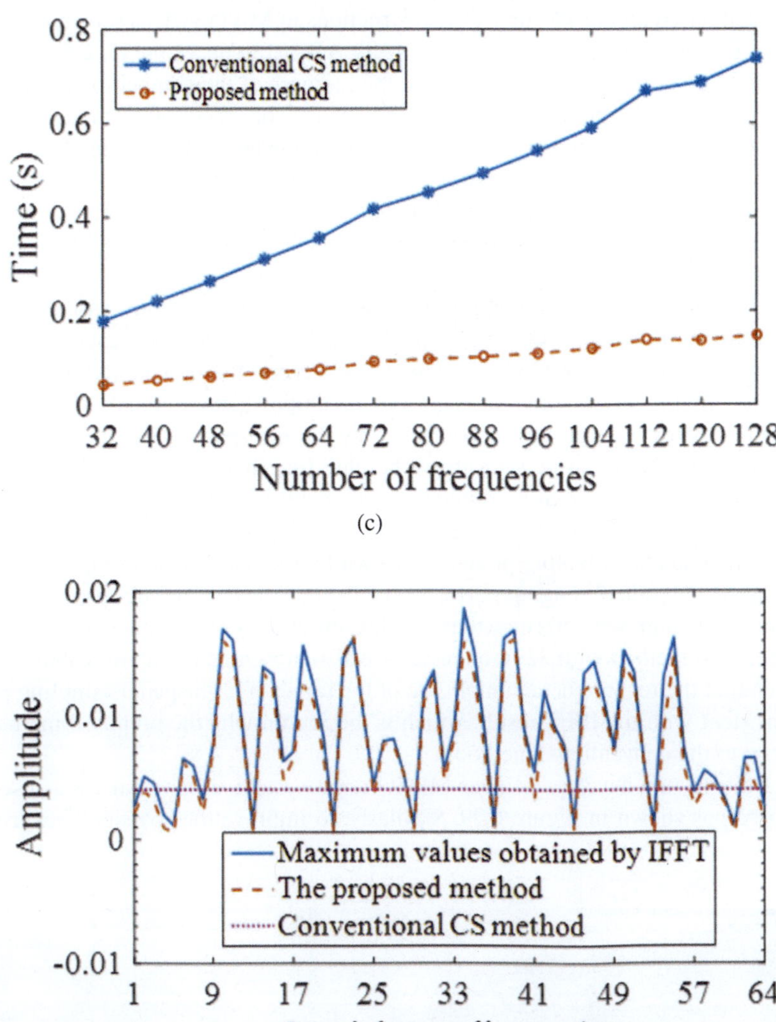

FIGURE 6.38   (Continued)

power of all the recovered coefficients of the proposed method and the conventional CS-based method. The ratio of target powers reveals the reconstruction accuracy of the positions of targets and their corresponding complex amplitudes. It can be seen that, even with simple summation fusion, the proposed method outperforms the conventional CS-based method in terms of high accuracy, which proves that the proposed imaging model is more suitable for the case of near-range targets and system phase/gain error. Moreover, based on the cross-correlation fusion method, the reconstruction accuracy can be further increased. To measure the computational complexity, the running time of each method against the number of selected frequencies is calculated based on the TIC and TOC instructions in MATLAB, as shown in Figure 6.38(c). The processing is conducted in MATLAB 2015b on a Core i5, 2.5 GHz, and 8-GB RAM PC. It can be seen that, with the increase of the number of frequencies, both methods have increased running time. Since the observation model is more suitable, the iteration number of the proposed imaging method is smaller than that of the conventional CS-based method. Although the complexity of the modified BOMP algorithm in a single iteration is higher than that of the OMP algorithm, the total computational cost of the proposed method is much smaller than that of the conventional CS-based method.

In the proposed method, the complex amplitudes of targets are assumed to be dependent on the spatial sampling points. In order to verify this assumption in practice, for the metallic sphere with a diameter of 7.5 cm, the maximum amplitude of the time-gated received signal at each spatial sampling point is picked and shown in Figure 6.38(d). This process is conducted by the zero-padded inverse fast Fourier transform (IFFT). It can be seen from the blue line that, at different spatial sampling points, the maximum amplitude of the signal is different. The conventional CS-based method assumes that the reflection coefficient of the sphere is unchanged for different spatial sampling points, as shown by the purple line in Figure 6.38(d). This causes degraded imaging performance of the conventional CS-based method. On the other hand, with 50 runs, the complex amplitudes of the sphere estimated by the proposed method with 32 frequencies are also presented in Figure 6.38(d). It can be seen that the reconstructed amplitude of the target at each spatial sampling point is consistent with the IFFT result, which is the reason why the proposed method is superior to the conventional method.

For the second developed linear MIMO radar system, a trihedral CR is used as the target, as shown in Figure 6.39. Similarly, to improve the sparsity, background

FIGURE 6.39 Measurement of a trihedral CR by the second designed linear MIMO radar system.

subtraction was conducted. Similar results to the first linear MIMO radar system can be obtained, as shown in Figure 6.40, where only 51 frequencies are randomly selected, the grid number is 41 × 41, and the iteration number of the OMP algorithm and the modified BOMP algorithm is 5. The advantage of the proposed method with only 1/32 data can be easily observed considering the strong artifacts in the image obtained by the conventional CS-based method.

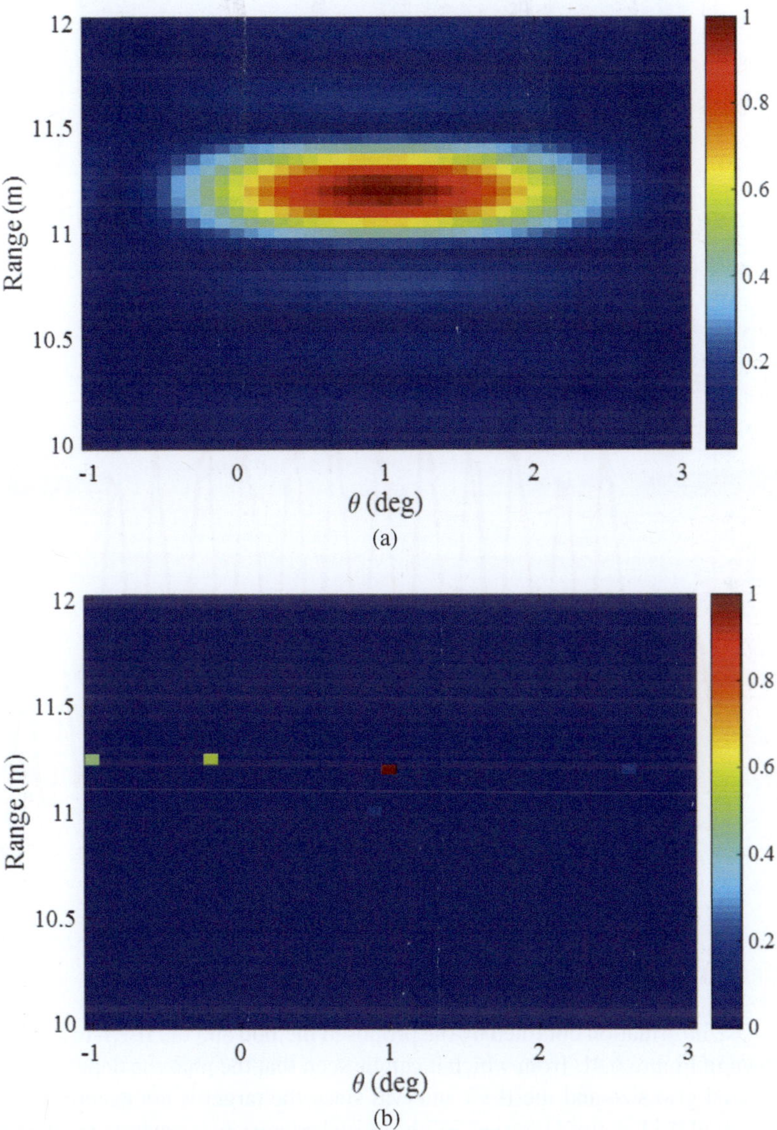

FIGURE 6.40   Imaging results of the trihedral CR obtained by: (a) the BP method with full data, (b) the conventional CS-based method with 1/32 data, (c) the proposed method with 1/32 data and the summation-based fusion, and (d) amplitude comparison between the proposed method and the IFFT-based method.

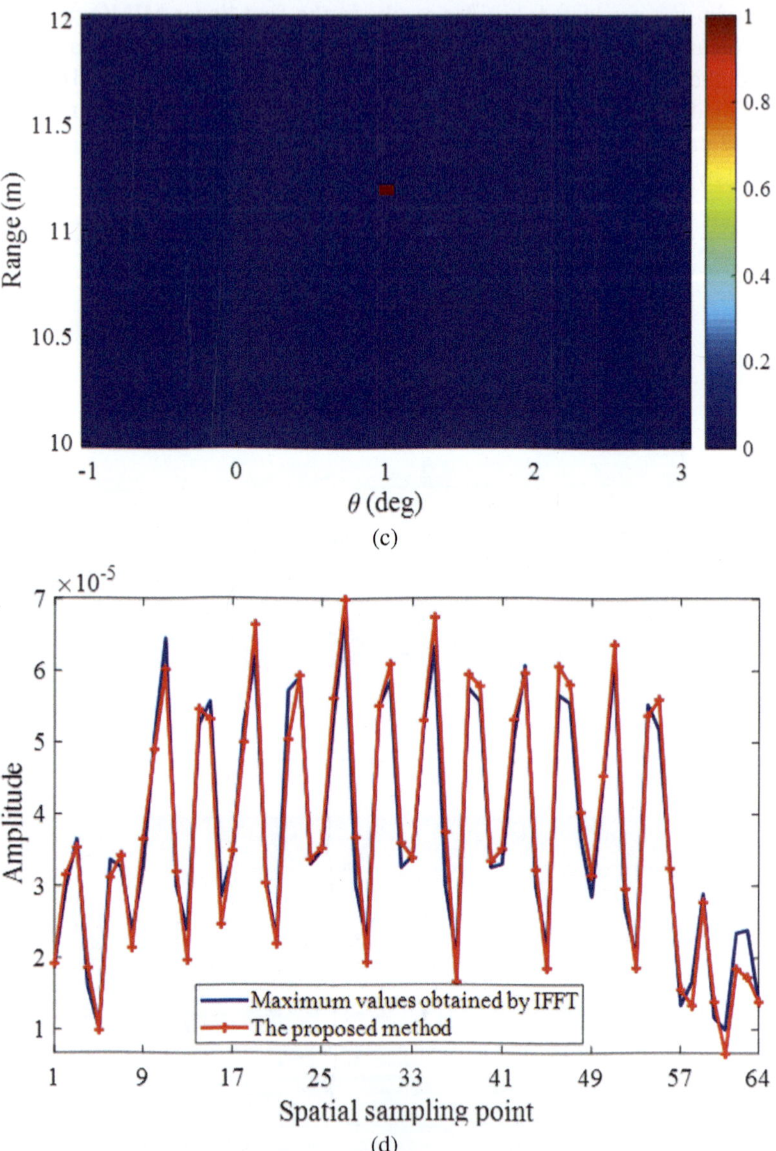

FIGURE 6.40 (Continued)

The phase information obtained by the proposed method and the IFFT-based method is shown in Figure 6.41, from which it can be seen that the phase is dependent on the discretized grid size and the IFFT interval since the target is not exactly located at the assumed grid or time interval. With the consideration of antenna position error, channel phase/gain difference, and time delay error, the received signal at the p-th spatial sampling point can be approximated by

$$s(p,q) = \sigma(x_0, y_0)[1 + G_p(f_q)]e^{-j\varphi_p(f_q)}e^{-j4\pi f_q \Delta R_p(x_0,y_0)/c}e^{-j4\pi f_q R_p(x_0,y_0)/c} \qquad (6.59)$$

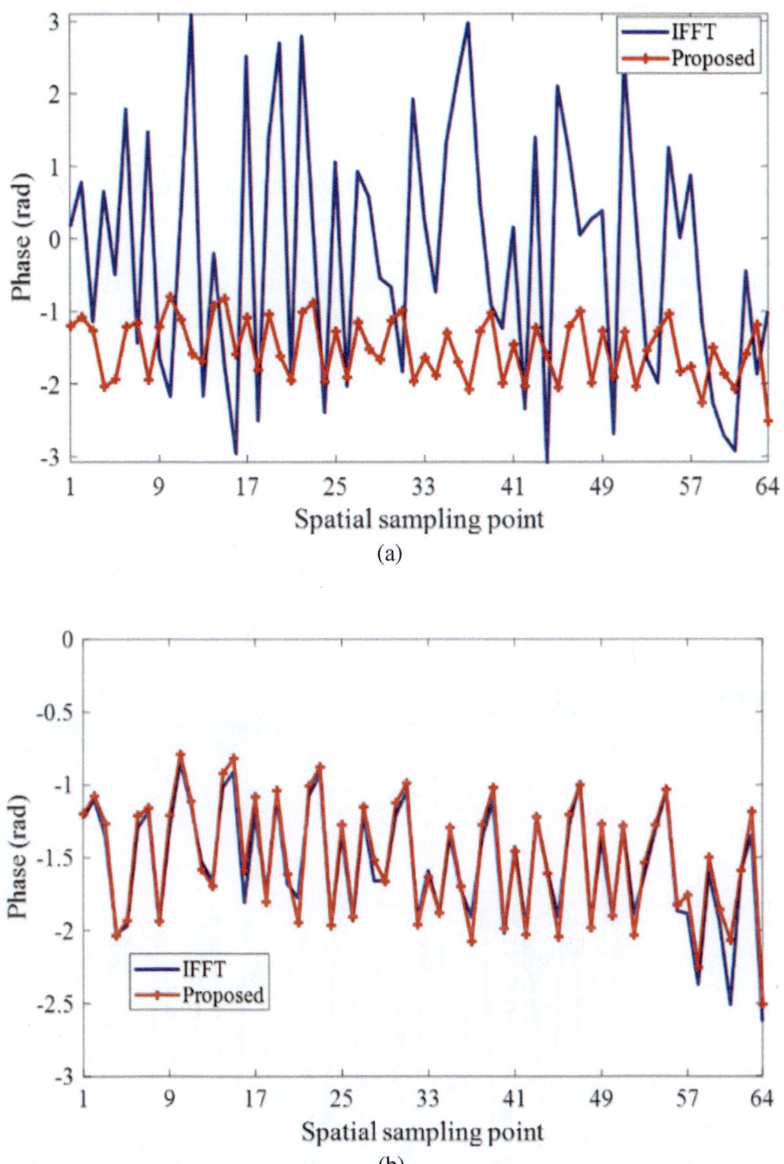

FIGURE 6.41   Estimated phases of the trihedral CR at different spatial sampling points obtained when: (a) the IFFT interval is different from the proposed method and (b) the IFFT interval is the same as the proposed method.

where G denotes gain, φ denotes phase, and ΔR is caused by the antenna position error and system delays. Ignoring the frequency dependency of the channel phase/gain difference, (6.59) can be simplified to

$$s(p,q) = \sigma(x_0, y_0)[1 + G_p]e^{-j\varphi_p} e^{-j4\pi f_q \Delta \tau_p^0} e^{-j4\pi f_q \tau_p^0} \tag{6.60}$$

Therefore, when the proposed method or IFFT is used, the obtained phase of the target at each spatial sampling point is given by

$$\varphi(p, x_0, y_0) = \varphi_p + 4\pi f_c (\Delta\tau_p^0 + \tau_p^0 - \tau) \qquad (6.61)$$

which is dependent on the value of τ. In real applications, in order to estimate the phase/gain difference for the calibration purpose, the interval of τ should be small enough to estimate τ_p.

It can be seen from Figure 6.41(a) that when the interval of IFFT is different from the grid size of the proposed method, their phase estimations are different. When the interval of FFT becomes the same as the proposed method, their phase estimations are close to each other. The advantage of the proposed method is that all the phase information in (6.61) is included in the calculations, while the conventional CS-based method simply assumes the phase of the target at different spatial sampling point is a constant.

Finally, different under-sampled frequencies are used to reconstruct the amplitude and phase of the trihedral CR at different antenna positions. Three different numbers of frequencies are selected and 50 time runs are carried out for each selection. The averaged results are presented in Figure 6.42. It can be observed that the reconstructed amplitudes and phases of the CR are consistent, which illustrates that the frequency dependency of channel phase/gain difference can indeed be ignored.

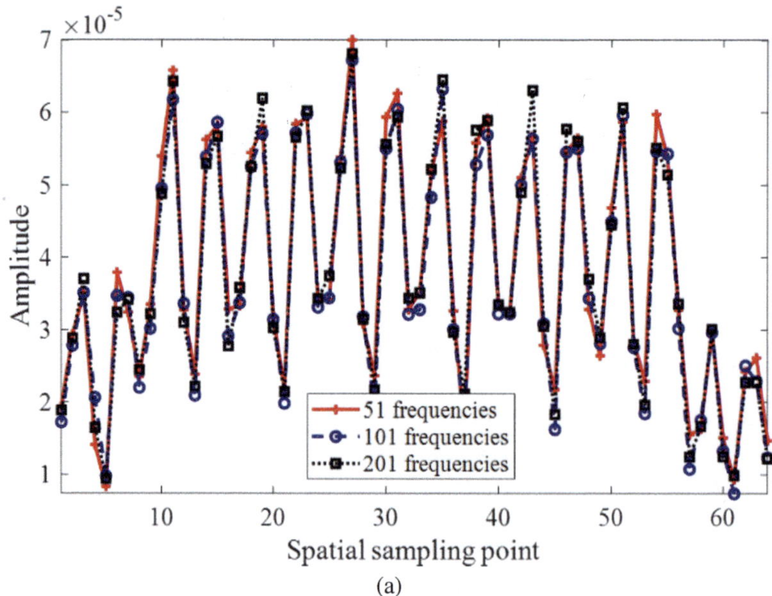

FIGURE 6.42   Amplitudes and phases obtained under different numbers of frequencies.

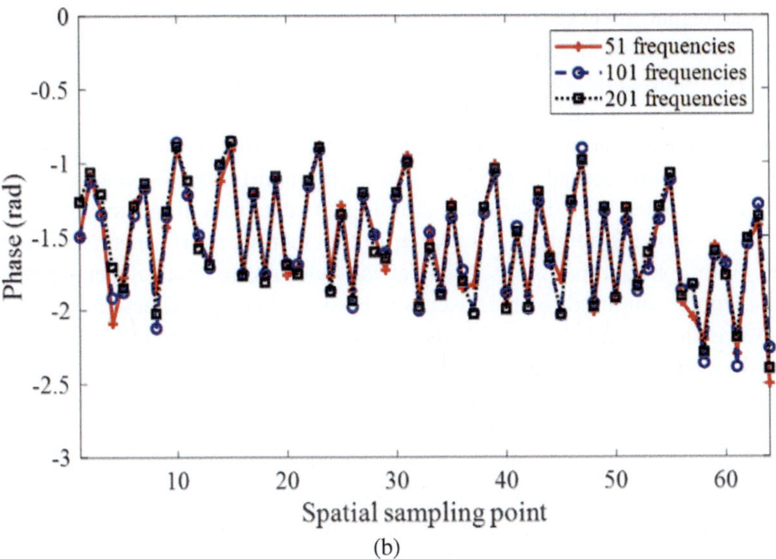

(b)

FIGURE 6.42 (Continued)

6.2.6 DISCUSSION

When targets are located at the near range or system phase/gain errors exist, the complex amplitudes of the targets are strongly dependent on the spatial sampling points. The observation model of the conventional CS-based imaging methods is no longer suitable, giving degraded imaging performance; that is, strong artifacts will be generated. The proposed azimuth-related measurement matrix is more suitable to image targets at near range or with system errors. The received signal in the proposed sensing matrix enjoys block sparsity, and its block partition is exactly known in advance. The advantages of conventional CS-based methods, that is, the capacity to get higher resolution with a reduced number of frequencies and spatial sampling points, are maintained by applying the theory of block CS. Based on the proposed modified BOMP algorithm, the complex amplitudes of targets with respect to different selected spatial sampling points can be separately reconstructed, resulting in multiple 2D-focused images. The proposed cross-correlation–based fusion method can further suppress artifacts, which is beneficial for some following processes, such as target identification. Experimental results obtained by the two designed SFCW-based linear MIMO array radar systems demonstrate that, compared to the BP method, the proposed method can achieve better imaging quality with undersampled data. In comparison with the conventional CS-based method, the proposed method produces fewer imaging artifacts, smaller reconstruction residual, and higher accuracy, while the computing time is also reduced.

Since multiple focused images can be obtained at different spatial sampling points, the potential applications of the proposed imaging method are fast displacement estimation. In such a case, the cross-correlation fusion method should not be used; the displacement of the target at each spatial points can be calculated by comparing the phases of different images. For example, for the first linear MIMO radar system, the data sampling rate is about 8 Hz. By using the proposed method, the data sampling rate is virtually increased to $8 \times 64 = 512$ Hz, which is sufficient for most displacement estimation applications. Actually, the proposed method can be easily extended for conventional GB-SAR imaging by modifying the sensing matrix [3]. Therefore, together with data under-sampling, the data sampling period of GB-SAR can also be virtually reduced. Moreover, in this section, short-range target imaging is the main consideration. For targets at far field, the azimuth dependency of the target complex amplitude is mainly caused by system errors. Although there are several conventional calibration methods for GB-SAR and MIMO radar, the proposed method can provide a focused image and system errors at the same time, which could be useful in practical applications. The problem of the proposed method in this section is indeed the same as the problem of the CS-based imaging method, that is, high computational complexity and low suitability for non-sparse observation scene with distributed targets. Also, the OMP algorithm needs to know the sparsity of the imaging scene or the noise level in advance, which may not be available in practical applications. Therefore, some user parameter-free sparse or block-sparse reconstruction algorithms should be applied.

6.3 TENSOR CS–BASED IMAGING METHOD

6.3.1 INTRODUCTION

From the imaging and displacement estimation results obtained by the designed cross-MIMO radar system in Section 6.1.2, it can be found that strong sidelobes in the azimuth and elevation directions make the result interpretation difficult. Also, the used time-domain BP-based imaging method for 3D high-resolution imaging is time consuming, especially for a large observation scene. Compared to other imaging algorithms, despite only being applicable in the far field, the far-field pseudo-polar image format algorithm has the great advantage of computational complexity, especially when the aperture length is comparable to the range resolution, based on which only 2D FFT is required to get the focused SAR image. To reduce the computation time of 3D imaging for the developed cross-MIMO radar system, according to the signal model in the far-field condition and by using a pseudo-polar spherical coordinate, the 3D FPFA algorithm is proposed in this section. Also, an adaptive amplitude windowing method, that is, the spatially variant apodization–based method, is adopted together with the proposed 3D FPFA algorithm to reduce the sidelobes without decreasing the spatial resolutions.

When the observation scene is sparse, that is, there are only several targets that have significant reflection coefficients, the use of a tensor compressive sensing (T-CS) technique allows the use of under-sampled raw data matrix and the optimization of data acquisition time and memory usage. The T-CS problem can be solved by some tensor-based sparse reconstruction algorithms, such as the Kronecker OMP (Kron-OMP) algorithm, the N-way block OMP (NBOMP) algorithm [29], and the 3D smooth L0 norm (SL0) algorithm [30]. However, for these algorithms, some user-defined parameters always need to be adjusted. In this section, a novel algorithm, the tensor-based iterative adaptive approach, which attempts to minimize a weighted least-squares cost function and does not need to fine-tune the user-defined parameters, is proposed for the effective and efficient reconstruction of the sparse scene with further improved imaging quality compared with the SVA-based FPFA method. Finally, the same data sets as those used in Section 6.1.2, where two trihedral CRs and a four-layer building acted as the targets, are processed to show the effectiveness and advantages of the proposed imaging algorithms.

6.3.2 3D PSEUDOPOLAR COORDINATE SYSTEM

According to the imaging geometry of the designed cross-MIMO radar shown in Figure 6.28, let $R_0 = \sqrt{x_0^2 + y_0^2 + z_0^2}$ denote the distance from the center of the radar system to the target, θ_0 denote the angle between Op and the y-z plane, and φ_0 denote the angle between Op and the x-y plane. It can be derived that $x_0 = R_0\sin\theta_0$ and $z_0 = R_0\sin\varphi_0$. Based on Taylor expansion, the distance between the m-n-th spatial sampling point and the target at the far field can be approximated by

$$R_{m,n}^0 = \sqrt{(x_m - x_0)^2 + y_0^2 + z_0^2} + \sqrt{x_0^2 + y_0^2 + (z_n - z_0)^2}$$
$$\simeq 2R_0 - \sin\theta_0 x_m - \sin\varphi_0 z_n \tag{6.62}$$

Therefore, the reflection coefficient of the target at (R, θ, φ) can be estimated by the frequency domain BP algorithm as

$$\tilde{\sigma}_{FDBP}(R,\theta,\varphi) \simeq \sum_{m=1}^{M}\sum_{n=1}^{N}\sum_{q=1}^{Q} s(m,n,q) e^{j2\pi f_q \frac{2R}{c}} e^{-j2\pi x_m \frac{\sin\theta}{\lambda_c}} e^{-j2\pi z_n \frac{\sin\varphi}{\lambda_c}} e^{-j\psi_{m,n,q}} \tag{6.63}$$

with

$$\psi_{m,n,q} = 2\pi f_q'(\sin\theta x_m + \sin\varphi z_n)/c \tag{6.64}$$

where $s(m, n, q)$ is the received signal corresponding to the m-n-th spatial sampling point and the q-th frequency, $f_q' = f_q - f_c$, f_c is the center frequency, and $\lambda_c = c/f_c$ is the wavelength. The absolute value of $\psi_{m,n,q}$ in (6.64) is bounded by

$$\left|\psi_{m,n,q}\right| = \left|2\pi f_q'(\sin\theta x_m + \sin\varphi z_n)/c\right| \leq \frac{\pi(\sin\theta L_x + \sin\varphi L_z)}{4\Delta R} \tag{6.65}$$

where $\Delta R = c/2B$ is the range resolution, B is the bandwidth, and L_x and L_z are the aperture lengths in the x and z directions, respectively. If $\psi_{m,n,q}$ can satisfy the condition in (6.66), the last exponential term of (6.63) can be ignored [31].

$$\left|\psi_{m,n,q}\right| \leq \frac{\pi(L_x \sin\theta_{max} + L_z \sin\varphi_{max})}{4\Delta R} \leq \frac{\pi}{2} \Rightarrow \frac{(L_x \sin\theta_{max} + L_z \sin\varphi_{max})}{2\Delta R} \leq 1 \tag{6.66}$$

For the designed cross-MIMO radar system, as $L_x = 90$ cm, $L_z = 75$ cm, $\Delta R = 30$ cm, and $\sin\theta_{max} \approx \sin\varphi_{max} \approx 1/2$, $L_x\sin\theta_{max} + L_z\sin\varphi_{max}$ is higher than $2\Delta R$. However, since the mean value of x_m (z_n) is zero and due to the factors $\sin\theta$ and $\sin\varphi$, ψ is always smaller than the value given in (6.66). Thus, even if the condition in (6.66) is slightly unsatisfied; that is, $L_x\sin\theta_{max} + L_z\sin\varphi_{max}$ is slightly higher than $2\Delta R$, the last exponential term in (6.63) can still be ignored without degrading the final imaging performance too much. In such a case, the reflection coefficient of the target at $(R,\ \theta,\ \varphi)$ can be approximately estimated by

$$\tilde{\sigma}_{FDBP}(R,\theta,\varphi) \simeq \sum_{m=1}^{M}\sum_{n=1}^{N}\sum_{q=1}^{Q} s(m,n,q)e^{j2\pi f_q \frac{2R}{c}} e^{-j2\pi x_m \frac{\sin\theta}{\lambda_c}} e^{-j2\pi z_n \frac{\sin\varphi}{\lambda_c}} \tag{6.67}$$

It can be observed that the exponential terms in (6.67) form the kernel of 3D Fourier transform. Therefore, a 3D pseudopolar spherical coordinate $(\alpha,\ \beta,\ \gamma)$ system is defined as

$$\alpha = -2R/c, \quad \beta = \sin\theta/\lambda_c, \quad \gamma = \sin\varphi/\lambda_c \tag{6.68}$$

Based on this pseudopolar coordinate, the received signal cube and the 3D FPFA algorithm can be formulated as the following multilinear expressions

$$\underline{S} = \underline{\Sigma} \times_1 F_1 \times_2 F_2 \times_3 F_3 \tag{6.69}$$

and

$$\underline{\tilde{\Sigma}} = \underline{S} \times_1 F_1^H \times_2 F_2^H \times_3 F_3^H = FFT_{3D}\left[\underline{S}\right] \tag{6.70}$$

where $(\cdot)^H$ denotes conjugate transpose; $\times_k$ denotes the mode-k tensor by matrix product [32]; $\underline{S}$ is the received signal cube; $\underline{\Sigma}$ is the reflection coefficient tensor of the observation scene; and F_1, F_2, and F_3 are the Fourier transform matrices in the range, azimuth, and elevation directions.

Compared to the BP imaging algorithm, the 3D FPFA imaging algorithm can significantly reduce the computational cost by using the fast Fourier transform. However, due to the intrinsic limitation of Fourier transform-based methods, high-level sidelobes will be generated by (6.70), which will introduce negative influences on the following displacement estimation. To suppress the sidelobes, amplitude windowing (a.k.a. amplitude apodization) methods can be used. For example, a Hanning function defined as

$$W_{Hann}(o) = 0.5 - 0.5\cos(2\pi o/O) = \sin^2(\pi o/O) \tag{6.71}$$

can effectively reduce the sidelobe level, where $o = 1, 2, \ldots, O$, and O is the number of sampling points in the frequency, azimuth, or elevation direction.

However, conventional window functions, including the Hanning window, will expand the main lobe, that is, will degrade the resolution. In this section, one of the adaptive apodization methods, the spatially variant apodization–based method [33–34], is used to suppress the sidelobes of the Fourier transforms in (6.70) without broadening the main lobe. SVA is a nonlinear operator based on the following "cosine on pedestal" weighting function.

$$W_{SVA}(o) = 0.5 - \xi(o)(2\pi o / O), \quad 0 \le \xi(o) \le 0.5 \tag{6.72}$$

where $\xi(o)$ is optimized to reduce the influences of other samples. If $\xi(o) = 0.5$, then the SVA function is changed to the Hanning function. With SVA, the compression in the range, azimuth, or elevation direction can be obtained by

$$\Sigma_{SVA}^a = F_a^{\ H}[S_a(f) \odot W_a] \tag{6.73}$$

where $a = 1$, 2, or 3, denoting the range, azimuth, and elevation directions. Given $\Sigma^a = F_a^H S_a(f)$, (6.73) gives the following results

$$\Sigma_{SVA}^a(o) = \Sigma^a(o) - \xi(o)[\Sigma^a(o-1) + \Sigma^a(o+1)] \tag{6.74}$$

Without considering the constraint in (6.72), the optimal $\xi(o)$ can be obtained from (6.74) as [33]

$$\xi_{opt}(o) = Re\left[\frac{\Sigma^a(o)}{\Sigma^a(o-1) + \Sigma^a(o+1)}\right] \tag{6.75}$$

Then, letting $\xi_{opt}(o)$ satisfy $0 \le \xi_{opt}(o) \le 0.5$, the following compression result can be obtained.

$$\Sigma_{SVA}^a(o) = \begin{cases} \Sigma^a(o), & \xi_{opt}(o) < 0 \\ \Sigma^a(o) - \xi_{opt}(o)[\Sigma^a(o-1) + \Sigma^a(o+1)], & 0 \le \xi_{opt}(o) \le 0.5 \\ \Sigma^a(o) - 0.5[\Sigma^a(o-1) + \Sigma^a(o+1)], & \xi_{opt}(o) > 0.5 \end{cases} \tag{6.76}$$

To summarize, the SVA-based FPFA method can provide the sidelobe-reduced and resolution-uninfluenced 3D image as

$$\widetilde{\underline{\Sigma}}_{SVA} = \underline{S}_{SVA} \times_1 F_1^H \times_2 F_2^H \times_3 F_3^H = FFT_{3D}\left[\underline{S} \odot \underline{W}\right] \tag{6.77}$$

For instance, given a target at 10 m, the range compression results obtained by IFFT, IFFT with Hanning, and IFFT with SVA are presented in Figure 6.43. It can be observed that, although it can effectively mitigate the sidelobes, the Hanning function will reduce the power and increase the width of main lobe. On the contrary, with a better sidelobe suppression performance, SVA will not reduce the range resolution, which is its advantage.

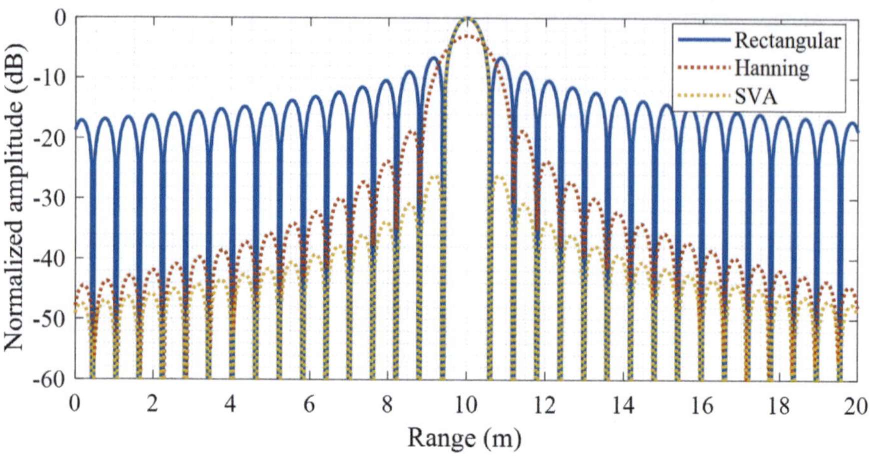

FIGURE 6.43   Range compression results obtained by different methods.

6.3.3   Tensor-Based IAA

When the imaging scene is sparse, that is, only a few strong targets exist, the T-CS-based method can be used for 3D imaging, as shown in (6.78), which can reduce the sidelobe level and achieve higher resolution than conventional Fourier transform-based imaging methods.

$$\widetilde{\underline{\Sigma}} = \min \left\| \underline{\Sigma} \right\|_0 \quad \text{s.t.} \quad \left\| \underline{S} - \underline{\Sigma} \times_1 F_1 \times_2 F_2 \times_3 F_3 \right\|_F \leq \varepsilon \tag{6.78}$$

where ε denotes the noise level, $\|\cdot\|_0$ denotes the number of nonzero elements in a vector or a tensor, and the Frobenius norm of a tensor $\underline{Y}$ is defined as

$$\left\| \underline{Y} \right\|_F \triangleq \sqrt{\sum_{i_1} \sum_{i_2} \sum_{i_3} \left| y_{i_1,i_2,i_3} \right|^2} \tag{6.79}$$

One advantage of the T-CS–based imaging method is that it can work well with under-sampled data, which can help to reduce the data acquisition time, the computation complexity, and the memory usage of (6.78). For the designed radar system, to reduce the spatial sampling points in the x and z directions, a data acquisition program is used to control the switches to randomly turn on several transmitter-receiver pairs of the cross-MIMO array. Since the frequencies of the SFCW signal cannot be randomly generated by the VNA, frequency under-sampling is realized in the data processing step. The random selection of frequencies by using a discrete frequency synthesizer driven by a set of random indication numbers can be conducted in practical applications. The under-sampled signal model, that is, the compressed measurement of the target reflection coefficient tensor, is shown in Figure 6.44, where $\underline{S}^{un}$ is the under-sampled signal cube, and F_1^{un}, F_2^{un}, and F_3^{un} are the under-sampled (partial) Fourier transform matrices.

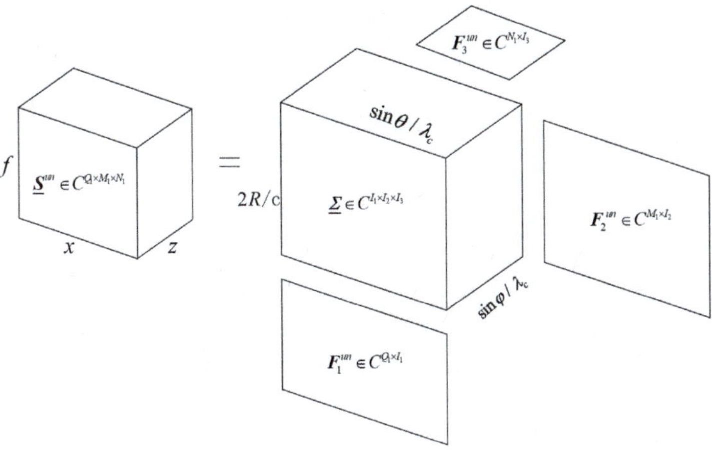

FIGURE 6.44 Compressive measurement of the target reflection coefficient tensor.

Then, 3D image of the observation scene can be obtained by solving the following minimization problem.

$$\widetilde{\underline{\Sigma}} = \min \left\| \underline{\Sigma} \right\|_0 \quad \text{s.t.} \quad \left\| \underline{S}^{un} - \underline{\Sigma} \times_1 F_1^{un} \times_2 F_2^{un} \times_3 F_3^{un} \right\|_F \leq \varepsilon \tag{6.80}$$

(6.80) can be solved by some efficient algorithms, such as the Kron-OMP algorithm, NBOMP algorithm, and 3D-SL0 algorithm [30]. However, Kron-OMP and NBOMP need to know the number of targets or the noise power, which is commonly unknown in practice. For the 3D-SL0 algorithm, several parameters need to be tuned, which requires a lot of effort in some cases. Therefore, based on the principle of the IAA algorithm [35] and its efficient 2D version [36], a non-parametric algorithm, T-IAA, is proposed to solve (6.80).

Conventionally, to fit classical CS theory, $\underline{\Sigma}$ is vectorized and estimated by the following 1D minimization problem.

$$\widetilde{\sigma} = \min \left\| \sigma \right\|_0 \quad \text{s.t.} \quad \left\| s^{un} - \Theta\sigma \right\|_2 \leq \varepsilon \tag{6.81}$$

where

$$\Theta = F_3^{un} \otimes F_2^{un} \otimes F_1^{un} \tag{6.82}$$

is the sensing matrix, $s^{un} = vec(\underline{S}^{un})$, $\sigma = vec(\underline{\Sigma})$, $\otimes$ denotes the Kronecker product, and $vec(\cdot)$ denotes the vectorization operation. It should be noted that, although (6.80) can be solved based on (6.81) by many classical vector-based sparse recovery algorithms [37], such as the convex relaxation algorithm, greedy algorithm, and non-convex optimization algorithm, it is too time consuming and needs too much memory. Thus, directly solving (6.80) without vectorization is preferred. IAA tries to solve (6.81) by minimizing the following weighted least-squares cost function.

$$\tilde{\sigma}_b = \min \left\| s^{un} - \sigma_b \psi_b \right\|^2_{Q_b^{-1}} \tag{6.83}$$

where σ_b is the b-th element of $\boldsymbol{\sigma}$, ψ_b is the b-th column of $\boldsymbol{\Theta}$, $\boldsymbol{R}$ is the covariance matrix of s^{un}, and

$$\left\| x \right\|^2_{Q_b^{-1}} \triangleq x^H Q_b^{-1} x, \quad Q_b = R - \left| \sigma_b \right|^2 \psi_b \psi_b^H \tag{6.84}$$

The minimization of (6.83) yields

$$\tilde{\sigma}_b = \psi_b^H R^{-1} s^{un} / \psi_b^H R^{-1} \psi_b \tag{6.85}$$

Due to the unknown $\boldsymbol{R}$, IAA operates iteratively to get (6.85). For most applications, convergence occurs after no more than 10–15 iterations [36]. The initialization of IAA is

$$\tilde{\sigma}_b^0 = \psi_b^H s^{un} / \psi_b^H \psi_b \tag{6.86}$$

and the o-th ($o = 1, 2, \ldots, O$) iteration is carried out by

$$\tilde{\sigma}_b^o = \psi_b^H (\boldsymbol{\Theta} \Pi^{o-1} \boldsymbol{\Theta}^H)^{-1} s^{un} / \psi_b^H (\boldsymbol{\Theta} \Pi^{o-1} \boldsymbol{\Theta}^H)^{-1} \psi_b \tag{6.87}$$

where $\Pi^{o-1} = diag\{ \left| \tilde{\sigma}^{o-1} \right|^2 \}$, and $(\cdot)^{-1}$ denotes the matrix inverse process.

Based on the Capon filtering property, the vector form of (6.87) can be approximated by

$$\tilde{\sigma}^o \simeq \Pi^{o-1} \boldsymbol{\Theta}^H (\boldsymbol{\Theta} \Pi^{o-1} \boldsymbol{\Theta}^H)^{-1} s^{un} = \Pi^{o-1} \boldsymbol{\Theta}^H u^{o-1} \tag{6.88}$$

where

$$u^{o-1} \simeq (\boldsymbol{\Theta} \Pi^{o-1} \boldsymbol{\Theta}^H)^{-1} s^{un} \tag{6.89}$$

can be obtained by the conjugate gradient (CG) algorithm, which can be used to reduce the computation complexity [36].

Assuming $\underline{A}^{o-1} = \left| \underline{\tilde{\Sigma}}^{o-1} \right|^2$, it can be derived that

$$\Pi^{o-1} = diag\{ vec(\underline{A}^{o-1}) \} \tag{6.90}$$

where $diag(\cdot)$ denotes the diagonalization operation.

Therefore, based on (6.82) and the properties of the Kronecker product and Hadamard product, it can be obtained that

$$\tilde{\sigma}^o \simeq vec(\underline{A}^{o-1} \odot [\underline{U}^{o-1} \times_1 (F_1^{un})^H \times_2 (F_2^{un})^H \times_3 (F_3^{un})^H]) \tag{6.91}$$

where $\underline{U}^{o-1}$ is the tensor form of u^{o-1} that can be obtained by the tensor version of the CG algorithm, as shown in Table 6.2, and $\odot$ denotes the Hadamard product.

Since $\boldsymbol{\sigma} = vec(\boldsymbol{\Sigma})$, the tensor-based IAA algorithm can be derived from (6.91), as shown in Table 6.3, in which the initialization step is obtained by making use of the minimum l_2-norm solution and $(\cdot)^{\dagger}$ denotes pseudo-inversion.

6.3.4 EXPERIMENTS AND RESULTS

In this section, experimental results are presented to illustrate the effectiveness of the proposed imaging algorithms, where the same data sets as those in Section 6.1.2 are used. The working frequency is from 4.75 to 5.25 GHz with 201 frequency steps, giving a range resolution of 30 cm and an ambiguous range of 60 m. It should be pointed out that, to improve the imaging quality, the antenna mutual coupling signal of the designed cross-MIMO radar system is measured in an anechoic chamber and then subtracted from the real measurement data. First, two trihedral corner reflectors with sizes 40 and 30 cm are used as the point targets. They are located at (0, 6.5, 0) and (1, 9.5, 0.5) m approximately, as shown in Figure 6.29. The radar images obtained by the BP algorithm, 3D FPFA algorithm, 3D FPFA algorithm together with the SVA approach and the T-IAA algorithm (for which $O = T = 10$) with 1/8

TABLE 6.2

Tensor-Based CG Algorithm

Input: $\underline{\boldsymbol{S}}^{un}$, $\boldsymbol{F}_1^{un}$, $\boldsymbol{F}_2^{un}$, $\boldsymbol{F}_3^{un}$, $\underline{\boldsymbol{A}}^{0-1}$. and the iteration number T.

1) Initialization: $\underline{\boldsymbol{U}}^0 = 0$, $\zeta_0 = 0$, $\underline{\boldsymbol{R}}_o = \underline{\boldsymbol{S}}^{un}$, $\boldsymbol{P}_0 = 0$, $\rho_0 = \left| \underline{\boldsymbol{R}}_0 \right|_F^2$, and $t = 1$;

2) Update:

$$\underline{\boldsymbol{P}}_t = \underline{\boldsymbol{R}}_{t-1} + \zeta_{t-1} \underline{\boldsymbol{P}}_{t-1}, \underline{\boldsymbol{T}}_1 = \underline{\boldsymbol{A}}^{0-1} \odot \left[\underline{\boldsymbol{P}}_t \times_1 \left(\boldsymbol{F}_1^{un} \right)^H \times_2 \left(\boldsymbol{F}_2^{un} \right)^H \times_3 \left(\boldsymbol{F}_3^{un} \right)^H \right],$$

$$\underline{\boldsymbol{T}}_2 = \underline{\boldsymbol{T}}_1 \times_1 \boldsymbol{F}_1^{un} \times_2 \boldsymbol{F}_2^{un} \times_3 \boldsymbol{F}_3^{un}, \eta_{t-1} = \rho_{t-1} / \sum\sum\sum \underline{\boldsymbol{P}}_t^* \odot \underline{\boldsymbol{T}}_2,$$

$$\underline{\boldsymbol{U}}_t = \underline{\boldsymbol{U}}_{t-1} + \eta_{t-1} \underline{\boldsymbol{P}}_t, \underline{\boldsymbol{R}}_t = \underline{\boldsymbol{R}}_{t-1} - \eta_{t-1} \underline{\boldsymbol{T}}_2, \rho_t = \left| \underline{\boldsymbol{R}}_t \right|_E^2, \zeta_t = \rho_t / \rho_{t-1}.$$

3) Iteration: $t \leftarrow t + 1$, if $t \le T$ go back to 2), otherwise, stop.

Output: $\underline{\boldsymbol{U}}^{0-1} = \underline{\boldsymbol{U}}^T$.

TABLE 6.3

Tensor-Based IAA Algorithm

Input: $\underline{\boldsymbol{S}}^{un}$, $\boldsymbol{F}_1^{un}$, $\boldsymbol{F}_2^{un}$, $\boldsymbol{F}_3^{un}$, and iteration number O and T.

1) Initialization: $\underline{\tilde{\boldsymbol{\Sigma}}}^0 = \underline{\boldsymbol{S}}^{un} \times_1 \left(\boldsymbol{F}_1^{un} \right)^{\dagger} \times_2 \left(\boldsymbol{F}_2^{un} \right)^{\dagger} \times_3 \left(\boldsymbol{F}_3^{un} \right)^{\dagger}, \underline{\boldsymbol{A}}^0 = \left| \underline{\tilde{\boldsymbol{\Sigma}}}^0 \right|^2$, and $o = 1$;

2) Estimation: with $\underline{\boldsymbol{S}}^{un}$, $\boldsymbol{F}_1^{un}$, $\boldsymbol{F}_2^{un}$, $\boldsymbol{F}_3^{un}$, and $\underline{\boldsymbol{A}}^{0-1}$ as inputs, use tensor-based CG algorithm to estimate $\underline{\boldsymbol{U}}^{0-1}$ with T iterations;

3) Update: $\underline{\tilde{\boldsymbol{\Sigma}}}^0 = \underline{\boldsymbol{A}}^{0-1} \odot \left[\underline{\boldsymbol{U}}^{0-1} \times_1 \left(\boldsymbol{F}_1^{un} \right)^H \times_2 \left(\boldsymbol{F}_2^{un} \right)^H \times_3 \left(\boldsymbol{F}_3^{un} \right)^H \right]$ and $\underline{\boldsymbol{A}}^0 = \left| \underline{\tilde{\boldsymbol{\Sigma}}}^0 \right|^2$;

4) Iteration: $o \leftarrow o + 1$, if $o \le O$ go back to 2), otherwise, stop.

Output: $\underline{\tilde{\boldsymbol{\Sigma}}} = \underline{\tilde{\boldsymbol{\Sigma}}}^o$.

---

data are shown in Figure 6.45, where the coordinate is the same as Figure 6.28 and the dynamic range is 20 dB.

It can be seen from Figure 6.45 that two CRs can both be effectively imaged by all the methods. Compared with the BP algorithm, the proposed 3D FPFA algorithm can achieve a similar result, as shown in Figure 6.45(b), but with a much-reduced computing time. Also, by using the SVA-based amplitude windowing method, high-level sidelobes in the azimuth and elevation directions can be suppressed without reducing the angular resolutions, as shown in Figure 6.45(c). With 64 randomly selected pairs of transceivers and 101 frequencies (1/8 data), the imaging result of the proposed T-IAA algorithm is shown in Figure 6.45(d), from which it can be seen that the sidelobes are much reduced and the imaging quality is further improved. Furthermore, measured by the TIC and TOC instructions in MATLAB and averaged by 50 trials, the running times of BP, 3D FPFA, 3D FPFA with SVA, and T-IAA are

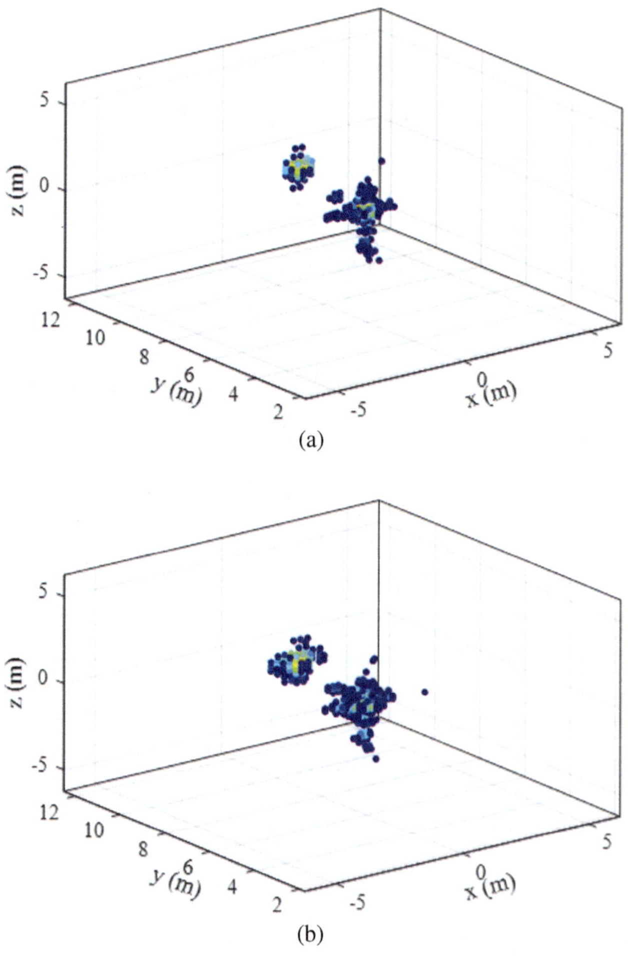

(a)

(b)

FIGURE 6.45 3D images obtained by (a) BP, (b) FPFA, (c) FPFA with SVA, and (d) T-IAA with 1/8 data.

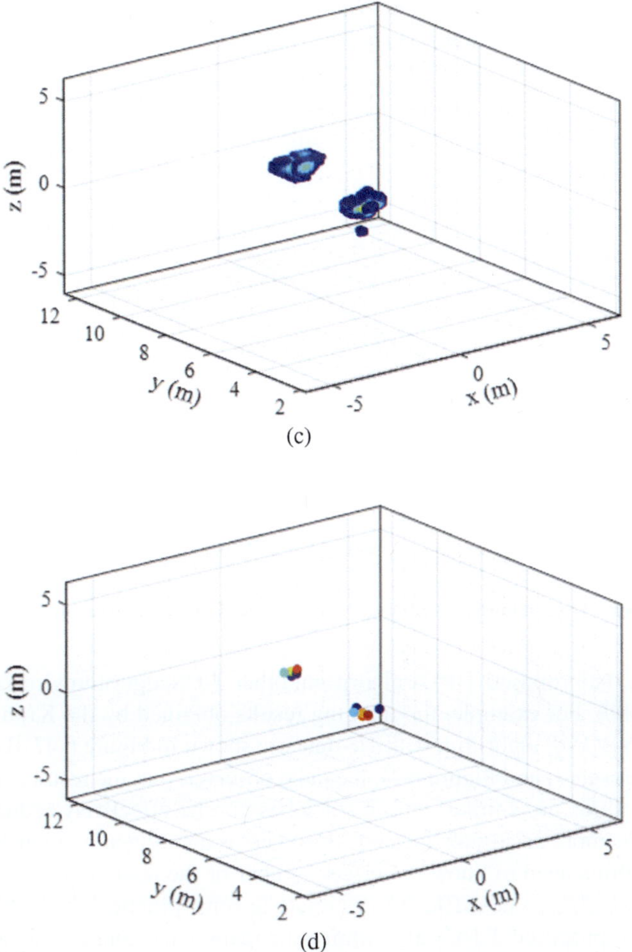

FIGURE 6.45 (Continued)

7.76, 0.12, 0.48, and 1.50 s, respectively. This indicates that the BP algorithm is the most time consuming, and all the proposed algorithms can reduce the computing cost. Also, since only 1/8 data is used by the T-IAA algorithm, its memory usage can also be reduced.

Then, one of the CRs, which is mounted on a linear scanner, as indicated by the red solid circle in Figure 6.29, is moved away from the radar system from 2 to 20 mm with a 2-mm step along the LOS direction. The displacement estimation results obtained by BP, FPFA, and T-IAA are shown in Figure 6.46. It can be seen that all these methods can achieve accurate displacement estimation. It should be noted that due to the influence of random selection, the joint T-IAA algorithm, which can be easily derived from the joint 2-D IAA algorithm proposed in [38], is used for the displacement estimation.

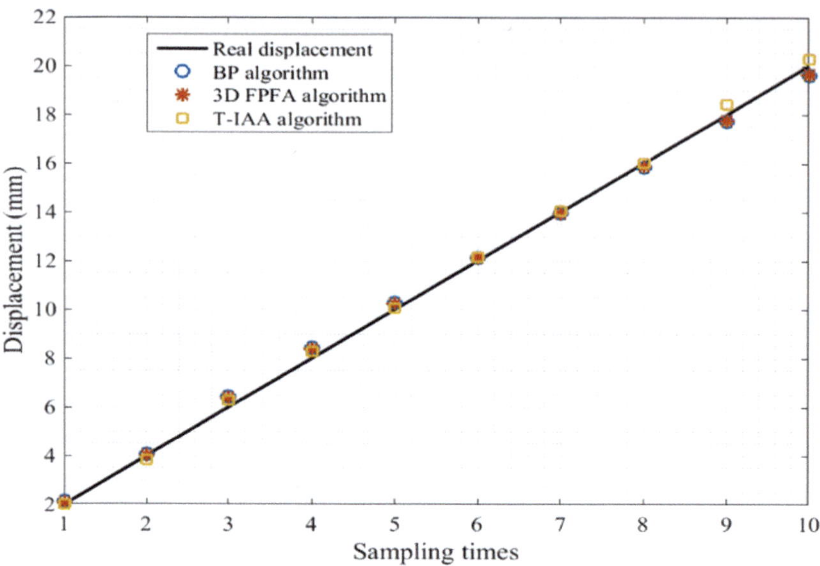

FIGURE 6.46   Displacement estimation of the bigger CR by different methods.

Apart from the proposed T-IAA algorithm, other T-CS algorithms can also be used to solve (6.80). For example, the imaging results obtained by the Kron-OMP algorithm and NBOMP algorithm with 1/8 data are shown in Figure 6.47. By comparing with Figure 6.45(d) and Figure 6.47, it can be observed that the results of these three T-CS algorithms are similar; strong sidelobes can be effectively reduced, and the target can be more accurately located. However, as mentioned previously, the OMP-type algorithms need to know the target number or the noise level to terminate the iterations, which may be difficult to estimate in some practical applications. On the contrary, the proposed T-IAA algorithm is nonparametric and can always converge within 10–15 iterations. Therefore, T-IAA is more suitable in practical applications.

Finally, a building with four layers, as shown in Figure 6.31, is used as the target. The purpose of this experiment is to validate the imaging performance of the designed system for a distributed target. The imaging results obtained by the BP algorithm and the 3D FPFA algorithm with the SVA approach are shown in Figure 6.48 with a dynamic range of 25 dB. For the BP algorithm, strong sidelobes make the imaging result difficult to interpret. However, by using the SVA approach, four layers and the main parts of the building can be clearly observed by the 3D FPFA algorithm. This result further demonstrates the effectiveness of the designed cross-MIMO radar system and the proposed imaging algorithm.

6.3.5   DISCUSSION

For the designed cross-MIMO radar system, based on the far-field approximation, a range-azimuth-elevation decoupled signal model is established. The 3D FPFA algorithm together with the SVA-based amplitude windowing method can generate high-resolution and sidelobe-reduced 3D radar images of the observation scene.

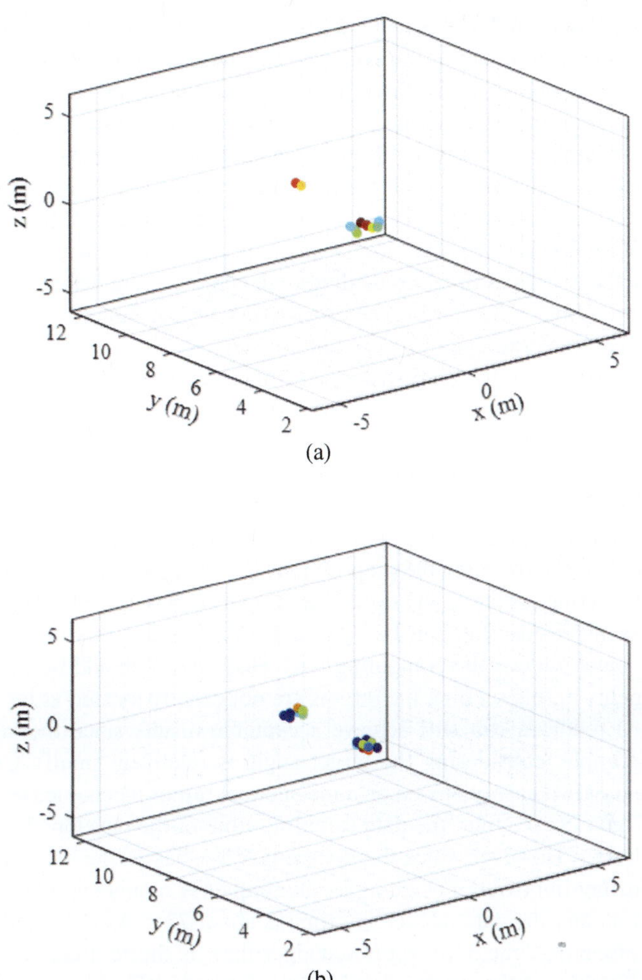

FIGURE 6.47 3D images obtained by (a) the Kron-OMP algorithm with 1/8 data and (b) the NBOMP algorithm with 1/8 data.

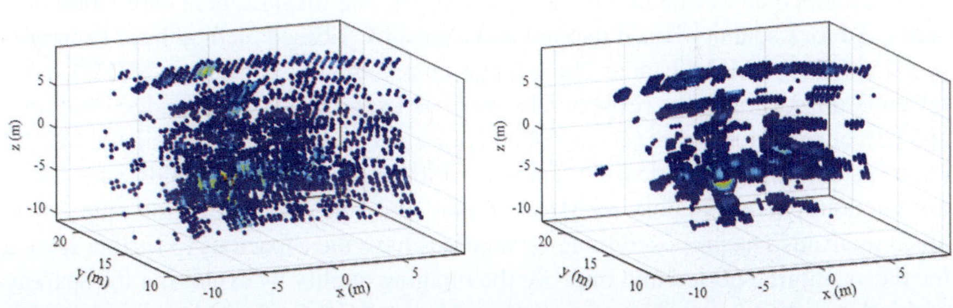

FIGURE 6.48 3D images of the building as a distributed target obtained by (a) the BP algorithm and (b) the 3D FPFA algorithm with the SVA approach.

Furthermore, based on T-CS theory and the proposed user-parameter–free T-IAA algorithm, under-sampling can be conducted to increase the data acquisition rate. The random selection of transmitters and receivers has been realized in the data acquisition stage rather than in the post-processing stage. Displacement estimation results show that the proposed algorithms can keep phase information and thus can be used for real applications, such as the monitoring of buildings, glaciers, dams, and landslides. In contrast to conventional GB-SAR or linear MIMO radar systems with only 2D imaging capacity, the 3D imaging capacity of the developed system provided by the proposed imaging methods can achieve more details and structures of the observed object. Therefore, the cross-MIMO radar system and the proposed algorithms have the potential to be used in practice.

The main problem of the proposed algorithm is that it can only be used for targets at far field. When the target is located at near range, a signal model error will be generated. Therefore, the range-azimuth-elevation coupled signal model will be more suitable. In such a case, conventional vector CS-based imaging should be used to provide more accurate estimation results, while the computing complexity and memory usage should be reduced. One of the potential solutions is the dimension-reduced CS imaging method. By using the proposed 3D FPFA algorithm with under-sampled data, the supporting region of the significant targets can be found. Then, the vector-based sparse reconstruction can be conducted only within this estimated region, which may much reduce the computing cost. However, it should be noted that, for the experiments in this section, the targets are not exactly located at far field, while the proposed methods can still achieve acceptable results since the signal model error is negligible (considering the target angle is relatively small). Compared to using a linear MIMO array moved on a mechanical rail or mechanically moving the transceiver on a 2D scanner, the data sampling time of the developed cross-MIMO system is shorter. However, since the system is VNA based, the data sampling rate should be further improved. For example, the frequency number is selected to be 201 in this section, and the data acquisition time is about 72 s. As indicated previously, the image refreshing time of the proposed algorithms is shorter than 72 s. Therefore, some modifications of the system should be performed.

6.4 SUMMARY

In this chapter, linear MIMO radar systems and cross-MIMO radar systems have been developed and evaluated for 2D/3D imaging and displacement estimation purposes. A block sparsity-based method and a tensor CS-based method have been proposed for linear MIMO radar short-range target imaging and cross-MIMO radar far-range target imaging, respectively. Various experiments with bridge, building, and corner reflectors as targets have been presented to validate the designed MIMO radar systems and the proposed imaging algorithms. Compared with GB-SAR, similar functions can be obtained by MIMO radar, while the data acquisition time can be much reduced. The proposed imaging methods have the capacities to further reduce the data sampling period and improve the imaging quality by exploiting the sparsity of the observation scene. Therefore, two problems of GB-SAR, the long data acquisition period and strong imaging artifacts, can be effectively solved by these advanced MIMO imaging radar techniques.

REFERENCES

[1] W.-Q. Wang and H. Shao, "MIMO antenna array design with polynomial factorization," Int. J. Antennas Propag., 2013.

[2] S. Mitra, K. Mondal, M. Tchobanou, et al., "General polynomial factorization-based design of sparse periodic linear arrays," IEEE Trans. Ultrason. Ferroelect. Freq. Control, vol. 57, no. 9, pp. 1952–1966, Sep. 2010.

[3] W. Feng, L. Yi, and M. Sato, "Near range radar imaging based on block sparsity and cross-correlation fusion algorithm," IEEE J. Sel. Topics Appl. Earth Observ. Remote Sens., vol. 11, no. 6, pp. 2079–2089, 2018.

[4] X. D. Zhuge "Short-range ultra-wideband imaging with multiple-input multiple-output arrays," Ph.D. dissertation, Delft University of Technology, Netherlands, 2010.

[5] Y. C. Eldar, P. Kuppinger, and H. Bolcskei, "Block-sparse signals: Uncertainty relations and efficient recovery," IEEE Trans. Signal Process., vol. 58, no. 6, pp. 3042–3054, Jun. 2010.

[6] G. Matrone, A. S. Savoia, G. Caliano, et al., "The delay multiply and sum beamforming algorithm in ultrasound b-mode medical imaging," IEEE Trans. Med. Imag., vol. 34, no. 4, pp. 940–949, Apr. 2015.

[7] S. Friedland, Q. Li, and D. Schonfeld, "Compressive sensing of sparse tensors," IEEE Trans. Ima. Process., vol. 23, no. 10, pp. 4438–4447, Oct. 2014.

[8] J. FortunyGuasch, "A fast and accurate far-field pseudopolar format radar imaging algorithm," IEEE Trans. Geosci. Remote Sens., vol. 47, no. 4, pp. 1187–1196, Apr. 2009.

[9] K. Han, Y. Wang, X. Chang, et al., "Generalized pseudopolar format algorithm for radar imaging with highly suboptimal aperture length," Sci. China Inf. Sci., vol. 58, no. 4, Apr. 2015.

[10] D. Tarchi, F. Oliveri, and P. F. Sammartino, "MIMO radar and ground based SAR imaging systems: Equivalent approaches for remote sensing," IEEE Trans. Geosci. Remote Sens., vol. 51, no. 1, pp. 425–435, Jan. 2013.

[11] J. Broussolle, V. Kyovtorov, M. Basso, et al., "MELISSA, a new class of ground based InSAR system: An example of application in support to the Costa Concordia emergency," ISPRS J. Photogramm. Remote Sens., vol. 91, pp. 50–58, May 2014.

[12] C. Hu, J. Wang, W. Tian, et al, "Design and imaging of ground-based Multiple-Input Multiple-Output Synthetic Aperture Radar (MIMO SAR) with non-collinear arrays," Sensors, vol. 17, no. 3, pp. 598–616, Mar. 2017.

[13] M. Pieraccini and L. Miccinesi, "An interferometric MIMO radar for bridge monitoring," IEEE Geosci. Remote Sens. Lett., vol. 16, no. 9, pp. 1383–1387, 2019.

[14] W. Wen-Qin, "Virtual antenna array analysis for MIMO synthetic aperture radars," Int. J. Antennas Propag, vol. 587276, pp. 1–10, 2012.

[15] C. Jakowatz, D. Wahl, P. Eichel, et al., Spotlight-mode synthetic aperture radar: A signal processing approach, Norwell: Kluwer Academic Publishers, 1996.

[16] W. Feng, G. Nico, J. Guo, et al., "Estimation of displacement vector by linear mimo arrays with reduced system error influences," in Proc. IEEE Int. Geosci. Remote Sens. Symp., Yokohama, Japan, pp. 346–349, Jul. 2019.

[17] A. Lyulyakin, I. Chernyak, and M. Sato, "Optimization of a sparse array antenna for 3D imaging in near range," IEICE Trans. Electron., vol. 102, no. 1, pp. 46–50, Jan. 2019.

[18] M. Pieraccini, M. Fratini, F. Parrini, et al., "Dynamic monitoring of bridges using a high-speed coherent radar," IEEE Trans. Geosci. Remote Sens., vol. 44, no. 11, pp. 3284–3288, Oct. 2006.

[19] X. Zhuge and A.G. Yarovoy, "A sparse aperture MIMO-SAR-based UWB imaging system for concealed weapon detection," IEEE Trans. Geosci. Remote Sens., vol. 49, no. 1, pp. 509–518, Jan. 2011.

[20] X. Zhuge, A. G. Yarovoy, T. Savelyev, et al., "Modified Kirchhoff migration for UWB MIMO array-based radar imaging," IEEE Trans. Geosci. Remote Sens., vol. 48, no. 6, pp. 2692–2703, Jun. 2010.

[21] A. Michelini, F. Coppi, A. Bicci, et al., "SPARX, a MIMO array for ground-based radar interferometry," Sensors, vol. 19, no. 2, p. 252, Feb. 2019.

[22] L. Yang, J. Zhou, L. Hu, et al., "A perturbation based approach for compressed sensing radar imaging," IEEE Antennas Wirel. Propag. Lett., vol. 16, pp. 87–90, Apr. 2016.

[23] J. Yang, X. Huang, J. Thompson, et al., "Compressed sensing radar imaging with compensation of observation position error," IEEE Trans. Geosci. Remote Sens., vol. 52, no. 8, pp. 4608–4620, Aug. 2014.

[24] M. Cetin, I. Stojanovic, O. Onhon, et al., "Sparsity-driven synthetic aperture radar imaging: Reconstruction, autofocusing, moving targets, and compressed sensing," IEEE Signal Process. Mag., vol. 31, no. 4, pp. 27–40, Jun. 2014.

[25] J. Yang, J. Thompson, X. Huang, et al., "Random-frequency SAR imaging based on compressed sensing," IEEE Trans. Geosci. Remote Sens., vol. 51, no.2, pp. 983–994, Feb. 2013.

[26] E. Elhamifar and R. Vidal, "Block-sparse recovery via convex optimization," IEEE Trans. Signal Process., vol. 60, no. 8, pp. 4094–4107, Aug. 2012.

[27] Z. Zhang and B. D. Rao, "Recovery of block sparse signals using the framework of block sparse Bayesian learning," in Proceeding IEEE International Conference Acoustics. Speech Signal Processing, 2012, pp. 3345–3348.

[28] L. Zhou, C. Huang, and Y. Su, "A fast back-projection algorithm based on cross correlation for GPR imaging," IEEE Geosci. Remote Sens. Lett., vol. 9, no. 2, pp. 228–232, Mar. 2012.

[29] C. F. Caiafa and A. Cichocki, "Multidimensional compressed sensing and their applications," Wiley Interdiscipl. Rev., Data Mining Knowl Discovery, vol. 3, no. 6, pp. 355–380, Oct. 2013.

[30] W. Qiu, J. Zhou, H. Zhao, et al., "Three-dimensional sparse turntable microwave imaging based on compressive sensing," IEEE Geosci. Remote Sens. Lett., vol. 12, no. 4, pp. 826–830, Apr. 2015.

[31] K. Han, Y. Wang, W. Tan, et al., "Efficient pseudopolar format algorithm for downlooking linear-array SAR 3-D imaging," IEEE Geosci. Remote Sens. Lett., vol. 12, no. 3, pp. 572–576, Mar. 2015.

[32] T. G. Kolda and B. W. Bader, "Tensor decomposition and applications," <cite lang="hmn">SIAM Rev., vol. 51, no. 3, pp. 455–500, 2009.

[33] H. C. Stankwitz, J. D. Rodney, and R. F. James, "Spatially variant apodization for sidelobe control in SAR imagery," IEEE National Radar Conference, pp. 132–137, Mar. 1994.

[34] H. C. Stankwitz, R. J. Dallaire, and J. R. Fienup, "Nonlinear apodization for sidelobe control in SAR imagery," IEEE Trans. Aerosp. Electron. Syst., vol. 31, no. 1, pp. 267–279, Jan. 1995.

[35] T. Yardibi, J. Li, P. Stoica, et al., "Source localization and sensing: A nonparametric iterative adaptive approach based on weighted least squares," IEEE Trans. Aerosp. Electron. Syst. vol. 46, no. 1, pp. 425–443, Jan. 2010.

[36] M. J. Jahromi and M. H. Kahaei, "Two-dimensional iterative adaptive approach for sparse matrix solution", Electron. Lett., vol. 50, no. 1, pp. 45–47, Jan. 2014.

[37] E. C. Marques, N. Maciel, L. Naviner, et al., "A review of sparse recovery algorithms," IEEE Access, vol. 7, pp. 1300–1322, Dec. 2018.

[38] W. Feng, Y. Guo, X. He, et al., "Jointly iterative adaptive approach based space time adaptive processing using MIMO radar," IEEE Access, vol. 6, pp. 26605–26616, Jun. 2018.

Index

0-9

1D phase unwrapping, 44, 45
2D displacement map, 82
2D map, 65, 80, 84
2D phase unwrapping, 45
3D spatial model, 53
3D terrain visualization, 53
3D visualization, 55, 119

A

angular resolution, 20, 28, 30, 223, 249, 284
antenna aperture, 2–3, 20–21
antenna radiation, 58–60
antenna radiation pattern, 60
Arato-zawa, 137–145, 214, 218
arc-scanning SAR (ArcSAR), 6
atmospheric delay, 99
atmospheric phase, 37–38, 44, 46, 85, 99–101
atmospheric phase compensation, 37, 46
atmospheric phase delay, 99–112
atmospheric phase screen (APS), 46, 85, 99–101, 103–112
atmospheric refractivity, 103
atmospheric turbulence, 108
azimuth resolution, 1–4, 6, 20–21, 26, 62, 221

B

back-projection algorithm, 30–31, 32
bandwidth, 1, 9, 20, 23–24, 26, 28, 31, 35, 62, 106, 130, 133, 220, 224–225, 232, 237–238, 252–253, 255, 277
block orthogonal matching pursuit (BOMP), 220, 261–263, 266, 268, 270–271, 275
block sparsity, 220, 256–276, 288
boxcar filter, 40, 43
bridge, 5, 9–11, 55–56, 246–248, 288

C

C-band, 130, 135
code division multiplexing (CDM), 8
coherence, 4, 6, 22, 37–44, 71, 74, 76–78, 81, 87, 104
coherence estimation, 40
coherence image, 38–41
coherent scatterer, 40
complex coherency, 77, 81

compressive sensing (CS), 10, 220–221, 253, 255–258, 260–263, 265–271, 274–276, 280–281, 286, 288
conjugate gradient (CG), 282, 283
continuous working mode, 37
co-registration, 37, 71, 238
corner reflector (CR), 81, 83–84, 87–96, 108, 245, 265, 283, 288
covariance, 101–103, 109–111, 282
covariance function, 102, 109
cross correlation, 8, 212–217, 220, 256, 264, 266, 269, 275
CS point, 80, 85

D

daisy chain, 38
data acquisition interval, 3, 76–78, 100
decorrelation, 4, 22, 38, 40, 42, 77, 104, 110
deformation, 44
Delaunay triangulation, 45, 146
DEM, 1, 4–5, 22, 36, 47, 57–58, 64, 80, 110, 146
deterministic approach, 101, 103
differential InSAR (DInSAR), 3, 21
differential phase, 99, 104–108
diffraction stacking, 115, 117, 119
digital elevation models (DEMs), 1, 4–5, 22, 36, 47, 57–58, 64, 80, 110, 146
disaster prevention, 56
disaster recovery, 55, 66
dispersion of amplitude (DA), 42–43, 104
displacement, 21–26, 31, 36–40, 43, 45–47, 220, 225, 237–239, 246, 248, 250–251, 253–254, 275–276, 278, 285–286, 288
displacement monitoring, 3–5, 11, 22, 77, 83, 87
dual ridged horn antenna, 126

E

early warnings, 56, 69, 77, 94, 97–98, 99, 214, 218
early warning system, 4, 11, 22, 56, 69, 77, 94, 97, 99, 214, 218
electric field, 47
electromagnetic signal, 46, 53
electromagnetic (EM) wave, 1, 3–4, 99, 100, 117, 134
empirical variogram, 110
environmental monitoring, 53, 65, 85
ESRI ArcGIS, 54

expectation operation, 39
extensometer, 94–96

F

far-field, 34, 36, 220, 225, 230, 276, 286
far-field pseudo-polar format algorithm (FPFA), 33, 36, 221, 245, 252, 276–279, 283–288
the fifth-generation Penn State/NCAR mesoscale model (MM5), 99
foliage penetration (FOPEN), 1
frequency-domain back projection (FDBP), 30–31, 33, 259–260
frequency modulated continuous waveform (FMCW), 22–26, 106

G

GB-SAR, 1–6, 9–12, 20–49, 53–98, 99–100, 103–108, 110, 115–135, 137–139, 144–148, 214, 220–222, 224–225, 227, 237, 244–245, 252, 255–256, 276, 288
GB-SAR illumination, 53, 65–66, 68, 72, 87
GB-SAR installation, 53, 55, 65–66, 69, 214
GB-SAR measurements, 12, 85, 94, 96
geocoding, 46–47, 57
geometric phase, 37
geostatistical approach, 109
geostatistics, 102–103, 108–109
GIS, 53, 55
global meteorological reanalysis data, 99
global positioning system (GPS), 11, 99–100
Goldstein filters, 43
grating lobe, 8–9, 225, 232, 234, 236, 255
ground-penetrating radar (GPR), 1

H

heterogeneous area, 40
horn antenna, 59

I

illumination area, 54, 58–64, 68, 71, 87, 97
image co-registration, 37, 71, 238
image resolution, 40
imaging algorithms, 26, 30, 33, 220, 237, 245, 252–253, 276–278, 283, 286, 288
incident plane, 48
incident wave, 47
integration path, 44
interferogram, 36–40, 42–43, 45, 73–74, 77, 83, 99, 103, 106, 110–111
interferogram filtering, 37, 43
interferogram network, 38
interferometric coherence, 6, 40–44, 81, 104
interferometric phase, 6, 37, 40, 44, 46, 71, 74–77, 81, 85, 99, 102, 106
interferometric SAR (InSAR), 3, 21, 99
intrinsic stationarity, 102
inverse SAR (ISAR), 3, 21
irrotational field, 44
iterative adaptive approach (IAA), 221, 277, 280–286, 288
Itoh condition, 44
Iwate-Miyagi Nairiku Earthqueake, 137

K

Kolmogorov turbulence theory, 101
Kriging, 109–112
Ku-band, 103–104, 106, 110
Kumamoto earthquake, 55

L

L_1-norm, 45
L_1-norm minimization problem, 45
landslide, 3–5, 10–11, 21–22, 40, 55, 137–138, 288
landslide area, 56, 64, 69, 83, 85, 92, 94
L-band, 126, 135
least squares-based unwrapping method, 45
LiDAR, 56–57, 69
LiDAR survey, 56–57
line of sight (LOS), 1, 20, 46, 246

M

maximum unambiguous displacement, 7
medium resolution imaging spectrometer, 99
migration, 115, 119
MIMO radar, 37
Minami-Aso, 40–41, 43, 55–56, 65–69, 73, 79–80, 83, 85, 87, 92, 95, 106
minimum cost flow (MCF), 45, 110
model-based approach, 103, 111
moderate resolution imaging spectroradiometer (MODIS), 99
monitoring site, 78, 95
moving average, 39
moving window, 39
multi-input multi-output (MIMO), 37, 220–288
multi-look filters, 43
multi-looking, 40
multiple-input-multiple-output (MIMO), 5–10
multispectral data, 100

N

natural distributed area, 40
near-real-time monitoring, 99–100
numerical weather prediction (NWP) models, 99
NWM, 100

O

ordinary least square (OLS), 106

P

path-following algorithms, 44
path-independent methods, 44
phase, 99–112
phase ambiguity, 36
phase center approximation (PCA), 20, 221–222, 228
phase difference, 36–37
phase disturbance, 40, 99
phase noise, 39, 43–44
phase noise filtering, 43
phase unwrapping, 37, 43–45, 73–74, 110
phase unwrapping algorithm, 44–45, 110
phase unwrapping problem, 44
phase wrapping, 38, 73, 77, 85
plane wave, 47
point spread functions (PSF), 220, 224–226, 229, 232–233, 235
polarimetric DInSAR (PolDInSAR), 3
polarimetric information, 5–6, 48, 115, 125
polarimetric InSAR (PolInSAR), 3
polarimetric SAR (PolSAR), 3, 21, 126
polarimetry, 47–48, 49
polynomial factorization (PF), 220, 223, 230
post-disaster, 56, 69
power-law distribution, 101
probability density function, 101
propagation phase term, 100

Q

quality-guided method, 44

R

radar interferometry, 3, 101
radar polarimetry, 48, 49
radar signal, 3–4, 12, 62
radiation pattern, 58–62, 117, 145, 224–228, 232, 234–236, 238–239, 241, 258
rainfall intensity, 139–144, 147–150, 163, 212–214, 216
rainy season, 82–94
random function, 101–102

range migration algorithm (RMA), 31–33, 35–36, 244
range resolution, 1, 20, 23, 25, 28, 35, 62, 237, 246, 248, 252–253, 276–277, 279, 283
raster, 56, 63–64
real-aperture-radar (RAR), 1–2, 20–21
redundant network, 38
reflector, 81–84, 87–96, 108
refractive index, 99–100
refractivity, 100–101, 103–106, 110
refractivity index, 100, 101, 103, 105, 110
relative humidity, 101
remote sensing, 48
residual APS, 108, 110–111
residue, 44
resolution, 40
RF, 102
rice paddy, 12, 126–137

S

SAR, 40
scattering matrix (SM), 47–49
scattering mechanism, 47, 126
scattering properties, 21, 36–37, 259
second-order stationarity, 102, 109
sidelobe, 8–9, 30–32, 36, 220–221, 225, 245, 248, 253, 255–256, 276, 278–280, 284, 285
simple Kriging, 109
single input and multiple output (SIMO), 8
single reference network, 38
SIR-C, 99
SNR, 8, 42
soft topography, 103–105
soil erosion, 137, 139, 143, 145
soil-loss, 144, 146–150, 163–164, 188, 212, 214
spaceborne SAR, 4–6, 22, 99–100
spatial baseline, 21, 36, 37, 40, 238
spatial coherence, 40
spatial dissemination, 53, 57, 65
spatial information, 62
spatially variant apodization (SVA), 221, 255, 276–277, 279, 283–284, 286–287
speckle, 40
speed of light, 1, 20, 22, 62, 243
spotlight mode, 3, 21
stability information, 56, 66
stacking SAR interferogram, 99
stationarity, 101–102, 109
steep topography, 103, 105–108
stepped frequency continuous waveform (SFCW), 22–26, 220, 243, 250–251, 275, 280

"stop-go" assumption, 2–3
strict stationarity, 101
stripmap mode, 2, 3, 20–21
structure-function, 101
synthetic aperture, 21
synthetic-aperture-radar, 1–3, 20, 115

T

target classification, 47
temporal baseline, 4, 12, 21–22, 36, 38, 40, 81
temporal coherence, 40
temporal decorrelation, 38, 42, 77
tensor compressive sensing (T-CS), 220–221, 276, 280, 286, 288
thermal coherence, 40
thermal noise, 4, 22, 26, 31, 37, 243
three-dimensional SAR image, 115
through-the-wall radar (TWR), 1
time division multiplexing (TDM), 8–9, 237, 255
time-domain back projection (TDBP), 31, 260–261, 264
time lag, 83, 212, 214
topographic phase, 5, 37
tree, 115, 117–124
troposphere, 99–100, 103
turbulent, 101, 103, 108
turbulent APS, 101, 108
turbulent eddie, 101
turbulent wind vortices, 101

U

unambiguous range, 23, 26, 237
uniform linear array (ULA), 220–221, 223–228, 237, 245, 248, 253–255
uniform planar array (UPA), 220, 227, 230–231, 233–236, 239, 253–254
unmanned aerial vehicle–borne (UAV) SAR, 4
unwrapped phase difference, 36, 44
unwrapping algorithm, 44–45, 110

V

variogram, 101–103, 109–110
vector network analyzer (VNA), 115, 116, 220, 238
vegetation, 115–126

W

water vapor, 77, 99–101
wavelength, 2–3, 20–21, 24, 28, 34, 37, 62, 130, 145, 239, 277
wave number, 47
wave vector, 47, 48
weather research and forecasting (WRF) model, 99
wrapped phase gradient, 44

X

X-band, 9, 99, 126
X-SAR, 99